Automatic

Sprinkler Systems

Handbook

Second Edition

Automatic Sprinkler Systems Handbook

Second Edition

Based on
Standard for the Installation of Sprinkler Systems, NFPA 13, 1985 Edition
Recommended Practice for the Inspection, Testing and
Maintenance of Sprinkler Systems, NFPA 13A, 1981 Edition
Standard for the Installation of Sprinkler Systems in
One- and Two-Family Dwellings and Mobile Homes, NFPA 13D, 1984 Edition

Edited by

Robert M. Hodnett

Extinguishing Systems Engineer

National Fire Protection Association, Inc.
Quincy, Massachusetts

This *Handbook* has not been processed in accordance with NFPA Regulations Governing Committee Projects. Therefore, the commentary in it shall not be considered the official position of NFPA or any of its committees and shall not be considered to be, nor relied upon as, a Formal Interpretation of the meaning or intent of any specific provision or provisions of NFPA 13, *Standard for the Installation of Sprinkler Systems*; NFPA 13A, *Recommended Practice for the Inspection, Testing and Maintenance of Sprinkler Systems*; or NFPA 13D, *Standard for the Installation of Sprinkler Systems in One- and Two-Family Dwellings and Mobile Homes.*

Project Manager: Jim Linville
Project Editors: Eileen Harrington and Ann E. Coughlin
Composition: Murial McArdle and Geoffrey S. Stevens
Illustrations: George Nichols
Cover Design: Danis Collett
Production: Donald McGonagle

NFPA No. 13AHB85
ISBN: 0-87765-306-2
Library of Congress No. 85-60090
Printed in the United States of America

Third Printing, May 1986

Contents

Installation of Sprinker Systems, NFPA 13

Inspection, Testing and Maintenance
of Sprinkler Systems, NFPA 13A

Installation of Sprinkler Systems in
One- and Two-Family Dwellings
and Mobile Homes, NFPA 13D

Foreword

NFPA 13, *Standard for the Installation of Sprinkler Systems*, is the oldest standard published by the National Fire Protection Association and, in some ways, predates the Association itself. The first edition of the sprinkler standard was published in 1896, the year in which the National Fire Protection Association was organized.

The Technical Committee responsible for development of the standard has, over the years, been made up of a group of highly qualified individuals respresenting a large number of interests. These have included fire service personnel, insurance company and association representatives, representatives of testing laboratories, manufacturers and installers of automatic sprinkler systems, representatives of users of automatic sprinkler systems, including industrial, commercial and governmental personnel, as well as independent consultants.

During the 1960s, the National Fire Protection Association's Board of Directors charged the Sprinkler Committee with development of a standard which would encourage innovative technology and economical system costs. This was intended to make the automatic sprinkler system available on a wider basis and to further help reduce our nation's fire losses. Major changes have included the introduction of hydraulic design methodology, new joining methods, and new types of pipe or tubing. As a result, the cost of automatic sprinkler installations has escalated at a considerably lesser rate than the various other mechanical systems in buildings. Alternate materials, new methods and devices have been — and continue to be — encouraged.

The life safety record of the automatic sprinkler system over its long history has been enviable in occupancies which have been protected. In May of 1973, the Committee, recognizing the need to reduce the annual life loss from fire in residential occupancies, established a subcommittee to deal with residential and light hazard occupancies. This subcommittee was responsible for the development of NFPA 13D, *Standard for the Installation of Sprinkler Systems in One- and Two-Family Dwellings and Mobile Homes*, which was first adopted in May 1975 and subsequently revised in 1980 and 1984 as a result of extensive fire tests. Activities in the residential sprinkler field have resulted in the development of a new residential sprinkler whose prime goal is life safety. This new fire protection tool now makes it possible to protect the person in the room of origin of the fire.

The sprinkler standard is not written as a legal document nor does it cover all situations which might be encountered. It must be used with engineering judgment and with the application of common sense. Assistance in the application of the standard can, of course, be found in the Formal Interpretations which have been issued by the Committee and which are collected and published in the *National Fire Codes*® Subscription Service. Additional information is also available through reviewing substantiation associated with the Technical Committee Reports and the Technical Committee Documentation developed under the Regulations Governing Committee Projects, which govern the standards-making system of the Association.

This *Handbook* is intended to assist in the application of NFPA 13, *Standard for the Installation of Sprinkler Systems*; NFPA 13A, *Recommended Practice for Inspection, Testing and Maintenance of Sprinkler Systems*; and NFPA 13D, *Standard for the Installation of Sprinkler Systems in One- and Two-Family Dwellings and Mobile Homes*. Contributors to the *Handbook* include members of the Technical Committee who have been responsible for the development of the individual documents who deserve our sincere thanks not only for their work on the *Handbook*, but also for their tireless efforts in Committee work.

Hopefully, this *Handbook* will help toward achieving easier and more accurate application of the standards and will contribute to the continuation of the enviable record of life safety and property protection provided by automatic sprinkler systems.

Chester W. Schirmer
Chairman
Technical Committee on
Automatic Sprinklers

Preface

The first automatic fire extinguishing system on record was patented in England in 1723 and consisted of a cask of water, a chamber of gunpowder, and a system of fuses. The first form of sprinkler system used in the United States, however, was the perforated pipe system, which was first installed about 1852.

The first automatic sprinkler was invented in 1864, but it wasn't until 1878 that Henry Parmelee invented a sprinkler that was used extensively in practice. Some 200,000 Parmelee sprinklers were installed throughout New England by the Providence Steam and Gas Pipe Company (later the Grinnell Company) between 1878 and 1882.

The first set of rules for the installation of automatic sprinklers was developed by John Wormald of the Mutual Fire Insurance Corporation of Manchester, England in 1885, based on an 1884 Factory Mutual Fire Insurance Company's study of the performance of sprinklers conducted by C. J. H. Woodbury of the Boston Manufacturers Mutual Fire Insurance Company and F. E. Cabot of the Boston Board of Fire Underwriters. In 1887, similar rules were prepared in the United States by the Factory Improvement Committee of the New England Insurance Exchange.

By 1895, the commercial growth and development of sprinkler systems was so rapid that a number of different installation rules had been adopted by various insurance organizations. Within a few hundred miles of Boston, nine radically different standards for the size of piping and sprinkler spacing were being used. To solve this problem, the following group of men met in Boston early in 1895:

Everett U. Crosby, Underwriters Bureau of New England
Uberto C. Crosby, New England Insurance Exchange
W. H. Stratton, Factory Insurance Association
John R. Freeman, Factory Mutual Fire Insurance Companies
Frederick Grinnell, Providence Steam and Gas Pipe Company
F. Eliott Cabot, Boston Board of Fire Underwriters

Subsequently, a small group of inspection bureau men met in New York in December of 1895 to draw up a set of uniform sprinkler rules. At a second meeting in March of 1896, these rules were completed and, at the same time, a committee was appointed to set up an association for the proper recognition of sprinklers. On November 6, 1896 this association, the National Fire Protection Association, came into being.

Thus, the first and only standard for the National Fire Protection Association in 1896 was the *Standard for the Installation of Sprinkler Systems*.

In 1900, the National Board of Fire Underwriters joined the NFPA and voted to adopt and assume the expense of publishing the NFPA standards. Thus, from 1901 to 1964, the NFPA *Standard for the Installation of Sprinkler Systems* was published by the National Board of Fire Underwriters, first as rules and requirements, later as regulations, and still later as a standard of the National Board of Fire Underwriters as recommended by the National Fire Protection Association.

When first printed in 1896, the Sprinkler Standard concerned itself principally with sprinkler pipe sizes, sprinkler spacing, and water supplies. Since 1900, the Sprinkler Standard has been subjected to considerable amplification and refinement. The 1985 Standard on which this second edition of the *Handbook* is based is the 51st Edition of NFPA 13.

In 1938, the Committee on Automatic Sprinklers began work on a *Recommended Practice for the Care and Maintenance of Sprinkler Systems* (NFPA 13A) because an improperly maintained sprinkler system may be worthless. The 1st Edition was adopted in 1939.

Recognizing the need to reduce the annual loss of life from fire in residential occupancies, the Committee on Automatic Sprinklers in 1973 began preparing a *Standard for the Installation of Sprinkler Systems in One- and Two-Family Dwellings and Mobile Homes* (NFPA 13D), which was first adopted in 1975.

It is hoped that this *Handbook* will assist fire protection practitioners in obtaining a better understanding of, and appreciation for, the requirements contained in NFPA 13 (1985), NFPA 13A (1981), and NFPA 13D (1984).

It should be pointed out, however, that the commentary on the text contained in this *Handbook* is the opinion of the Editor and Technical Consultants. The commentary does not necessarily reflect the official position of NFPA nor the Committee on Automatic Sprinklers.

Arthur E. Cote, P.E.
Assistant Vice President
Standards

Acknowledgments

The editor of this Second Edition of the *Automatic Sprinkler Systems Handbook* acknowledges with gratitude the efforts of the many persons and organizations who contributed to its preparation and development. The following persons deserve special acknowledgment for their review of the materials contained herein:

Robert L. Retelle
Richard E. Hughey
John J. Walsh
Russell P. Fleming
Denison Featherstonhaugh
Robert E. Duke
Jack A. Wood
Rolf Jensen
Chester W. Schirmer

Much of the material in this Second Edition of the *Automatic Sprinkler Systems Handbook* is based on the First Edition. The persons who researched and wrote that material were Robert L. Retelle, Richard E. Hughey, John J. Walsh, Russell P. Fleming, Denison Featherstonhaugh, Robert E. Duke, Wayne Ault, Jack A. Wood, Rolf Jensen and Chester W. Schirmer. The editor would like to give special thanks to Richard E. Hughey who was inadvertently omitted from the list of authors in the First Edition.

Appreciation is expressed to the NFPA staff members who attended to the countless details that went into the preparation of this *Handbook*, especially Arthur E. Cote, Assistant Vice President, Standards, for his review of the final version.

Finally, the editor wishes to express his appreciation to Eileen Harrington for her invaluable assistance.

Robert M. Hodnett, P.E.
Editor

NOTES:

The text, illustrations, and captions to photographs and drawings that make up the commentary on various sections of this *Handbook* are printed in color and in a column narrower than the standard text. The photographs, most of which are part of the commentary, are printed in black for clarity. The text of the standards and the recommended practice is printed in black.

The Formal Interpretations included in this *Handbook* were issued as a result of questions raised on specific editions of the standard or recommended practice. They apply to all previous and subsequent editions in which the text remains substantially unchanged. Formal Interpretations are not part of the standards or recommended practice and therefore are printed in color to the full column width.

NFPA 13

Standard for the Installation of

Sprinkler Systems

1985 Edition

NOTICE: An asterisk (*) following the number or letter designating a subdivision indicates explanatory material on that subdivision in Appendix A. Material from Appendices A, B, and C is integrated with the text and is identified by the letter A, B, or C preceding the subdivision number to which it relates.

Information on referenced publications can be found in Chapter 10 and Appendix D.

1
GENERAL INFORMATION

1-1 Scope. This standard provides the minimum requirements for the design and installation of automatic sprinkler systems and of exposure protection sprinkler systems, including the character and adequacy of water supplies and the selection of sprinklers, piping, valves and all materials and accessories; but not including the installation of private fire service mains and their appurtenances, the installation of fire pumps, the construction and installation of gravity and pressure tanks and towers.

NOTE: Consult other NFPA standards for additional requirements relating to water supplies.

It is important to understand this is an installation standard for sprinkler systems having automatic and open sprinklers and using water as the extinguishing medium. Also, it is a standard of minimum requirements. While it is possible to achieve improved performance, for example, by providing supervisory signaling systems or by providing additional protection devices and systems, this standard establishes minimum requirements for sprinkler systems that are not dependent upon these features.

The *Standard for the Installation of Sprinkler Systems* establishes the water supply needs and the types of water supply sources that are acceptable. It does not, however, contain installation or performance requirements for those water supply sources. Other NFPA standards provide requirements for pumps, and storage tanks.

Other NFPA standards and recommended practices utilizing fixed piped systems for application of water and water based extinguishing media for fire extinguishment and control include:

NFPA 11, *Standard for Low Expansion Foam and Combined Agent Systems*

NFPA 11A, *Standard for Medium and High Expansion Foam Systems*

NFPA 13D, *Standard for the Installation of Sprinkler Systems in One- and Two-Family Dwellings and Mobile Homes*

NFPA 14, *Standard for Standpipe and Hose Systems.*

NFPA 15, *Standard for Water Spray Fixed Systems for Fire Protection*

NFPA 16, *Standard for the Installation of Deluge Foam-Water Sprinkler Systems and Foam-Water Spray Systems*
    NFPA 16A, *Recommended Practice for the Installation of Closed-Head Foam-Water Sprinkler Systems.*

1-2 Purpose.   The purpose of this standard is to provide a reasonable degree of protection for life and property from fire through installation requirements for sprinkler systems based upon sound engineering principles, test data, and field experience. The standard endeavors to continue the excellent record that has been established by standard sprinkler systems and meet the needs of changing technology. Nothing in this standard is intended to restrict new technologies or alternate arrangements, providing the level of safety prescribed by the standard is not lowered.

NOTE: A sprinkler system is a specialized fire protection system and requires knowledgeable and experienced design and installation.

It is difficult to quantify the level of protection for life and property safety that sprinkler systems should provide. It would be unreasonable to expect, for example, that individuals in a fire exposure area of a building with automatic sprinklers might not be injured in the event of a rapidly spreading flammable liquid fire or that property losses can be restricted to specific dollar values or percentages of floor areas. The best that reasonably can be expected is that, by complying with this standard, protection of life and property will be greatly improved. According to NFPA data collected for over eighty years, there has never been a multiple (three or more) loss of life due to fire in a building with a properly operating sprinkler system.

Requirements are based upon proven engineering principles, test data and field experience that has been analyzed, evaluated and agreed upon by knowledgeable people. The aim of this standard is to continue to strive for innovative, progressive, and economically feasible measures that promote life and property safety.

The last sentence is a policy statement that overrides any possibility that developments in sprinkler technology, not currently recognized in the standard, but which provide comparable or improved levels of protection, will be prohibited by the standard.

1-3* Definitions.

Approved.   Acceptable to the "authority having jurisdiction."

NOTE: The National Fire Protection Association does not approve, inspect or certify any installations, procedures, equipment, or materials nor does it approve or evaluate testing laboratories. In determining the acceptability of installations or procedures, equipment or materials, the authority having jurisdiction may base acceptance on compliance with NFPA or other appropriate standards. In the absence of such standards, said authority may require evidence of proper installation, procedure or use. The authority having jurisdiction may also refer to the listings or labeling practices of an organization concerned with product evaluations which is in a position to determine compliance with appropriate standards for the current production of listed items.

"Approved," which means acceptable to the authority having jurisdiction, is different from "listed." An item that is approved is not necessarily listed (see definition of listed). Most items that are installed in a sprinkler system such as alarm valves, dry-pipe valves, sprinklers, hangers, etc. would be both listed and approved. Items such as drain valves that are not listed would be approved.

Authority Having Jurisdiction.   The "authority having jurisdiction" is the organization, office or individual responsible for "approving" equipment, an installation or a procedure.

NOTE: The phrase "authority having jurisdiction" is used in NFPA documents in a broad manner since jurisdictions and "approval" agencies vary as do their responsibilities. Where public safety is primary, the "authority having jurisdiction" may be a federal, state, local or other regional department or individual such as a fire chief, fire marshal, chief of a fire prevention bureau, labor department, health department, building official, electrical inspector, or others having statutory authority. For insurance purposes, an insurance inspection department, rating bureau, or other insurance company representative may be the "authority having jurisdiction." In many circumstances the property owner or his designated agent assumes the role of the "authority having jurisdiction"; at government installations, the commanding officer or departmental official may be the "authority having jurisdiction."

Dwelling Unit.   One or more rooms arranged for the use of one or more individuals living together as in a single housekeeping unit normally having cooking, living, sanitary and sleeping facilities.

For purposes of this standard, dwelling unit includes hotel rooms, dormitory rooms, apartments, condominiums, sleeping rooms in nursing homes and similar living units.

Listed.   Equipment or materials included in a list published by an organization acceptable to the "authority having jurisdiction" and concerned with product evaluation, that maintains periodic inspection of production of listed equipment or materials and whose listing states either that the equipment or material meets appropriate standards or has been tested and found suitable for use in a specified manner.

NOTE: The means for identifying listed equipment may vary for each organization concerned with product evaluation, some of which do not recognize equipment as listed unless it is also labeled. The "authority having jurisdiction" should utilize the system employed by the listing organization to identify a listed product.

One prominent testing inspection agency laboratory uses the designation "classified" for pipe. Material with this designation meets the intent of "listed."

Shall.   Indicates a mandatory requirement.

Should.   Indicates a recommendation or that which is advised but not required.

Sprinkler System.   A sprinkler system, for fire protection purposes, is an integrated system of underground and overhead piping designed in accordance with fire protection engineering standards. The installation includes one or more automatic water supplies. The portion of the sprinkler system above ground is a network of specially sized or hyraulically designed piping installed in a building, structure or area, generally overhead, and to which sprinklers are attached in a systematic pattern. The valve controlling each system riser is located in the system riser or its supply piping. Each sprinkler system riser includes a device for actuating an alarm when the system is in operation. The system is usually activated by heat from a fire and discharges water over the fire area.

NOTE: The design and installation of water supply facilities such as gravity tanks, fire pumps, reservoirs or pressure tanks, and underground piping are covered by the following NFPA standards: NFPA 22, *Water Tanks for Private Fire Protection*; NFPA 20, *Installation of Centrifugal Fire Pumps*, and NFPA 24, *Installation of Private Fire Service Mains and Their Appurtenances*.

It is important to note that the sprinkler system definition includes water supplies and underground piping. While tanks, pumps and underground piping installation and design requirements are covered by other standards, when they are used in connection with overhead sprinkler piping, they become an integral part of the sprinkler system. They are so important to sprinkler system performance that they must be treated as parts of the system.

Systems can be designed according to predeterminded pipe sizes conforming to pipe schedules in Chapter 3 or hydraulically designed in accordance with Chapter 7. Pipe schedule systems have stood the test of time and were the only pipe sizing method allowed until the 1966 Edition of NFPA 13. Systems of hydraulic design are more often less costly than pipe schedule systems and are considered by many to be more efficient because the most economical sizes of pipe are used. The standard therefore provides the designers the option of either designing according to the pipe schedules or hydraulically designing, as long as field experience continues to be satisfactory.

A-1-3   A sprinkler system is considered to have a single system riser control valve.

Standard.   A document containing only mandatory provisions, using the word "shall" to indicate requirements. Explanatory material may be included only in the form of "fine print" notes, in footnotes, or in an appendix.

1-4 Other Publications.   A selected list of other publications related to the installation of sprinkler systems is published at the end of this standard.

1-5 Maintenance.

1-5.1*   A sprinkler system installed under this standard shall be properly maintained for efficient service. The owner is responsible for the condition of the sprinkler system and shall use due diligence in keeping the system in good operating condition.

A-1-5.1 Impairments.   Before shutting off a section of the fire service system to make sprinkler system connections, notify the authority having jurisdiction, plan the work carefully, and assemble all materials to enable completion in shortest possible time. Work started on connections should be rushed to completion without interruption, and protection restored as promptly as possible. During the impairment, provide emergency hose lines, additional fire pails and extinguishers, and maintain extra watch service in the areas affected.

When changes involve shutting off water from any considerable number of sprinklers for more than a few hours, temporary water supply connections should be made to sprinkler systems so that reasonable protection can be maintained. In adding to old systems or revamping them, protection should be restored each night so far as possible. The members of the private fire brigade as well as public fire department should be notified as to conditions.

While it may appear to be beyond the scope of this standard, proper maintenance of sprinkler systems is important for effective sprinkler system performance at the time fire occurs. It is mandated, therefore, that the owner maintain and service the system properly. A system not maintained in accordance with instructions is not in compliance with this standard.

1-5.2   The installing contractor shall provide the owner with:

(a) Instruction charts describing operation and proper maintenance of sprinkler devices.

(b) Publication entitled NFPA 13A, *Inspection, Testing and Maintenance of Sprinkler Systems.*

It is quite important that the requirements of this section are followed to ensure that the system will be operable as long as is required. There are many properly maintained sprinkler systems that are still protecting lives and property even though they were installed fifty or more years ago.

NFPA 13A contains extensive material on the care and maintenance of automatic sprinklers and sprinkler system components. It also has sections on flushing, impairments and fire codes.

1-6 Classification of Sprinkler Systems.

1-6.1   This standard covers automatic sprinkler systems of the types listed below, also systems of outside sprinklers for protection against exposure fires covered specifically in Chapter 6. Manually operated deluge systems, used for certain special hazard conditions, are not specifically covered in this standard but certain provisions of this standard will be found applicable.

Wet-Pipe Systems (*See Section 5-1.*)

Dry-Pipe Systems (*See Section 5-2.*)

Pre-Action Systems (*See Section 5-3.*)

Deluge Systems (*See Section 5-3.*)

Combined Dry-Pipe and Pre-Action Systems (*See Section 5-4.*)

Sprinkler Systems — Special Types. Special purpose systems employing departures from the requirements of this standard, such as special water supplies and reduced pipe sizing, shall be installed in accordance with their listing.

1-7 Classification of Occupancies.

1-7.1*   Occupancy classifications for this standard relate to sprinkler installations and their water supplies only. They are not intended to be a general classification of occupancy hazards.

A-1-7.1   Occupancy examples in the listings as shown in the various hazard classifications are intended to represent the norm for those occupancy types. Unusual or abnormal fuel loadings or combustible characteristics and susceptibility for changes in these characteristics, for a particular occupancy, are considerations that should be weighed in the selection and classification.

The Light Hazard classification is intended to encompass residential occupancies; however, this is not intended to preclude the use of listed residential sprinklers in residential occupancies or residential portions of other occupancies.

The occupancies listed under each category are meant to be general. Some of the occupancies listed could be a more or less hazardous classification depending on the combustibility of the contents.

It is recognized that interior arrangements of a given occupancy can vary markedly, changing the fire challenge to sprinkler systems. Each occupancy example, therefore, must be considered as to whether or not it is within the norm for the particular nature of a business or operation. If it is not, then the proper occupancy classification must be selected by comparing quantity, combustibility, and heat release potential of the contents of the building or portion of building being studied with other occupancy classifications.

1-7.2 Light Hazard Occupancies.

1-7.2.1* Light Hazard. Occupancies or portions of other occupancies where the quantity and/or combustibility of contents is low and fires with relatively low rates of heat release are expected.

A-1-7.2.1 Light Hazard Occupancies include occupancies having conditions similar to:

Churches
Clubs
Eaves and overhangs, if combustible construction with
    no combustibles beneath
Educational
Hospitals
Institutional
Libraries, except large stack rooms
Museums
Nursing or Convalescent Homes
Office, including Data Processing
Residential
Restaurant seating areas
Theaters and Auditoriums excluding stages
    and prosceniums
Unused attics

Light Hazard Occupancies, as defined in this section, offer the least challenge to sprinkler systems. The nature of occupancies in this classification results in the low fire fuel contribution. Generally there is no processing, manufacturing, or large storage accumulation. Fixtures and furniture remain in fairly fixed arrangements in this group of occupancies. Mainly these are institutional, educational, religious, residential and office type properties.

1-7.3 Ordinary Hazard Occupancies.

1-7.3.1* Ordinary Hazard (Group 1).   Occupancies or portions of other occupancies where combustibility is low, quantity of combustibles is moderate, stock piles of combustibles do not exceed 8 ft (2.4 m) and fires with moderate rates of heat release are expected.

A-1-7.3.1   Ordinary Hazard Occupancies (Group 1) include occupancies having conditions similar to:

Automobile parking garages
Bakeries
Beverage manufacturing
Canneries
Dairy products manufacturing and processing
Electronic plants
Glass and glass products manufacturing
Laundries
Restaurant service areas

Ordinary Hazard Occupancies include a wide range of occupancy types. Because of the divergent water supply demands put upon sprinkler systems in this classification, it is divided into three groups.

Group 1 occupancies offer the lowest sprinkler system challenge of the Ordinary Hazard Cccupancies, and are comprised mostly of light manufacturing and servicing industries. Use of flammable and combustible liquids or gases is either very limited or so arranged that it presents low challenge to sprinklers. Stock piles of combustible commodities having Group 1 characteristics that are not within the scope of NFPA 231, *Indoor General Storage*, and not over 8 ft (2.4 m) high are Group 1 Ordinary Occupancies.

1-7.3.2* Ordinary Hazard (Group 2).   Occupancies or portions of other occupancies where quantity and combustibility of contents is moderate, stock piles do not exceed 12 ft (3.7 m) and fires with moderate rate of heat release are expected.

A-1-7.3.2   Ordinary Hazard Occupancies (Group 2) include occupancies having conditions similar to:

Cereal mills
Chemical Plants — Ordinary
Cold storage warehouses
Confectionery products
Distilleries
Leather goods manufacturing
Libraries-large stack room areas
Machine shops
Metal working
Mercantiles
Printing and publishing
Textile manufacturing
Tobacco products manufacturing
Wood product assembly

Ordinary Hazard Group 2 occupancies cover the intermediate range of the Ordinary Hazard Classes. Most of the Ordinary Hazard Occupancies fall into this group. They represent the norm or the average occupancy for manufacturing and processing industries.

When attempting to classify occupancies that decidedly are not Light Hazard or Extra Hazard, the Group 2 classification should be considered first. Those that do not compare well with the examples of Group 1 and 3 occupancies are most probably Group 2 classes.

Stock piles of combustibles for Group 2 are the same as in Group 1, except the height limitation is 12 ft (3.6 m) rather than 8 ft (2.4 m).

1-7.3.3* Ordinary Hazard (Group 3).   Occupancies or portions of other occupancies where quantity and/or combustibility of contents is high, and fires of high rate of heat release are expected.

A-1-7.3.3   Ordinary Hazard Occupancies (Group 3) include occupancies having conditions similar to:

Feed mills
Paper and pulp mills
Paper process plants
Piers and wharves
Repair garages
Tire manufacturing

Warehouses (having moderate to higher combustibility
of content, such as paper, household furniture, paint,
general storage, whiskey, etc.)[1]
Wood machining

When hazards in those buildings or portions of buildings of this occupancy
group are severe, the authority having jurisdiction should be consulted for
special rulings regarding water supplies, types of equipment, pipe sizes, types of
sprinklers and sprinkler spacing.

The occupancies in Group 3 border on being Extra Hazard
Classes. These occupancies can typically include dust or residue
problems or include materials with higher rates of heat release than
contemplated in the other two groups. They are usually the
troublesome occupancies of the Ordinary Classes with which fire
experience has shown a need for strengthened water supplies.

1-7.4* Extra Hazard Occupancies.

A-1-7.4   New installations protecting extra hazard occupancies should be
hydraulically designed where standards giving design criteria are available.

1-7.4.1*   Extra hazard occupancies or portions of other occupancies where
quantity and combustibility of contents is very high, and flammable and
combustible liquids, dust, lint or other materials are present introducing the
probability of rapidly developing fires with high rates of heat release.

A-1-7.4.1   Extra Hazard Occupancies (Group 1) include occupancies having
conditions similar to:

Combustible Hydraulic Fluid use areas
Die Casting
Metal Extruding
Plywood and particle board manufacturing
Printing [using inks with below 100°F (37.8°C)
    flash points]
Rubber reclaiming, compounding, drying, milling,
    vulcanizing
Saw mills
Textile picking, opening, blending, garnetting, carding,
    combining of cotton, synthetics, wood shoddy or burlap
Upholstering with plastic foams

---

[1]For high-piled storage as defined in 4-1.3.8, see Appendix D for separately published NFPA
standards relating to water supply requirements, particularly NFPA 231, *Indoor General Storage*, and
NFPA 231C, *Rack Storage of Materials*.

Extra Hazard Occupancies (Group 2) include occupancies having conditions similar to:

Asphalt saturating
Flammable liquids spraying
Flow coating
Mobile Home or modular building assemblies (where
    finished enclosure is present and has combustible
    interiors)
Open Oil quenching
Solvent cleaning
Varnish and paint dipping

1-7.4.2   Extra hazard occupancies involve a wide range of variables which may produce severe fires. The following shall be used to evaluate the severity of extra hazard occupancies:

Extra Hazard (Group 1) include occupancies described in 1-7.4.1 with little or no flammable or combustible liquids.

Extra Hazard (Group 2) include occupancies described in 1-7.4.1 with moderate to substantial amounts of flammable or combustible liquids or where shielding of combustibles is extensive.

Extra Hazard Occupancies represent the occupancy fire conditions covered by this standard which provide more severe challenges to sprinkler protection. The Extra Hazard Group 1 Occupancies include those in which hydraulic machinery or systems with flammable or combustible hydraulic fluids under pressure are present. Ruptures and leaks in piping or fittings have resulted in fine spray discharge of such liquids, and intense fires result. Those occupancies with process machinery that use flammable or combustible fluids in closed systems are Extra Hazard Group 1. Also in this Group are occupancies that have dust and lint in suspension or that contain moderate amounts of combustible cellular foam materials.

The Extra Hazard Group 2 occupancies have more than small amounts of flammable or combustible liquids, usually in open systems where rapid evaporation can occur when subjected to high temperatures.

This occupancy classification also applies when ceiling sprinklers are obstructed by occupancy conditions, not structural conditions, and water discharged by sprinklers may not reach the burning material because of the shielding.

The Extra Hazard Occupancy examples were classified on the basis of actual field experience with sprinkler system operations in those kinds of occupancies shown in the examples.

1-8* Design and Installation.

A-1-8 Sprinkler Systems in Buildings Subject to Flood.

When sprinkler systems are installed in buildings subject to recurring floods, the location of control valves, alarm devices, dry-pipe valves, pumps, compressors, power and fuel supplies should be such that system operation will be uninterrupted by high water.

1-8.1 Devices and Materials.

1-8.1.1   Only new sprinklers shall be employed in the installation of sprinkler systems.

The sprinkler is one component of the system that must be depended upon to operate efficiently and effectively when fire occurs. Consequently, it should not leak or rupture, or operate for any reason other than to extinguish fire. For these reasons and because of the relatively low cost of sprinklers themselves, new sprinklers are required.

1-8.1.2*   When a sprinkler system is installed, only approved materials and devices shall be used.

A-1-8.1.2   Under special conditions used equipment may be reused by the original owner, subject to the approval of the authority having jurisdiction. Second-hand alarm valves, retarding chambers, circuit closers, water motor alarms, dry-pipe valves, quick-opening devices and other devices may be used as replacement equipment in existing systems subject to the approval of the authority having jurisdiction.

The key word here is "approved" (acceptable to the authority having jurisdiction). Some components of sprinkler systems are required to be "listed" (*See Section 1-3, Definitions*). This is a quality control that ensures that materials such as pipe, fittings, sprinklers, hangers, and other accessory equipment meet a certain level for performance and reliability. When listed materials are not specifically required by the standard, it means that they are not evaluated by Underwriters Laboratories Inc., Factory Mutual, or other agencies with similar testing and follow-up programs. In that event, the materials must be "approved" (acceptable to the authority having jurisdiction), meaning the authority having jurisdiction would require materials and devices equivalent to materials as specified in standards such as the ANSI/ASTM standard for piping as shown in Table 3-1.1.1, or for fittings as shown in Table 3-13.1.1. If "listed" material is not required by specific reference and "approved" material is required, it will be left to the authority having jurisdiction to determine the suitability of the product or component. An example of the latter is the nonindicating valve referred to in 3-14.1.1. In that case the nonindicating valve is not "listed" or required by this standard to meet the equivalency of a specified ANSI/ASTM standard. The authority having jurisdiction must, in this case, decide if it is to be considered "approved."

Formal Interpretation

Question: In the second sentence, are the words "and other devices" intended to include equipments comprising the water supply, such as a fire pump assembly complete with driver and controls, a water tank, and the heating systems for the valve room, pump room and water tank?

Answer: No. While the scope of NFPA 13 includes the character and adequacy of water supplies, "equipments" of the nature described are within the scopes of the NFPA 20 and 22 Standards.

1-8.1.3   Sprinkler systems shall be designed for a maximum working pressure of 175 psi (12.1 bars).

Exception: Higher design pressures may be used when all system components are rated for pressures higher than 175 psi (12.1 bars).

Sprinklers are currently rated for a 175 psi (12.1 bars) working pressure. Until and unless all components that make up sprinkler systems are approved for working pressures higher than 175 psi (12.1 bars), systems shall be designed for not more than 175 psi (12.1 bars) working pressure.

The exception merely removes the 175 psi (12.1 bars) restriction in the event that higher pressure rated sprinklers are made available subsequent to the publication of the 1985 edition of the standard.

1-8.1.3.1   Interior system components subject to pressure shall be designed for a working pressure not less than 175 psi (12.1 bars).

1-9* Working Plans.

A-1-9   Preliminary layouts should be submitted for review to the authority having jurisdiction before any equipment is installed or remodeled in order to avoid error or subsequent misunderstanding. Any material deviation from approved plans will require permission of the authority having jurisdiction.

Preliminary layouts should show:

(a) Name of owner and occupant

(b) Location, including street address

(c) Point of compass

(d) Construction and occupancy of each building

NOTE: Date on special hazards should be submitted as they may require special rulings.

(e) Building height in feet

(f) If it is proposed to use a city main as a supply, whether the main is dead-end or circulating, size of the main and pressure in psi; and if dead-end, direction and distance to nearest circulating main

(g) Distance from nearest pumping station or reservoir

(h) In cases where reliable up-to-date information is not available, a water-flow test of the city main should be conducted in accordance with Appendix B-2-1.1. (The preliminary plan should specify who conducted the test, date and time, the location of the hydrants where flow was taken and where static and residual pressure readings were recorded, the size of main supplying these hydrants, and the results of the test, giving size and number of open hydrant butts flowed; also data covering minimum pressure in connection with city main should be included.)

(i) Data covering waterworks systems in small towns in order to expedite the review of plans

(j) Fire walls, fire doors, unprotected window openings, large unprotected floor openings, and blind spaces

(k) Distance to and construction and occupancy of exposing buildings — e.g., lumber yards, brick mercantiles, fire-resistive office buildings, etc.

(l) Spacing of sprinklers, number of sprinklers in each story or fire area and total number of sprinklers, number of sprinklers on each riser and on each system by floors, total area protected by each system on each floor, total number of sprinklers on each dry-pipe system or pre-action or deluge system and if extension to present equipment, number of sprinklers on riser per floor, sprinklers already installed

(m) Capacities of dry-pipe systems with the bulk pipe included (*see Table A-5-2.3*), and if an extension is made to an existing dry-pipe system. The total capacity of the existing and also extended portion of the system

(n) Weight or class, and size and material of any proposed underground pipe

(o) Whether property is located in a flood area requiring consideration in the design of sprinkler system

(p) Name and address of party submitting the layout.

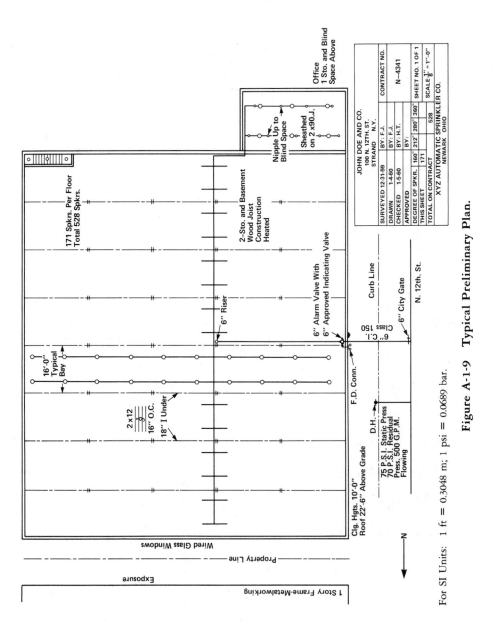

Figure A-1-9   Typical Preliminary Plan.

For SI Units: 1 ft = 0.3048 m; 1 psi = 0.0689 bar.

1-9.1   Working plans shall be submitted for approval to the authority having jurisdiction before any equipment is installed or remodeled. Deviation from approved plans will require permission of the authority having jurisdiction.

Plans are prepared primarily for the mechanics who do the on-the-job installation. They also serve to protect the interest of the owner, who usually is not knowledgeable in the art of sprinkler system installations and consequently will rely upon others to check the plans for conformance to this standard. The owner determines to which "authorities having jurisdiction" — be they code enforcers, consultants, insurers, or even the owner himself — the plans should be submitted for examination and approval.

When the installation is completed, the installation is checked against the plans to determine compliance with approved plans and the sprinkler standard.

The property owner should retain working plans and specifications for reference. In the event alterations are undertaken sometime in the future, a good deal of expense and time saving, especially for systems of hydraulic design, can result if they are available at that time.

The symbols commonly used in plan drawings can be found in NFPA 172, *Fire Protection Symbols for Architectural and Engineering Drawings.*

1-9.2*   Working plans shall be drawn to an indicated scale, on sheets of uniform size, with plan of each floor, made so that they can be easily duplicated, and shall show the following data:

(a) Name of owner and occupant

(b) Location, including street address

(c) Point of compass

(d) Ceiling construction

(e) Full height cross section

(f) Location of fire walls

(g) Location of partitions

(h) Occupancy of each area or room

(i) Location and size of concealed spaces and closets. (*See 4-4.3 to 4-4.17 inclusive, except 4-4.5 and 4-4.6.*)

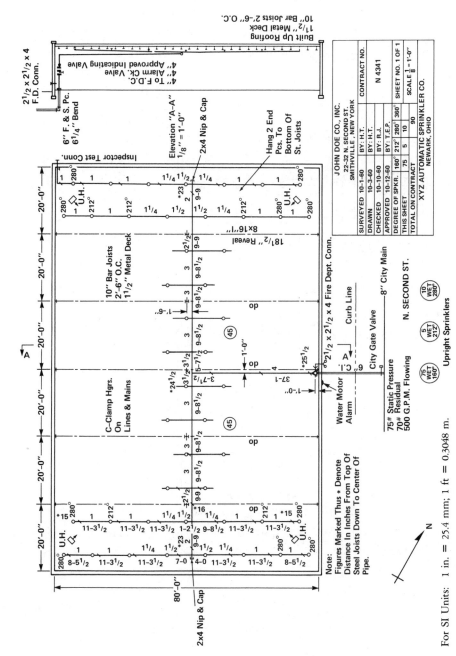

Figure A-1-9.2(A)   Typical Working Plans.

For SI Units: 1 in. = 25.4 mm; 1 ft = 0.3048 m.

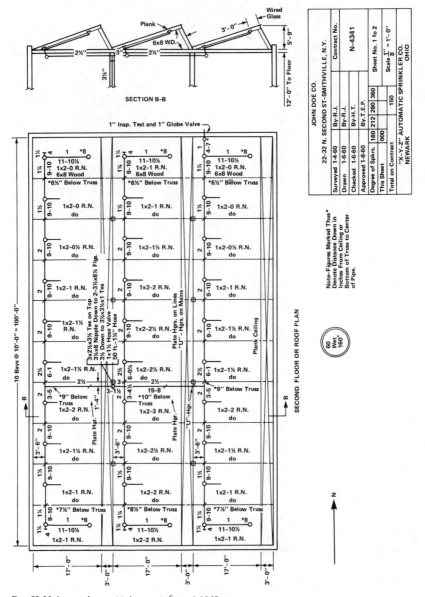

For SI Units: 1 in. = 25.4 mm; 1 ft = 0.3048 m.

Figure A-1-9.2(B)   Typical Working Plans (cont.)

(j) Any questionable small enclosures in which no sprinklers are to be installed

(k) Size of city main in street, pressure and whether dead-end or circulating and, if dead-end, direction and distance to nearest circulating main, city main test results (*see B-2-1*)

(l) Other sources of water supply, with pressure or elevation

(m) Make, type and nominal orifice size of sprinkler

(n) Temperature rating and location of high temperature sprinklers

(o) Total area protected by each system on each floor

(p) Number of sprinklers on each riser per floor

(q) Make, type, model and size of alarm or dry-pipe valve

(r) Make, type, model and size of pre-action or deluge valve

(s) Kind and location of alarm bells

(t) Total number of sprinklers on each dry-pipe system, pre-action system, combined dry-pipe/pre-action system or deluge system

(u) Approximate capacity in gallons of each dry-pipe system

(v) Pipe type and schedule of wall thickness

(w) Nominal pipe size and cutting lengths of pipe (or center to center dimensions)

NOTE: Where typical branch lines prevail, it will be necessary to size only one line.

(x) Location and size of riser nipples

(y) Type of fittings and joints and location of all welds and bends

(z) Type and locations of hangers and sleeves

(aa) All control valves, check valves, drain pipes and test pipes

(bb) Size and location of hand hose, hose outlets and related equipment

(cc) Underground pipe size, length, location, weight, material, point of connection to city main; the type of valves, meters and valve pits; and the depth that top of the pipe is laid below grade

(dd) Provision for flushing (*see 3-8.2*)

(ee) When the equipment is to be installed as an addition to an existing system enough of the existing system shall be indicated on the plans to make all conditions clear

(ff) For hydraulically designed systems, the material to be included on the hydraulic data nameplate

(gg) Name and address of contractor.

Formal Interpretation

Question: Is it the intent of 1-9.2(z) that hanger locations be dimensioned?

Answer: No, but hanger locations are required to be in accordance with 3-15.5 and 3-15.6.

1-10 Approval of Sprinkler Systems.

1-10.1   The installer shall perform all required acceptance tests (*see Section 1-11*), complete the Contractor's Material and Test Certificate(s) (*see Section 1-12*), and forward the certificate(s) to the authority having jurisdiction prior to asking for approval of the installation.

Material and test certificates, one for aboveground installations and one for underground installations, are shown in Section 1-12. Occasionally the aboveground and underground work is performed by separate contractors. To accommodate that event a certificate is provided for each circumstance.

The certificates are acknowledgments that materials used and tests made are in accordance with the requirements of this standard. They also provide a record of the test results that can be used for comparison with tests that will or may take place in the future.

1-10.2   When the authority having jurisdiction desires to be present during the conduct of acceptance tests, the installer shall give advance notification of the time and date the testing will be performed.

Occasionally the authority having jurisdiction wishes to be present at the time performance tests are conducted. It is the responsibility of the authority having jurisdiction to inform the installer that he desires to be present. This requirement then obligates the installer to comply with such a request and to give the authority having jurisdiction advance notice. Failure to do so may make it necessary to re-run tests.

1-11 Acceptance Tests.

1-11.1* Flushing of Underground Connections.

A-1-11.1

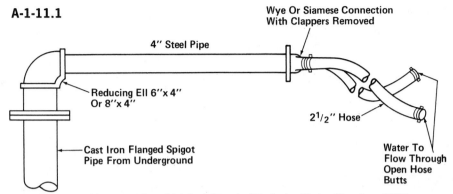

Employing Horizontal Run Of 4-Inch Pipe And Reducing Fitting Near Base Of Riser.

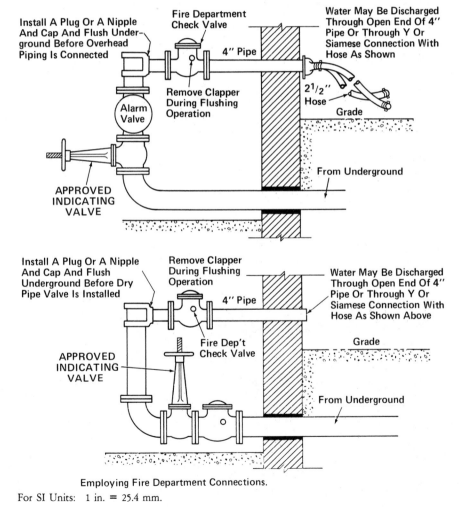

Employing Fire Department Connections.

For SI Units:　1 in. = 25.4 mm.

Figure A-1-11.1　Methods of Flushing Water Supply Connections.

1-11.1.1   Underground mains and lead-in connections to system risers shall be flushed before connection is made to sprinkler piping in order to remove foreign materials which may have entered the underground piping during the course of the installation. For all systems, the flushing operation shall be continued until water is clear.

Experience has shown that stones, gravel, blocks of wood, bottles, work tools, work clothes and other objects have been found in piping when flushing was performed. Also, objects in underground piping quite remote from the sprinkler installation, that otherwise would remain stationary, will sometimes be transported into sprinkler system piping when sprinkler systems operate. Sprinkler systems can draw greater flows than normal for domestic or process uses. Fire department pumpers when taking suction from hydrants for pumping into sprinkler systems may compound the problem by increasing the velocity of water flow in underground piping.

Because of the inherent nature of sprinkler system design in which pipe sizes diminish from the point of connection of underground piping, objects that move from underground piping and enter sprinkler system risers will lodge at the point in the system where they may totally obstruct passage of water.

1-11.1.2*   Underground mains and lead-in connections shall be flushed at a flow rate not less than indicated in Table 1-11.1.2 or at the hydraulically calculated water demand rate of the system, whichever is greater.

Table 1-11.1.2

Pipe Size	Flow Rate	L/min.
4 in.	400 gpm	1514
5 in.	600 gpm	2271
6 in.	750 gpm	2839
8 in.	1000 gpm	3785
10 in.	1500 gpm	5678
12 in.	2000 gpm	7570

Exception No. 1:   When the water supply will not produce the stipulated flow rate, connections to a hydraulically designed system may be flushed at the demand rate of the system, including hose streams if hose or hydrants or both are supplied from that connection.

Exception No. 2:   For pipe schedule systems, when the water supply will not produce the stipulated flow rate, the maximum flow rate available shall be used.

Based upon a study by Factory Mutual Research it has been determined that the size of particles that will move upward in piped water streams can be determined if the specific gravity of the particle and velocity of the water stream are known. For example: granite, 2 inches in diameter, will move upward in piping if the water stream velocity is 5.64 feet per second (1.72 m/s). All of the flow rates shown in Table 1-11.1.2 will develop velocities equal to or greater than 5.64 fps (1.72 m/s). The flow rates in the table have been considered adequate for sprinkler system installations in which pipe sizing was based upon the pipe schedules, as shown in Chapter 3. With the advent of hydraulically calculated system design the number of sprinklers expected to operate in a fire situation at a given density (gpm/ft^2) [(L/min)/m^2] can result in flows that exceed the rates in the table. For this reason, if the hydraulically calculated water demand exceeds the values in Table 1-11.1.2, the higher values should be used to ensure proper flushing.

Exception No. 1: In the case that the demand flow of a hydraulically calculated system is less than that in Table 1-11.1.2 for any given pipe size, the calculated rate, including rates designed for hose or hydrants, may be used for flushing. The reason for this exception is that, if only the design flow rate is anticipated and it is less than those of the table, it is not logical to require that this flow in Table 1-11.1.2 be required.

Exception No. 2: The reason for not requiring the table flow rates when they are not available is the same as explained in Exception No. 1.

It should be recognized that overall field experience is such that actual flushing is normally done without a measurement of the flow rate. Flushing is normally accomplished at the maximum flow rate available from the water supply.

A-1-11.1.2 Underground mains and lead-in connections to system risers should be flushed through hydrants at dead ends of the system or through accessible aboveground flushing outlets allowing the water to run until clear. If water is supplied from more than one source or from a looped system, divisional valves should be closed to produce a high velocity flow through each single line. The flows specified in Table 1-11.1.2 will produce a velocity of at least 6 ft/sec (1.8 m/s) which is necessary for cleaning the pipe and for lifting foreign material to an aboveground flushing outlet.

1-11.1.3 Provision shall be made for the disposal of water issuing from test outlets to avoid property damage.

1-11.2 Hydrostatic Tests.

1-11.2.1* All new systems including yard piping shall be hydrostatically tested at not less than 200 psi (13.8 bars) pressure for 2 hours, or at 50 psi (3.4 bars) in excess of the maximum pressure, when the maximum pressure to be maintained in the system is in excess of 150 psi (10.3 bars).

The test pressure shall be read from a gage located at the low elevation point of the individual system or portion of the system being tested.

Exception: At seasons of the year that will not permit testing with water an interim test may be conducted with air pressure of at least 40 psi (2.8 bars) allowed to stand for 24 hours. The standard hydrostatic test shall be conducted when weather permits.

A-1-11.2.1 Example. A sprinkler system has for its water supply a connection to a public water service main. A 100 psi (6.9 bars) rated pump is installed in the connection. With a maximum normal public water supply of 70 psi (4.8 bars) at the low elevation point of the individual system or portion of the system being tested and a 120 psi (8.3 bars) pump (churn) pressure, the hydrostatic test pressure is 70 + 120 + 50 or 240 psi (16.5 bars).

Systems that have been modified or repaired to any appreciable extent should be hydrostatically tested at not less than 50 psi (3.4 bars) in excess of normal static pressure for 2 hours.

To reduce the possibility of serious water damage in case of a break, pressure may be maintained by a small pump, the main controlling gate meanwhile being kept shut during test.

Formal Interpretation

Question: Is hydro-pneumatic the substitute or equal of hydrostatic procedure according to Testing Laboratory or NFPA Standards?

Answer: No. The NFPA Standard for Installation of Automatic Sprinkler Systems requires hydrostatic pressure testing of new systems under 1-11.2. During seasons in which freezing may occur, interim testing of dry pipe systems with air is required with the customary hydrostatic testing required when weather permits in accordance with 1-11.2.1.

Hydro-pneumatic procedures are associated with flushing of foreign materials that may obstruct waterways in piping and not with the pressure tests required for leakage testing in new systems.

Formal Interpretation

Question: Is it the intent of 1-11.2.1 and 1-11.2.2 that the sprinkler pipings at the top of a system fed by a riser 105 ft in height be pressurized to 200 psi for 2 hours when the sprinkler piping and riser are considered as "the individual system or zone"?

Answer: No.

Formal Interpretation

Question: In determining hydrostatic test pressure, when a booster pump is installed, should the shut off pressure or normal operating pressure be considered in determining hydrostatic test pressure? If a relief valve is installed, should the relief valve setting be considered in determining the hydrostatic pressure test?

Answer: The hydrostatic pressure should, in the case of new installations, be based upon the maximum pressure exerted upon the new system piping and components. When a booster pump is provided, the maximum pressure is shut off pressure or "churn" pressure without consideration of relief valve pressure settings.

All new systems are tested hydrostatically to at least 200 psi (13.8 bars). This standard is set to ensure pipe joints are made up to withstand that pressure without coming apart or leaking. It is primarily a workmanship test and not a materials performance test. However, damaged materials (cracked fittings, leaky sprinklers, etc.) are routinely discovered during the hydrostatic test. As noted previously, all materials used must be rated for at least 175 psi (12.1 bars) working pressure, the maximum pressure at which the system is normally maintained.

The measurement of the hydrostatic test pressure is taken at the lowest elevation within the system or portion of the system being tested. It is not considered necessary to test at the high point of the system (which, due to static head, would increase the test pressure significantly) due to the fact that application of pressure typically occurs at the lower elevation, and these high pressures would not be anticipated at the higher elevations within the system.

The air test for dry-pipe systems in 1-11.3.2 is permitted, by the Exception to 1-11.2.1, as an interim substitution for the hydrostatic test when there is a danger of freezing water in the system. The standard hydrostatic test must be made, however, when it is possible to do so without danger of freezing.

The measure of success for inside sprinkler piping under hydrostatic test is that there is no visible leakage.

1-11.2.2* Permissible Leakage.   The inside sprinkler piping shall be installed in such a manner that there will be no visible leakage when the system is subjected to the hydrostatic pressure test. Refer to NFPA 24, *Installation of Private Fire Service Mains and Their Appurtenances*, for permissible leakage in underground piping. The amount of leakage shall be measured by pumping from a calibrated container.

A-1-11.2.2   Valves isolating the section to be tested may not be "drop tight." When such leakage is suspected, test blanks of the type recommended in 1-11.2.5 should be used in a manner that includes the valve in the section being tested.

1-11.2.3 Fire Department Connection.   Piping between the check valve in the fire department inlet pipe and the outside connection shall be tested the same as the balance of the system.

This section is intended as a reminder that the fire department pumping connection is required to be tested. The fact that the piping from the check valve to the hose connections is not normally subjected to water pressure may lead the contractor to believe testing of that portion is not required.

1-11.2.4 Corrosive Chemicals.   Brine or other corrosive chemicals shall not be used for testing systems.

1-11.2.5 Test Blanks.   Whenever a test blank is used it shall be of the self-indicating type. Test blanks shall have red painted lugs protruding beyond the flange in such a way as to clearly indicate their presence. The installer shall have all test blanks numbered so as to keep track of their use and assure their removal after the work is completed.

It is extremely important that all test blanks are removed. A test blank without a protruding lug is virtually impossible to detect by visual inspection.

1-11.3 Test of Dry-Pipe Systems.

1-11.3.1 Differential Dry-Pipe Valves.   The clapper of a differential type dry-pipe valve shall be held off its seat during any test in excess of 50 psi (3.4 bars) to prevent damaging the valve.

Due to the design of differential dry-pipe valves in which flexible gaskets are used to seal water from entering the system, it has been found that pressures in excess of 50 psi (3.4 bars) may cause damage to the mechanism of the valve.

1-11.3.2 Air Test.   In dry-pipe systems an air pressure of 40 psi (2.8 bars) shall be pumped up, allowed to stand 24 hours, and all leaks which allow a loss of pressure over 1½ psi (0.1 bar) for the 24 hours shall be stopped.

Formal Interpretation

Question: If a new dry-pipe system is hydrostatically tested in accordance with 1-11.2.1 is it also required to conduct an air test of the system in accordance with 1-11.3.2?

Answer: Yes, both tests are required. The hydrostatic test is for visible signs of water leakage providing watertight integrity, while the air test, limiting air pressure loss to 1½ psi minimum over a twenty-four hour period, is for proof of airtight integrity. One test does not guarantee that the system will perform satisfactorily when subjected to the other test.

The air test is a test for airtight integrity of the system. Under normal conditions dry-pipe systems are subjected to lower pressures than those when the system has tripped and filled with water. Since dry systems are subjected to pressures from both air and water they must be pressure tested by both air and water. Air and water characteristic differences make it impossible to judge system integrity based upon one and only one of these tests. The 200 psi water and 40 psi air tests are considered the extreme pressures for normal use in systems.

1-11.3.3 Operating Test of Dry-Pipe Valve.   A working test of the dry-pipe valve alone and with quick opening device, if installed, shall be made before acceptance by opening the system test pipe. Trip and water delivery times shall be measured from time inspector's test pipe is opened and shall be recorded using the Contractor's Material and Test Certificate for Aboveground Piping.

Full operational tests of dry-pipe valves are required. The Contractor's Material and Test Certificate has provisions for recording the time it takes for the valve to trip after opening the system test valve, the air and water pressures before the test, the air pressure at trip point, the delivery time for water to reach the test outlet, and whether or not alarms operated properly. The valve should operate in accordance with the manufacturer's specifications for the trip point. Delivery timing for water to reach the test outlet should be in accordance with 5-4.5.

1-11.4 Tests of Drainage Facilities.   Tests of drainage facilities shall be made while the control valve is wide open. The main drain valve shall be opened and remain open until the system pressure stabilizes. (*See 2-9.1.*)

1-12 Contractor's Material and Test Certificates.

CONTRACTOR'S MATERIAL & TEST CERTIFICATE FOR **A**BOVEGROUND PIPING

PROCEDURE

Upon completion of work, inspection and tests shall be made by the contractor's representative and witnessed by an owner's representative. All defects shall be corrected and system left in service before contractor's personnel finally leave the job.

A certificate shall be filled out and signed by both representatives. Copies shall be prepared for approving authorities, owners and contractor. It is understood the owner's representative's signature in no way prejudices any claim against contractor for faulty material, poor workmanship, or failure to comply with approving authority's requirements or local ordinances.

PROPERTY NAME	DATE

PROPERTY ADDRESS

PLANS	ACCEPTED BY APPROVING AUTHORITY('S) NAMES ADDRESS INSTALLATION CONFORMS TO ACCEPTED PLANS ☐ YES ☐ NO EQUIPMENT USED IS APPROVED ☐ YES ☐ NO IF NO, EXPLAIN DEVIATIONS
INSTRUCTIONS	HAS PERSON IN CHARGE OF FIRE EQUIPMENT BEEN INSTRUCTED AS TO LOCATION OF CONTROL VALVES AND CARE AND MAINTENANCE OF THIS NEW EQUIPMENT ☐ YES ☐ NO IF NO, EXPLAIN HAVE COPIES OF APPROPRIATE INSTRUCTIONS AND CARE AND MAINTENANCE CHARTS AND NFPA 13A BEEN LEFT ON PREMISES ☐ YES ☐ NO IF NO, EXPLAIN
LOCATION OF SYSTEM	SUPPLIES BLDGS.

	MAKE	MODEL	YEAR OF MANUFACTURE	ORIFICE SIZE	QUANTITY	TEMPERATURE RATING
SPRINKLERS						

PIPE AND FITTINGS	PIPE CONFORMS TO _____ STANDARD ☐ YES ☐ NO FITTINGS CONFORM TO _____ STANDARD ☐ YES ☐ NO IF NO, EXPLAIN

	ALARM DEVICE			MAXIMUM TIME TO OPERATE THROUGH TEST PIPE	
ALARM VALVE OR FLOW INDICATOR	TYPE	MAKE	MODEL	MIN.	SEC.

	DRY VALVE			Q.O.D.		
	MAKE	MODEL	SERIAL NO.	MAKE	MODEL	SERIAL NO.

DRY PIPE OPERATING TEST		TIME TO TRIP THRU TEST PIPE*		WATER PRESSURE	AIR PRESSURE	TRIP POINT AIR PRESSURE	TIME WATER REACHED TEST OUTLET*		ALARM OPERATED PROPERLY	
		MIN.	SEC.	PSI	PSI	PSI	MIN.	SEC.	YES	NO
	Without Q.O.D.									
	With Q.O.D.									
	IF NO, EXPLAIN									

*MEASURED FROM TIME INSPECTOR'S TEST PIPE IS OPENED.

85A (10-80)   PRINTED IN USA   (OVER)

Contractor's Material & Test Certificate for Aboveground Piping

DELUGE & PREACTION VALVES	OPERATION ☐ PNEUMATIC ☐ ELECTRIC ☐ HYDRAULIC								

	OPERATION ☐ PNEUMATIC ☐ ELECTRIC ☐ HYDRAULIC
DELUGE & PREACTION VALVES	PIPING SUPERVISED ☐ YES ☐ NO DETECTING MEDIA SUPERVISED ☐ YES ☐ NO
	DOES VALVE OPERATE FROM THE MANUAL TRIP AND/OR REMOTE CONTROL STATIONS ☐ YES ☐ NO
	IS THERE AN ACCESSIBLE FACILITY IN EACH CIRCUIT FOR TESTING IF NO, EXPLAIN ☐ YES ☐ NO

	MAKE	MODEL	DOES EACH CIRCUIT OPERATE SUPERVISION LOSS ALARM		DOES EACH CIRCUIT OPERATE VALVE RELEASE		MAXIMUM TIME TO OPERATE RELEASE	
			YES	NO	YES	NO	MIN.	SEC.

TEST DESCRIPTION	<u>HYDROSTATIC:</u> Hydrostatic tests shall be made at not less than 200 psi (13.6 bars) for two hours or 50 psi (3.4 bars) above static pressure in excess of 150 psi (10.2 bars) for two hours. Differential dry-pipe valve clappers shall be left open during test to prevent damage. All aboveground piping leakage shall be stopped. <u>FLUSHING:</u> Flow the required rate until water is clear as indicated by no collection of foreign material in burlap bags at outlets such as hydrants and blow-offs. Flush at flows not less than 400 GPM (1514 L/min) for 4-inch pipe, 600 GPM (2271 L/min) for 5-inch pipe, 750 GPM (2839 L/min) for 6-inch pipe, 1000 GPM (3785 L/min) for 8-inch pipe, 1500 GPM (5678 L/min) for 10-inch pipe and 2000 GPM (7570 L/min) for 12-inch pipe. When supply cannot produce stipulated flow rates, obtain maximum available. <u>PNEUMATIC:</u> Establish 40 psi (2.7 bars) air pressure and measure drop which shall not exceed 1-½ psi (0.1 bars) in 24 hours. Test pressure tanks at normal water level and air pressure and measure drop which shall not exceed 1-½ psi (0.1 bars) in 24 hours.

TESTS	ALL PIPING HYDROSTATICALLY TESTED AT _____ PSI FOR _____ HRS. IF NO, STATE REASON
	DRY PIPING PNEUMATICALLY TESTED ☐ YES ☐ NO
	EQUIPMENT OPERATES PROPERLY ☐ YES ☐ NO

	DRAIN TEST	READING OF GAGE LOCATED NEAR WATER SUPPLY TEST PIPE: STATIC PRESSURE: _____ PSI	RESIDUAL PRESSURE WITH VALVE IN TEST PIPE OPEN WIDE _____ PSI

Underground mains and lead in connections to system risers flushed before connection made to sprinkler piping.	
VERIFIED BY COPY OF THE U FORM NO. 85B ☐ YES ☐ NO	OTHER EXPLAIN
FLUSHED BY INSTALLER OF UNDER-GROUND SPRINKLER PIPING ☐ YES ☐ NO	

BLANK TESTING GASKETS	NUMBER USED	LOCATIONS	NUMBER REMOVED

WELDING	WELDED PIPING ☐ YES ☐ NO
	IF YES ...
	DO YOU CERTIFY AS THE SPRINKLER CONTRACTOR THAT WELDING PROCEDURES COMPLY WITH THE REQUIREMENTS OF AT LEAST AWS D10.9, LEVEL AR-3 ☐ YES ☐ NO
	DO YOU CERTIFY THAT THE WELDING WAS PERFORMED BY WELDERS QUALIFIED IN COMPLIANCE WITH THE REQUIREMENTS OF AT LEAST AWS D10.9, LEVEL AR-3 ☐ YES ☐ NO
	DO YOU CERTIFY THAT WELDING WAS CARRIED OUT IN COMPLIANCE WITH A DOCUMENTED QUALITY CONTROL PROCEDURE TO INSURE THAT ALL DISCS ARE RETRIEVED, THAT OPENINGS IN PIPING ARE SMOOTH, THAT SLAG AND OTHER WELDING RESIDUE ARE REMOVED, AND THAT THE INTERNAL DIAMETERS OF PIPING ARE NOT PENETRATED ☐ YES ☐ NO

HYDRAULIC DATA NAMEPLATE	NAMEPLATE PROVIDED ☐ YES ☐ NO IF NO, EXPLAIN

REMARKS	DATE LEFT IN SERVICE WITH ALL CONTROL VALVES OPEN:

SIGNATURES	NAME OF SPRINKLER CONTRACTOR
	TESTS WITNESSED BY

	FOR PROPERTY OWNER (SIGNED)	TITLE	DATE
	FOR SPRINKLER CONTRACTOR (SIGNED)	TITLE	DATE

ADDITIONAL EXPLANATION AND NOTES

Contractor's Material & Test Certificate for Aboveground Piping

CONTRACTOR'S MATERIAL & TEST CERTIFICATE FOR U**NDERGROUND PIPING**

PROCEDURE

Upon completion of work, inspection and tests shall be made by the contractor's representative and witnessed by an owner's representative. All defects shall be corrected and system left in service before contractor's personnel finally leave the job.

A certificate shall be filled out and signed by both representatives. Copies shall be prepared for approving authorities, owners and contractor. It is understood the owner's representative's signature in no way prejudices any claim against contractor for faulty material, poor workmanship, or failure to comply with approving authority's requirements or local ordinances.

PROPERTY NAME	DATE

PROPERTY ADDRESS

	ACCEPTED BY APPROVING AUTHORITY('S) NAMES	
	ADDRESS	
PLANS	INSTALLATION CONFORMS TO ACCEPTED PLANS	☐YES ☐NO
	EQUIPMENT USED IS APPROVED	☐YES ☐NO
	IF NO, STATE DEVIATIONS	
	HAS PERSON IN CHARGE OF FIRE EQUIPMENT BEEN INSTRUCTED AS TO LOCATION OF CONTROL VALVES AND CARE AND MAINTENANCE OF THIS NEW EQUIPMENT IF NO, EXPLAIN	☐YES ☐NO
INSTRUCTIONS	HAVE COPIES OF APPROPRIATE INSTRUCTIONS AND CARE AND MAINTENANCE CHARTS BEEN LEFT ON PREMISES IF NO, EXPLAIN	☐YES ☐NO
LOCATION	SUPPLIES BLDGS.	

	PIPE TYPES AND CLASS	TYPE JOINT	
UNDERGROUND PIPES AND JOINTS	PIPE CONFORMS TO _____ STANDARD		☐YES ☐NO
	FITTINGS CONFORM TO _____ STANDARD		☐YES ☐NO
	IF NO, EXPLAIN		
	JOINTS NEEDING ANCHORAGE CLAMPED, STRAPPED, OR BLOCKED IN		☐YES ☐NO
	ACCORDANCE WITH _____ STANDARD IF NO, EXPLAIN		

TEST DESCRIPTION	FLUSHING. Flow the required rate until water is clear as indicated by no collection of foreign material in burlap bags at outlets such as hydrants and blow-offs. Flush at flow not less than 400 GPM (1514 L/min) for 4-inch pipe, 600 GPM (2271 L/min) for 5-inch pipe, 750 GPM (2839 L/min) for 6-inch pipe, 1000 GPM (3785 L/min) for 8-inch pipe, 1500 GPM (5678 L/min) for 10-inch pipe and 2000 GPM (7570 L/min) for 12-inch pipe. When supply cannot produce stipulated flow rates, obtain maximum available. HYDROSTATIC. Hydrostatic tests shall be made at not less than 200 psi (13.8 bars) for two hours or 50 psi (3.4 bars) above static pressure in excess of 150 psi (10.3 bars) for two hours. LEAKAGE. New pipe laid with rubber gasketed joints shall, if the workmanship is satisfactory, have little or no leakage at the joints. The amount of leakage at the joints shall not exceed 2 qts. per hr. (1.89 L/h) per 100 joints irrespective of pipe diameter. The leakage shall be distributed over all joints. If such leakage occurs at a few joints the installation shall be considered unsatisfactory and necessary repairs made. The amount of allowable leakage specified above may be increased by 1 fl oz per in. valve diameter per hour (30 mL/25 mm/h) for each metal seated valve isolating the test section. If dry barrel hydrants are tested with the main valve open, so the hydrants are under pressure, an additional 5 oz per minute (150 mL/min) leakage is permitted for each hydrant.

	NEW UNDERGROUND PIPING FLUSHED ACCORDING TO _____ STANDARD		☐YES ☐NO
	BY (COMPANY) IF NO, EXPLAIN		
	HOW FLUSHING FLOW WAS OBTAINED	THROUGH WHAT TYPE OPENING	
FLUSHING TESTS	☐PUBLIC WATER ☐TANK OR RESERVOIR ☐FIRE PUMP	☐HYDRANT BUTT.	☐OPEN PIPE
	LEAD-INS FLUSHED ACCORDING TO _____ STANDARD		☐YES ☐NO
	BY (COMPANY) IF NO, EXPLAIN		
	HOW FLUSHING FLOW WAS OBTAINED	THROUGH WHAT TYPE OPENING	
	☐PUBLIC WATER ☐TANK OR RESERVOIR ☐FIRE PUMP	☐ Y CONN. TO FLANGE & SPIGOT	☐OPEN PIPE

85B(10-80)   PRINTED IN USA   (OVER)

Contractor's Material & Test Certificate for Underground Piping

HYDROSTATIC TEST	ALL NEW UNDERGROUND PIPING HYDROSTATICALLY TESTED AT _____ PSI FOR _____ HOURS	JOINTS COVERED ☐ YES ☐ NO		
LEAKAGE TEST	TOTAL AMOUNT OF LEAKAGE MEASURED _____ GALS. _____ HOURS ALLOWABLE LEAKAGE _____ GALS. _____ HOURS			
HYDRANTS	NUMBER INSTALLED TYPE AND MAKE		ALL OPERATE SATISFACTORILY ☐ YES ☐ NO	
CONTROL VALVES	WATER CONTROL VALVES LEFT WIDE OPEN IF NO, STATE REASON	☐ YES ☐ NO		
	HOSE THREADS OF FIRE DEPARTMENT CONNECTIONS AND HYDRANTS INTERCHANGEABLE WITH THOSE OF FIRE DEPARTMENT ANSWERING ALARM	☐ YES ☐ NO		
REMARKS	DATE LEFT IN SERVICE			
SIGNATURES	NAME OF INSTALLING CONTRACTOR			
	TESTS WITNESSED BY			
	FOR PROPERTY OWNER (SIGNED)	TITLE		DATE
	FOR INSTALLING CONTRACTOR (SIGNED)	TITLE		DATE

ADDITIONAL EXPLANATION AND NOTES

85B BACK

Contractor's Material & Test Certificate for Underground Piping

1-13 Operation of Sprinkler System Control Valves by Contractors.
When work on a sprinkler system requires that a contractor operate a valve controlling water supplies to a sprinkler system, the contractor shall inform the owner so that the owner may follow the normal valve supervision procedure.

1-14 Units. Metric units of measurement in this standard are in accordance with the modernized metric system known as the International System of Units (SI). Two units (liter and bar), outside of but recognized by SI, are commonly used in international fire protection. These units are listed in Table 1-14 with conversion factors.

Table 1-14

Name of Unit	Unit Symbol	Conversion Factor
liter	L	1 gal = 3.785 L
liter per minute per square meter	$(L/min)/m^2$	1 gpm/ft^2 = 40.746 (L/min)/m^2
millimeter per minute	1 mm/min	1 gpm/ft^2 = 40.746 mm/min
cubic decimeter	dm^3	1 gal = 3.785 dm^3
pascal	Pa	1 psi = 6894.757 Pa
bar	bar	1 psi = 0.0689 bar
bar	bar	1 bar = 10^5 Pa

For additional conversions and information see ASTM E380, *Standard for Metric Practice.*

1-14.1 If a value for measurement as given in this standard is followed by an equivalent value in other units, the first stated is to be regarded as the requirement. A given equivalent value may be approximate.

When and if the United States converts to SI, some of the values will probably be rounded off to the nearest whole number. A value such as 0.15 gpm/ft^2 would become 6.0(L/min)/m^2 or 6.0 mm/min instead of 6.1119(L/min)/m^2 or 6.1119 mm/min.

1-14.2 The conversion procedure for the SI units has been to multiply the quantity by the conversion factor and then round the result to the appropriate number of significant digits.

2
WATER SUPPLIES

2-1* General Provisions. Every automatic sprinkler system shall have at least one automatic water supply.

An "automatic" supply is one that is not dependent on any manual operation such as making connections, operating valves or starting pumps to supply water at the time of a fire.

A-2-1 Water supplies should have adequate pressure, capacity and reliability.

The water supply needed for various occupancies, including Extra Hazard Occupancies, is determined by evaluating the number of sprinklers which may be expected to operate from any one fire plus quantities needed simultaneously for hose streams.

Determination of the water supply needed for Extra Hazard Occupancies will require special consideration of the four factors: (1) area of sprinkler operation, (2) density of discharge, (3) required time of discharge, and (4) amount of water needed simultaneously for hose streams.

When the occupancy presents a possibility of intense fires requiring extra heavy discharge, this may be obtained by an increase in the pressure and volume of the water supply, by a closer spacing of sprinklers, by the use of larger pipe sizing, or by a combination of these methods. In such cases, consideration should be given to hydraulically designed systems. (*See Chapter 7.*)

When separately published standards on various subjects contain specific provisions for water supplies, these should be consulted. (*See Appendix D for availability of standards.*)

B-2 Water Supplies.

B-2-1 Testing of Water Supply.

B-2-1.1 To determine the value of public water as a supply for automatic sprinkler systems, it is generally necessary to make a flow test to determine how much water can be discharged at a residual pressure at rate sufficient to give the required residual pressure under the roof (with the volume flow hydraulically translated to the base of the riser) — i.e., a pressure head represented by the height of the building plus the required residual pressure.

Fire flow tests are the subject of NFPA 291, *Recommended Practice for Fire Flow Testing and Marking of Hydrants*, and are also covered in "Fire Flow Tests — Discharge Tables for Circular Outlets — Friction Losses in Pipes" published by NFPA. Also see A-2-3.1.1.

B-2-1.2   The proper method of making such test is to use two hydrants in the vicinity of the property. The static pressure should be measured on the hydrant in front of or nearest to the property and the water allowed to flow from the hydrant next nearest the property, preferably the one farthest from the source of supply if main is fed only one way. The residual pressure will be that indicated at the hydrant where water is not flowing.

B-2-1.3   Referring to Figure B-2, the method of conducting the flow tests is as follows:

1. Attach gage to hydrant (A) and obtain static pressure.

2. Either attach second gage to hydrant (B) or use pitot tube at outlet. Have hydrant (B) opened wide and read pressure at both hydrants.

3. Use the pressure at (B) to compute the gallons flowing and read the gage on (A) to determine the residual pressure or that which will be available on the top line of sprinklers in the property.

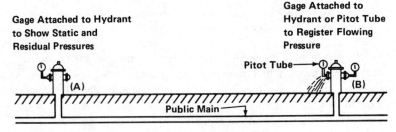

Figure B-2   Method of Conducting Flow Tests.

B-2-1.4   Water pressure in pounds for a given height in feet equals height multiplied by 0.434.

B-2-1.5   In making flow tests, whether from hydrants or from nozzles attached to hose, always measure the size of the orifice. While hydrant outlets are usually 2½ in. they are sometimes smaller and occasionally larger. The Underwriter's play pipe is 1⅛ in. and 1¾ in. with tip removed, but occasionally nozzles will be 1 in. or 1¼ in. and with the tip removed the opening may be only 1½ in.

B-2-1.6   The pitot tube should be held approximately one-half the diameter of the hydrant or nozzle opening away from the opening. It should be held in the center of the stream, except that in using hydrant outlets the stream should be explored to get the average pressure.

2-2* Water Supply Requirements for Sprinkler Systems.

A-2-2   The water supply requirement for sprinkler protection is determined by the number of sprinklers expected to operate in event of fire. The primary factors affecting the number of sprinklers which might open are:

1. Occupancy

2. Combustibility of contents

3. Area shielded from proper distribution of water

4. Height of stock piles

5. Combustibility of construction (ceilings and blind spaces)

6. Ceiling heights and draft conditions

7. Horizontal and vertical cutoffs

8. Wet or dry sprinkler system

9. High water pressure

10. Housekeeping

11. Temperature rating of sprinklers

12. Water flow alarm and response thereto.

Many of these items may be interdependent, but all are individual factors which, if adverse, affect the ability of sprinklers to control fire in its incipient stage. The proper design of systems, including density and operating areas, normally considers Items 1, 2, 4, 8, 9, 11 and 12. However, the potential of operating all sprinklers in a room or compartment is very real when some of the other serious factors, such as 3, 5, 6, 7 and 10, are considered.

B-2-2 Interconnection of Water Supplies.

B-2-2.1   All main water supplies should be connected with the sprinkler system at the base of riser, except that where a gravity or pressure tank or both constitute the only automatic source of water supply, permission may be given to connect the tank or tanks with the sprinkler system at the top of the riser.

B-2-2.2   Where a gravity tank and a pressure tank are connected to a common riser, approved means should be provided to prevent residual air pressure in the pressure tank (after water has been drained from it) from holding the gravity tank check valve closed, a condition known as air lock. Under normal conditions, air lock may be conveniently prevented in new equipment by connecting the gravity tank and pressure tank discharge pipes together 45 ft (13.7 m) or more below the bottom of the gravity tank and placing the gravity tank check valve at the level of this connection.

2-2.1 Water Supply Requirement Tables.

2-2.1.1*   Water supply requirement tables shall be used in determining the minimum water supply requirements for light, ordinary and extra hazard occupancies. Occupancy classification shall be determined from Section 1-7.

This section covers the minimum requirements and minimum volume and pressure for normal or average occupancies within the types classified. Many occupancies may have a need for increased requirements, especially when occupancy hazards are above normal or a very high degree of reliability is desired.

(a) Table 2-2.1(A) is used to determine the minimum volume of water and pressure normally required for a pipe schedule sprinkler system. THE TABLE IS TO BE USED ONLY WITH EXPERIENCED JUDGMENT.

(b) Table 2-2.1(B) is used to determine the minimum volume of water and pressure normally required for a hydraulically designed sprinkler system.

Formal Interpretation

Question: When using Table 2-2.1(B) for calculating the minimum water supply for sprinklers only for buildings having individual room areas less than 1500 sq ft and all rooms constructed in accordance with the requirements set forth in 2-2.1.2.8, can areas of sprinkler operation less than 1500 sq ft be used?

Answer: Yes. The minimum water supply requirement for the above described conditions is then calculated on the basis of the area of the largest room; however, the sprinkler discharge density for 1500 sq ft must be used (see 2-2.1.2.9) provided it is not a dry pipe system (see 2-2.1.2.10) and there are no unsprinklered combustible spaces (see 2-2.1.2.11).

Table 2-2.1(A)   Guide to Water Supply Requirements for Pipe Schedule Sprinkler Systems

Occupancy Classification	Residual Pressure Required (see Note 1)	Acceptable Flow at Base of Riser (see Note 2)	Duration in Minutes (see Note 4)
Light Hazard	15 psi	500-750 gpm (see Note 3)	30-60
Ordinary Hazard (Group 1)	15 psi or higher	700-1000 gpm	60-90
Ordinary Hazard (Group 2)	15 psi or higher	850-1500 gpm	60-90
Ordinary Hazard (Group 3)	Pressure and flow requirements for sprinklers and hose streams to be determined by authority having jurisdiction.		60-120
High-Piled Storage (see 4-1.3.9)	Pressure and flow requirements for sprinklers and hose streams to be determined by authority having jurisdiction. See Chapter 7 and NFPA 231 and NFPA 231C.		
High-Rise Buildings	Pressure and flow requirements for sprinklers and hose streams to be determined by authority having jurisdiction. See Chapter 8.		
Extra Hazard	Pressure and flow requirements for sprinklers and hose streams to be determined by authority having jurisdiction.		

For SI Units: 1 psi = 0.0689 bar; 1 gpm = 3.785 L/min.

Notes:

1. The pressure required at the base of the sprinkler riser(s) is defined as the residual pressure required at the elevation of the highest sprinkler plus the pressure required to reach this elevation.

2. The lower figure is the minimum flow including hose streams ordinarily acceptable for pipe schedule sprinkler systems. The higher flow should normally suffice for all cases under each group.

3. The requirement may be reduced to 250 gpm if building area is limited by size or compartmentation or if building (including roof) is noncombustible construction.

4. The lower duration figure is ordinarily acceptable where remote station water flow alarm service or equivalent is provided. The higher duration figure should normally suffice for all cases under each group.

Table 2-2.1(B)   Table and Design Curves for Determining Density, Area of Sprinkler Operation and Water Supply Requirements for Hydraulically Designed Sprinkler Systems

Minimum Water Supply Requirements

Hazard Classification	Sprinklers Only — gpm	Inside Hose — gpm	Total Combined Inside and Outside Hose — gpm	Duration in Minutes
Light	See 2-2.1.2.1	0, 50 or 100	100	30
Ord. — Gp. 1	See 2-2.1.2.1	0, 50 or 100	250	60-90
Ord. — Gp. 2	See 2-2.1.2.1	0, 50 or 100	250	60-90
Ord. — Gp. 3	See 2-2.1.2.1	0, 50 or 100	500	60-120
Ex. Haz. — Gp. 1	See 2-2.1.2.1	0, 50 or 100	500	90-120
Ex. Haz. — Gp. 2	See 2-2.1.2.1	0, 50 or 100	1000	120

For SI Units: 1 gpm=3.785 L/min.

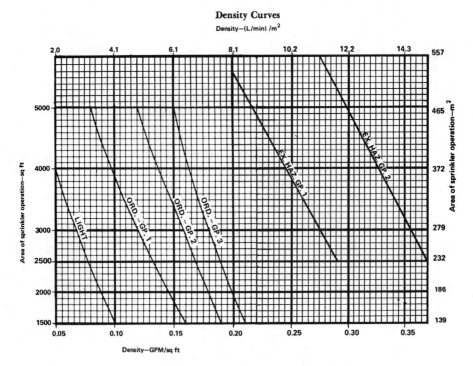

Density Curves

For SI Units: 1 sq ft = 0.0920 m²; 1 gpm/sq ft = 40.746 (L/min)/m².

Figure 2-2.1(B)

When hydraulically designing a sprinkler system, any point on the design curve is satisfactory for an adequate sprinkler system. At the lower end of the curve, the density and therefore the required pressure are greater but the total water required is less than at the upper end. Designs at the lower end of the curve are usually more economical and provide a greater degree of fire control. Where pressures available are lower or a larger number of sprinklers are expected to open, it might be advantageous to design for a larger water supply and lower pressure.

A-2-2.1.1   For occupancies which have the potential for fast-spreading fire due to the presence of lint, combustible residue, combustible hydraulic fluids under high pressure with ignition sources nearby, etc., the minimum area of operation should encompass the entire area likely to be involved in such a fire.

2-2.1.2   The following shall be used in applying Table 2-2.1(B).

2-2.1.2.1   The densities and areas provided in Figure 2-2.1(B) are based on the use of standard response, standard orifice (½ in.) and large orifice (¹⁷⁄₃₂ in.) sprinklers. For use of other types of sprinklers see 4-1.1.3.

The densities and areas shown in Figure 2-2.1(B) should not be used for extra large orifice and large-drop sprinklers. See Chapter 9 for the requirements for large-drop sprinklers.

2-2.1.2.2   The water supply requirement for sprinklers only shall be calculated from the density curves in Figure 2-2.1(B). System piping shall be calculated to satisfy a single point on the appropriate design curve. It is not necessary to meet all points on the selected curve.

Formal Interpretation

Question: When applying Sections 2-2.1.2.2 and 2-2.1.2.8 and Table 2-2.1(B) to the hydraulic design of a room covering less than 1500 sq ft in area, is it the intent of the standard to impose a minimum water flow demand requirement for sprinklers only of 150 US gpm (i.e., 0.1 × 1500 = 150) and if so, should this minimum water flow be carried through the sprinkler piping from the area of application back to the alarm valve header?

Answer: No.

It is *not* necessary to meet any specific point on the curve, nor is it necessary to meet more than one or all of the points on the curve. Any design which satisfies any point (density and area) from Figure 2-2.1(B) meets the standard. However, there are modifications to the selected densities and areas which may be made, and some which must be made, in accordance with 2-2.1.2.3 through 2-2.1.2.13. The higher densities associated with the smaller areas of application tend to limit fire size. Large-scale fire tests indicate substantial reductions in fire damage using sprinklers with higher densities.

2-2.1.2.3   When inside hose stations are planned or are required by other standards, a water allowance of 50 gpm (189 L/min) for a one hose station installation [100 gpm (378 L/min) for a two or more station installation] shall be added to the sprinkler requirement at the point of connection to the system at the residual pressure required by the sprinkler system design.

Inside hose must be considered as water used from a hydraulically calculated system. Because any diversion of water from calculated systems is critical to the calculated area, the one or more hose stations nearest the remote area should be carried in the hydraulic calculations.

2-2.1.2.4   Water supply demands for ceiling sprinklers and in-rack sprinklers for rack storage systems shall be combined and hydraulically balanced at the common supply point.

In addition to inside hose stations, in-rack sprinklers can also divert water normally supplied to ceiling sprinklers. Therefore, the total combined supply for both ceiling sprinklers and in-rack sprinklers serving the same area should be included and balanced at the common supply point.

2-2.1.2.5   Water allowance for outside hose shall be added to the sprinkler and inside hose requirement at the connection to the city water main, or at a yard hydrant, whichever is closer to the system riser.

2-2.1.2.6   The lower duration figure is ordinarily acceptable where remote station water flow alarm service or equivalent is provided.

2-2.1.2.7   When pumps, gravity tanks or pressure tanks supply sprinklers only, requirements for inside and outside hose need not be considered in determining the size of such pumps or tanks.

2-2.1.2.8   The water supply requirement for sprinklers only shall be based upon the area of the sprinkler operation selected from Table 2-2.1(B) or upon the area of the largest room, at the discretion of the designer. Such a room shall be enclosed with construction having a fire resistance rating equal to the water supply duration indicated in Table 2-2.1(B) with minimum protection of openings as follows:

(a) Light Hazard — automatic or self-closing doors.

Exception: When openings are not protected, calculations shall include the sprinklers in the room plus two sprinklers in the communicating space nearest each such unprotected opening unless the communicating space has only one sprinkler, in which case calculations shall be extended to the operation of that sprinkler.

Formal Interpretation

Question: In a building classified as light hazard (such as an apartment building), where all walls are at least 30-minute fire rated, but without self-closing doors, would it be correct to base the calculations on the largest room and adjacent communicating rooms even though the combined area of these rooms total less than 1500 sq ft?

Answer: Yes. The Exception to 2-2.1.2.8(a) would apply when doors are not self-closing. The density for 1500 sq ft should be used although the calculated area may be smaller.

Formal Interpretation

Question: Is it the intent of the Committee to permit the omission from the hydraulic calculations, sprinkler discharge in closets and washrooms (not protected by self-closing or automatic doors), when the Exception to 2-2.1.2.8(a) is used to determine the area of operation to be used?

Answer: No. The Exception to 2-2.1.2.8(a) says communicating space and does not differentiate between closets, washrooms or other rooms. Section 7-4.3.1.6 allows the omission of sprinklers in closets, washrooms and similar small compartments when the areas in Table 2-2.1(B) are being calculated.

(b) Ordinary and Extra Hazard — automatic or self-closing doors with appropriate fire resistance ratings for the enclosure.

Exception: The water supply for dwelling units protected by residential sprinklers shall be in accordance with 7-4.4 in wet systems only.

2-2.1.2.9   For areas of sprinkler operation less than 1500 sq ft (139 m²) used for light and ordinary hazard occupancies, the density for 1500 sq ft (139 m²) shall be used. For areas of sprinkler operation less than 2500 sq ft (232 m²)for extra hazard occupancies (Groups 1 and 2), the density for 2500 sq ft (232 m²) shall be used.

A minimum density is needed for fire control. The density is not reduced by compartmentation but the total water supply can be reduced by using the required higher density for a smaller area of sprinkler operation.

2-2.1.2.10   For dry-pipe systems, increase area of sprinkler operation by 30 percent without revising density.

Formal Interpretation

Question: For a dry-pipe sprinkler system in an ordinary hazard occupancy having unsprinklered combustible concealed spaces, would 2-2.1.2.10 and 2-2.1.2.11 permit a density based on less than 3000 sq ft area of sprinkler operation provided the 30 percent increase in the area of sprinkler operation required by 2-2.1.2.10 brings the area of sprinkler operation to over 3000 sq ft as required by 2-2.1.2.11?

Answer: Yes.

Dry-pipe systems introduce an element of delay in water delivery which may allow an incipient fire to spread. The fire continues to burn without control until water reaches opened sprinklers. This usually causes more sprinklers to open than would be expected with a wet-pipe system.

Because of this delay the area of operation, but not the density, is increased by 30 percent in order that water supply for the additional operating area be available. (*See also A-2-2.*)

2-2.1.2.11*   For construction having unsprinklered combustible concealed spaces (as described in 4-4.4) the minimum area of sprinkler operation shall be 3000 sq ft (279 m^2).

Exception: Combustible concealed spaces filled entirely with noncombustible insulation.

A-2-2.1.2.11   This section is included to compensate for possible delay in operation of sprinklers from fires in combustible concealed spaces found in wood frame, brick veneer, and ordinary construction.

Formal Interpretation

Question 1: Is it the intent of Section 2-2.1.2.11 to impose the 3000 square foot design area where sprinklers have been omitted in accordance with Section 4-4.4.1, regardless of the solid joist or beam depth; or, are these areas regarded as joist space voids which do not require 3000 square feet to be used?

Answer: Yes, regardless of the solid joist or beam depth.

Question 2: Would this also apply in part to walls?

Answer: No.

Formal Interpretation

Question: Is it the intent of 2-2.1.2.11 to impose the 3000 sq ft design area in a fully sprinklered wood frame construction building even though there exists no unsprinklered combustible concealed spaces?

Answer: No. Section 2-2.1.2.11 applies only when you have unsprinklered combustible concealed spaces.

Formal Interpretation

Question: In a building where a ceiling is attached directly to floor or roof joists (solid joists or TJI truss joist) construction which forms an unsprinklered combustible concealed space as described in Section 4-4.4.1, is it the intent of the Committee to require a hydraulically calculated minimum area of sprinkler operation of 3,000 sq ft (279 m^2) per Section 2-2.1.2.11?

Answer: Yes.

Formal Interpretation

Question 1: Section 2-2.1.2.11. Can unsprinklered combustible concealed spaces as described in 4-4.4.1 be permitted, provided the below ceiling design area is increased to a minimum of 3000 square feet?

Answer: No. Permitted unsprinklered combustible concealed spaces as provided in 4-4.4.1 are not contingent upon use of hydraulically designed systems. However, when hydraulically designed systems are installed and such conditions exist the minimum design area of sprinkler operation is 3000 square feet.

Question 2: Section 3-8.4. Would a filtered and chlorine-treated swimming pool be classified as a raw source of water thus requiring return bends *to avoid* accumulation of sediment *in drop nipples*?

Answer: No.

2-2.1.2.12   For hazard classifications other than those indicated see appropriate NFPA standards for design criteria.

2-2.1.2.13 When high temperature sprinklers are used for extra hazard occupancies (Groups 1 and 2) the area of sprinkler operation may be reduced by 25 percent without revising the density but not to less than 2000 sq ft (185.8 m²).

Large-scale fire tests indicate substantial reduction in fire damage using sprinklers with higher temperature ratings. That is the reason for allowing the 25 percent reduction in area.

2-2.1.3 When other NFPA standards have developed sprinkler system design criteria, they shall take precedence.

Other NFPA standards and recommended practices that prescribe density and area of operation are NFPA 231, *Standard for Indoor General Storage*; NFPA 231C, *Standard for Rack Storage of Materials*; NFPA 231D, *Standard for Storage of Rubber Tires*; NFPA 231E, *Recommended Practice for the Storage of Baled Cotton*; NFPA 231F, *Standard for the Storage of Roll Paper*, and NFPA 409, *Standard on Aircraft Hangars*.

2-3 Connections to Water Works Systems.

2-3.1 Acceptability.

2-3.1.1* A connection to a reliable water works system shall be an acceptable water supply source. The volume and pressure of a public water supply shall be determined from water flow test data.

A-2-3.1.1 Care should be taken in making water tests to be used in designing or evaluating the capability of sprinkler systems. The water supply tested should be representative of the supply that may be available at the time of a fire. For example, testing of public water supplies should be done at times of normal demand on the system. Public water supplies are likely to fluctuate widely from season to season and even within a 24-hour period. Allowance should be made for seasonal or daily fluctuations, for drought conditions, for possibility of interruption by flood, or for ice conditions in winter. Testing of water supplies also normally used for industrial use should be done while water is being drawn for industrial use. The range of industrial-use demand should be taken into account.

Future changes in water supplies should be considered. For example, a large, established urban supply is not likely to change greatly within a few years. However, the supply in a growing suburban industrial park may deteriorate quite rapidly as greater numbers of plants draw more water.

See other information about performing flow tests in B-2-1.

2-3.1.2 Meters.   Meters are not recommended for use on sprinkler systems; however, where required by other authorities, they shall be of approved type.

Meters are not considered necessary on fire protection connections and are to be discouraged due to the friction loss they introduce as well as due to the additional cost associated with them and with regard to automatic sprinkler systems. The arrangements indicated in Figure B-2-3.1 provide for proper metering of water used for other than fire protection purposes. Where there is a concern for possible leakage or loss from the fire protection lines, a detector check valve can be substituted for the check valve indicated.

2-3.2* Capacity.   The connection and arrangement of underground supply piping shall be capable of supplying the volume as required in Table 2-2.1(A) or 2-2.1(B). Pipe size shall be at least as large as the system riser. (*See NFPA 24, Standard for the Installation of Private Fire Service Mains and Their Appurtenances.*)

Exception: Unlined cast or ductile iron shall not be less than 4 in. in size.

A-2-3.2   In private underground piping systems for buildings of other than light hazard occupancy, any dead-end pipe which supplies both sprinklers and hydrants should be not less than 8 in. in size. Also see NFPA 24, *Standard for the Installation of Private Fire Service Mains and Their Appurtenances.*

In order to provide adequate water capacity, yard systems which supply one or more sprinkler systems, in addition to one or more hydrants, may be sized hydraulically to comply with 2-2.1.2.5. The capacity should be calculated to supply the most hydraulically demanding system, and the closest hydrant to that system. When underground systems are not hydraulically calculated, larger pipe sizing such as a minimum of 6 inches for looped and a minimum of 8 inches for dead-end underground mains may be necessary.

B-2-3 Special Provisions.

B-2-3.1 Domestic Connections.   Connections for domestic water service should be made on the water supply side of the check valve in the water supply main so that the use of the fire department connection will not subject the domestic water system to high pressure. If the domestic consumption will significantly reduce the sprinkler water supply, an increase in the size of the pipe supplying both the domestic and sprinkler water may be justified. Circulation of water in sprinkler pipes is objectionable, owing to increased corrosion, deposits of sediment, and condensation drip from pipes.

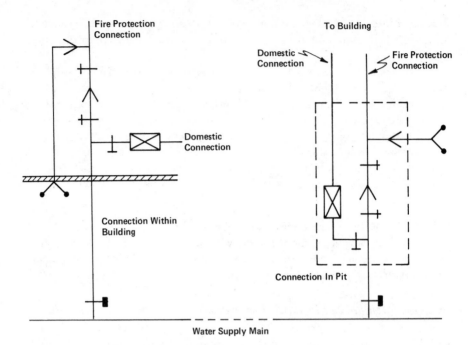

Figure B-2-3.1   Connection for Domestic Water.

B-2-3.2 Water Hammer.   When connections are made from water mains subject to severe water hammer [especially when pressure is in excess of 100 lbs per sq in. (6.9 bars)], it may be desirable to provide either a relief valve, properly connected to a drain, or an air chamber in the connection. If an air chamber is used, it should be located close to where the pipe comes through wall and on the supply side of all other valves and so located as to take the full force of water hammer. Air chambers should have a capacity of not less than 4 cu ft (0.09 m³), should be controlled by an approved indicating valve, and should be provided with a drain at the bottom, also an air vent with control valve and plug to permit inspection.

B-2-3.3 Penstocks, Flumes, etc.   Water supply connections from penstocks, flumes, rivers or lakes should be arranged to avoid mud and sediment, and should be provided with approved double removable screens or approved strainers installed in an approved manner.

B-2-3.4 Pressure Regulating Valves.   Pressure regulating valves should not be used except with permission of the authority having jurisdiction.

2-4 Gravity Tanks.

2-4.1 Acceptability.   An elevated tank sized in accordance with Table 2-2.1(A) or 2-2.1(B) shall be an acceptable water supply source. (*See NFPA 22, Water Tanks for Private Fire Protection.*)

2-4.2 Capacity and Elevation.   The capacity and elevation of the tank and the arangement of the underground supply piping shall provide the volume and pressure required by Table 2-2.1(A) or 2-2.1(B) designs.

2-5 Pumps.

2-5.1* Acceptability.   A single automatically controlled fire pump sized in accordance with Table 2-2.1(A) or 2-2.1(B) supplied under positive head shall be an acceptable water supply source. (*See NFPA 20, Installation of Centrifugal Fire Pumps.*)

A-2-5.1   An automatically controlled vertical turbine pump taking suction from a reservoir, pond, lake, river or well complies with 2-5.1.

A vertical turbine pump is a centrifugal fire pump which operates in a vertical position. In that position the first impeller, or bowl, of the pump is always submerged; therefore, the first pump impeller has positive suction and the pump complies with 2-5.1. This type of pump, with its first impeller submerged, does not actually take "suction from," but rather pumps from an impounded water source.

2-5.2* Supervision.   When a single fire pump constitutes the sole sprinkler supply, it shall be provided with supervisory service from an approved central station, proprietary, remote station system or equivalent.

A-2-5.2   See sections dealing with sprinkler equipment supervisory and water flow alarm services in NFPA 71, *Standard for Central Station Signaling Systems,* NFPA 72A, *Standard for Local Protective Signaling Systems,* NFPA 72B, *Standard for Auxiliary Protective Signaling Systems,* NFPA 72C, *Standard for Remote Station Protective Signaling Systems for Fire Alarm and Supervisory Service,* or NFPA 72D, *Standard for Proprietary Protective Signaling Systems.*

In order to improve the reliability of a single fire pump to the level of acceptance as sole supply for sprinkler systems, supervisory service to a constantly attended location is required. This supervision should include a pump-running indication as well as pump power or supervision of conditions which would render the pump inoperative.

2-6 Pressure Tanks.

2-6.1 Acceptability.

2-6.1.1   A pressure tank sized in accordance with Table 2-2.1(A) or 2-2.1(B) is an acceptable water supply source. (*See NFPA 22, Water Tanks for Private Fire Protection.*)

2-6.1.2   Pressure tanks shall be provided with an approved means for automatically maintaining the required air pressure. When a pressure tank is the sole water supply there shall also be provided an approved trouble alarm to indicate low air pressure and low water level with the alarm supplied from an electrical branch circuit independent of the air compressor.

As in the case with a single pump, pressure tanks which are the sole water supply should be supervised for reliability. Trouble alarms which indicate low air pressure and low water levels should be received at a constantly attended location.

2-6.1.3   Pressure tanks shall not be used to supply other than sprinklers and hand hose attached to sprinkler piping.

2-6.2 Capacity.

2-6.2.1   The size of the pressure tank required shall be in accordance with Table 2-2.1(A) or 2-2.1(B) and shall include the extra capacity needed to fill dry-pipe systems when installed. Minimum requirements when pressure tanks are not the sole water supply source shall be as indicated in 2-6.2.2, 2-6.2.3, and 2-6.2.4.

2-6.2.2 Light Hazard Occupancy.   Amount of available water, not less than 2,000 gal (7570 L).

2-6.2.3 Ordinary Hazard Occupancy.   Amount of available water, not less than 3,000 gal (11 355 L) for Groups 1 and 2. For Group 3, refer to authority having jurisdiction.

2-6.2.4 Extra Hazard and Woodworker Occupancies.   Refer to authority having jurisdiction.

2-6.2.5   For high-rise buildings, see Chapter 8.

2-6.3* Water Level and Air Pressure.   Unless otherwise approved by the authority having jurisdiction, the pressure tank shall be kept two-thirds full of water, and an air pressure of at least 75 psi (5.2 bars) by the gage shall be maintained. When the bottom of the tank is located below the highest sprinklers served, the air pressure by the gage shall be at least 75 psi (5.2 bars) plus three times the pressure caused by the column of water in the sprinkler system above the tank bottom.

A-2-6.3   The air pressure to be carried and the proper proportion of air in the tank may be determined from the following formulas, in which,

P = Air pressure carried in pressure tank.
A = Proportion of air in tank.
H = Height of highest sprinkler above tank bottom.

When tank is placed above highest sprinkler

$$P = \frac{30}{A} - 15.$$

A = ⅓ then P = 90 − 15 = 75 lbs per sq in.
A = ½ then P = 60 − 15 = 45 lbs per sq in.
A = ⅔ then P = 45 − 15 = 30 lbs per sq in.

When tank is below level of the highest sprinkler

$$P = \frac{30}{A} - 15 + \frac{0.434H}{A}$$

A = ⅓ then P = 75 + 1.30H.
A = ½ then P = 45 + 0.87H.
A = ⅔ then P = 30 + 0.65H.

The respective air pressures above are calculated to ensure that the last water will leave the tank at a pressure of 15 psi when the base of the tank is on a level with the highest sprinkler, or at such additional pressure as is equivalent to a head corresponding to the distance between the base of the tank and the highest sprinkler when the latter is above the tank.

The final pressure required at the pressure tank for systems designed from Table 2-2.1(B) will normally be higher than the 15 psi anticipated in the previous paragraph. The following formula should be used to determine the tank pressure and ratio of air to water in hydraulically designed systems.

$$P_i = \frac{P_f + 15}{A} - 15$$

where
P_i = Tank pressure
P_f = Pressure required from hydraulic calculations
A = Proportion of air

Example: Hydraulic calculations indicate 75 psi is required to supply the system. What tank pressure will be required?

$$P_i = \frac{75 + 15}{.5} - 15$$

$$P_i = 180 - 15 = 165 \text{ psi}$$

For SI Units: 1 ft = 0.3048 m; 1 psi = 0.0689 bar.

In this case the tank would be filled with 50 percent air and 50 percent water and the tank pressure would be 165 psi. If the pressure is too high the amount of air carried in the tank will have to be increased.

Location of Pressure Tanks.   Pressure tanks should be located above the top level of sprinklers but may be located in the basement or elsewhere.

See B-2-2.2 for the proper location of a pressure tank used in conjunction with a gravity tank in order to prevent air lock.

When water supplies other than circulating public systems are used, care should be taken that components (piping, pumps or tanks) are reliably installed and maintained. Even though the capacities of private systems may be adequate, the reliability of these emergency supply systems may not prove equal to public systems without regular periodic testing, maintenance and supervision. Public systems are "tested" and "supervised" continually by domestic usage of water, while private systems sometimes sit idle for long periods of time.

2-7 Fire Department Connections.

2-7.1*   A fire department connection shall be provided as described in this section.

Exception: *When permission of the authority having jurisdiction has been obtained for its omission.*

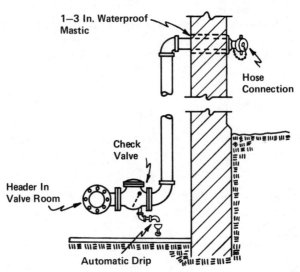

For SI Units: 1 in. = 25.4 mm.

Figure 2-7.1   Fire Department Connection.

Seldom, if ever, should fire department connections be omitted from sprinkler systems. They are an important supplement to normal water supplies, they allow the fire department to apply water directly over the fire, and in most cases they allow the fire department to bypass a closed control valve. When omissions are considered, they are normally for very small buildings which are completely without interior partitions and which have windows and doors providing easy access for manual fire fighting as a supplement to normal sprinkler operation. Even when gravity tanks or pressure tanks are the sole source of supply, fire department connections should be included. This allows fire departments to pump from tanker trucks directly into the system, thereby either augmenting the supply from tanks prior to it being depleted or adding to it. (*See commentary to 2-7.3.4.*)

A-2-7.1   The fire department connection should be located not less than 18 in. (457 mm) and not more than 5 ft (1.5 m) above the level of the adjacent grade or access level.

2-7.2* Size.   Pipe size shall be not less than 4 in. for fire engine connections and not less than 6 in. for fire boat connections, except that 3-in. pipe may be used to connect a single hose connection to a 3-in. or smaller riser.

A-2-7.2   For hydraulically designed sprinkler systems, the size of the fire department connection should be sufficient to supply the sprinkler water demand developed from Table 2-2.1(B).

2-7.3* Arrangement.   (*See 3-14.2.5 and 3-14.2.6.*)

A-2-7.3   Fire department connections should be located and arranged so that hose lines can be readily and conveniently attached without interference from nearby objects including buildings, fences, posts or other fire department connections. When a hydrant is not available, other water supply sources such as a natural body of water, a tank or reservoir should be utilized.

2-7.3.1   The fire department connection shall be made on the system side of a check valve in the water supply piping.

Each separate water supply connection for sprinkler systems, whether automatic or manual, should be made on the system side of a check valve. Only in this way can pressure differentials between municipal, fire pump, pressure tank, or fire department connections effectively supply the sprinklers. The highest pressure supply will come in first until its capacity is reached. As capacity is reached the pressure will drop until the next highest pressure supply comes in. The check valve prevents the higher pressure supply from backing up or circulating into lower pressure supplies. (See 3-14.2, A-3-14.2, and commentary.) For example:

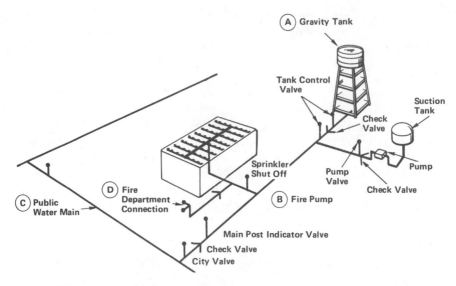

Figure 2.1.   Alternative water supplies to sprinkler system.

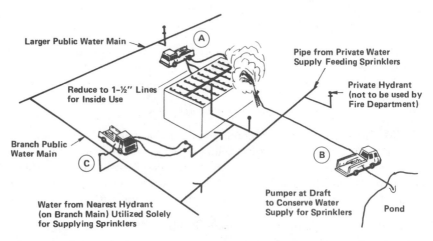

Figure 2.2.   Fire department water supply connections to sprinkler system.

Sprinklers in the example are normally supplied by a public main pressure of 50 psi static. Upon arrival at a fire in this building, the fire officer orders a pumper to connect to the hydrant across the street and to pump at 100 psi into the yard fire department connection, thus adding supplementary pressure to the sprinkler system. If the connection were not on the system side of a check valve, the pumped water would circulate back into the same main to the suction side of the pumper, and no supplementary pressure could be added.

2-7.3.2   On wet-pipe systems with a single riser the connection shall be made on the system side of approved indicating, check, and alarm valves to the riser, unless the system is supplied by a fire department pumper connection in the yard. (*See 3-14.2.6.*)

Formal Interpretation

Section 2-7.3.2 indicates that the fire deparment connection shall be made on the system side of an approved indicating valve, unless the sprinklers are supplied by the fire department pumper connection in the yard.

Question: Is it the intent of paragraph 2-7.3.2 to permit a post indicator valve to be present, in line, between the fire department connection and the main riser? (Please refer to the following diagram for clarity on this matter):

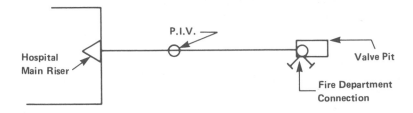

Answer: Yes.

Formal Interpretation

Question: In a sprinkler system having the fire department connection in the yard, may the system indicating valve be of the OS and Y type located on the system riser inside the building?

Answer: Yes.

2-7.3.3   On dry-pipe systems with a single riser the connection shall be made between the approved indicating valve and the dry-pipe valve, unless the system is supplied by a fire department pumper connection in the yard.

2-7.3.4   On systems with two or more risers the connection shall be made on the system side of all shutoff valves controlling other water supplies, but on the supply side of the riser shutoff valves so that, with any one riser off, the connection will feed the remaining sprinklers, unless the sprinklers are supplied by a fire department pumper connection in the yard.

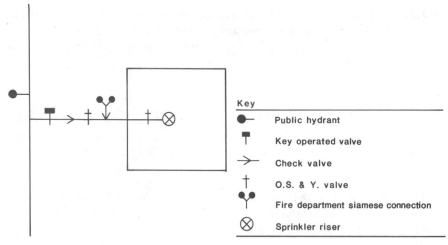

Key

●—   Public hydrant

Key operated valve

→   Check valve

†   O.S. & Y. valve

Fire department siamese connection

⊗   Sprinkler riser

Figure 2.3.   Fire department yard connection to a system with one or more multiple risers.

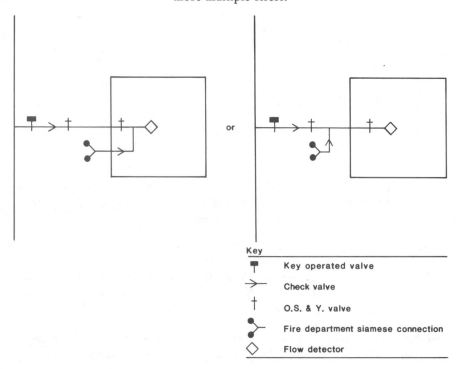

or

Key

Key operated valve

→   Check valve

†   O.S. & Y. valve

Fire department siamese connection

◇   Flow detector

Figure 2.4.   Fire department connections to a dry-pipe sprinkler system having a single riser. Either method is satisfactory according to the standard.

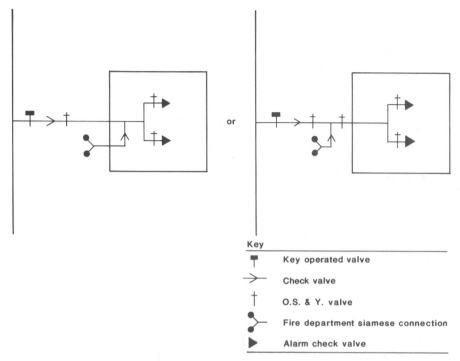

Key

⊤	Key operated valve
⟩	Check valve
†	O.S. & Y. valve
	Fire department siamese connection
▶	Alarm check valve

Figure 2.5.   Alternative fire department connections to systems having two or more risers. Either method is satisfactory according to the standard.

Formal Interpretation

Question: Are control valves required on the individual risers shown in Figure 1?

Answer: Yes.

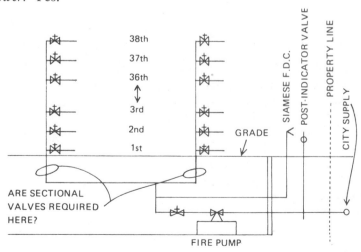

Figure 1

Formal Interpretation

Question 1: Does Section 2-7.3.4 require that a fire department pumper connection be provided on each riser in a multiple riser system?

Answer: No.

Question 2: Is a fire department pumper connection installed at the city connection valve pit satisfactory to comply with Section 2-7.3.4?

Answer: Yes.

Fire department connections have long presented difficulty between the concept of protection and the feasibility of installation. On the one hand, the concept would say that fire departments should be able to pump into any system whether the final control valve is open or closed, thereby assisting low pressure water supplies, or supplementing water supplies from other sources such as tanks or impounded storage. However, when multiple risers are in a single building system or when single risers are fed directly from a private yard-main supply, it may not be feasible to run individual fire department connections to the system side of each riser control valve. It would also be difficult to identify these multiple fire department connections for effective use. Therefore, the section's exceptions, drawings, and formal interpretations are intended to clarify that the standard is taking a reasonable approach in this area.

2-7.3.5   Fire department connections shall not be connected on the suction side of booster pumps.

According to 2-7.3.1 through 2-7.3.4 the fire department connection is made on the system side of a check valve. It should be clear then that the fire department connection should not be on the suction side of a pump. A check valve is not permitted on the suction side of a pump according to NFPA 20, *Standard for the Installation of Centrifugal Fire Pumps*. Therefore, a fire department connection on the suction side could not be made on the system side of a check. In addition, full efficiency and reliability cannot be obtained for a suction side connection because of losses through pump and valves, and the inability to supplement the system if discharge side control valves are closed. It is important to note that pumping into the suction side of a fire pump or booster pump will increase the discharge pressure of the pump, frequently beyond the pressure limits of the system. The fire department connection should be made on the system side of a pump, and on the system side of the pump discharge check valve and control valves.

2-7.3.6 Fire department connections to sprinkler systems shall be designated by a sign having raised letters at least 1 in. (25 mm) in size cast on plate or fitting reading for service designated: viz. — "AUTOSPKR.," "OPEN SPKR." or "AUTOSPKR. and STANDPIPE."

2-7.4 Valves.

2-7.4.1 An approved check valve shall be installed in each fire department connection, located as near as practicable to the point where it joins the system.

2-7.4.2 There shall be no shutoff valve in the fire department connection.

Formal Interpretation

Question: What is the Official Interpretation of NFPA 13, Section 2-7.4.2 "There shall be no shutoff valve in the fire department connection"?

Answer: A shutoff valve shall not be installed in the connection between the fire department siamese and the underground or yard system (Standards 13 and 24). This is an emergency checked connection for water supply to hydrants and sprinkler systems connected to the underground or yard system. Sectional valves are recommended in large yard systems (Standard 24, paragraph 3-5.1) and post indicator valves, wall post indicator valves or inside control valves are recommended for sprinkler systems with two or more risers (Standard 13, paragraph 2-7.3.4). This also means that single riser building connections to a large yard system will be valved.

Section 2-7.4.2 is needed only because it is usually considered good practice to install a shutoff valve on both sides of a check valve to facilitate repair. The check valve in the fire department connection is considered to be an exception to that general good practice and no shutoff valve is permitted at that point in the emergency connection.

2-7.5 Drainage. The piping between the check valve and the outside hose coupling shall be equipped with an approved automatic drip.

In the event the check valve in the siamese connection leaks, it is the purpose of the automatic drip to drain this water to a safe location and maintain the piping between the check valve and the hose couplings without water collection. Without the automatic drip, such collected water might freeze and prevent the fire department from being able to pump into the system under fire conditions.

2-7.6 Hose Connections.

2-7.6.1   The fire department connection(s) shall be internal threaded swivel fitting(s) having the NH standard thread, at least one of which shall be 2.5-7.5 NH standard thread, as specified in NFPA 1963, *Screw Threads and Gaskets for Fire Hose Connections.*

Exception: When local fire department connections do not conform to NFPA 1963, the authority having jurisdiction shall designate the connection to be used.

All hose coupling threads in sprinkler systems, and threads for hydrants on yard mains supplying sprinkler systems, should match those of the first responding fire department. If fire department hose couplings are of the type without threads, the sprinkler system connections should be compatible.

2-7.6.2   Hose connections shall be equipped with listed plugs or caps.

2-8 Arrangement of Water Supply Connections.

2-8.1 Connection Between Underground and Aboveground Piping.
The connection between the system piping and underground piping shall be made with a suitable transition piece and shall be properly strapped or fastened by approved devices. The transition piece shall be protected against possible damage from corrosive agents, solvent attack, or mechanical damage.

2-8.2* Connection Passing Through or Under Foundation Walls.
When system piping pierces a foundation wall below grade or is located under the foundation wall, clearance shall be provided to prevent breakage of the piping due to building settlement.

A-2-8.2   When the system riser is close to an outside wall, underground fittings of proper length should be used in order to avoid pipe joints located in or under the wall. When the connection passes through the foundation wall below grade, a 1- to 3-in. (25- to 76-mm) clearance should be provided around the pipe and the clear space filled with asphalt mastic or similar flexible waterproofing material. (*Also see Appendix B-3-1.*)

2-9 Water Supply Test Pipes and Gages.

2-9.1* Test Pipes.   Test pipes, which may also be used as drain pipes, shall be provided at locations that will permit flow tests to be made to determine whether water supplies and connections are in order. Such test pipes shall be not less than the sizes specified in 3-11.2 and equipped with a shutoff valve. They shall be so installed that the valve may be opened wide for a sufficient time to assure a proper test without causing water damage. (*See 3-11.2 and 3-11.4.*)

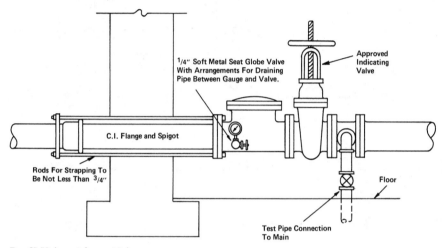

1/4" Soft Metal Seat Globe Valve
With Arrangements For Draining
Pipe Between Gauge and Valve.

Approved
Indicating
Valve

C.I. Flange and Spigot

Rods For Strapping To
Be Not Less Than 3/4"

Floor

Test Pipe Connection
To Main

For SI Units: 1 in. = 25.4 mm.

Figure 2-9.1   Water Supply Connection with Test Pipe.

Test or drain pipes are used for both implied purposes. They are used to drain systems when repairs are necessary or when dry-pipe valves have tripped. They are also used to indicate that water supply is available at the system riser and on the system side of all check valves, control valves, and underground piping.

Test or drain pipes are normally smaller than all other water supply piping. Therefore, they are not capable of the same large scale flow tests discussed in Section B-2-1. However, when the system demand is small a drain test can be used; the test pressures and flow at the test pipe represent exactly the water supply available for the amount of water flowed. These tests should not be extrapolated to larger flows or system demands because of the unknown friction values of check valves under low flow conditions which are included in the drain tests. Two-inch drain tests, as they are known, normally consist of opening the main 2-inch drain wide until the pressure stabilizes. Records are kept of these tests to detect possible deterioration of water supplies or to detect valves which may have been closed in the overall water supply system.

A-2-9.1

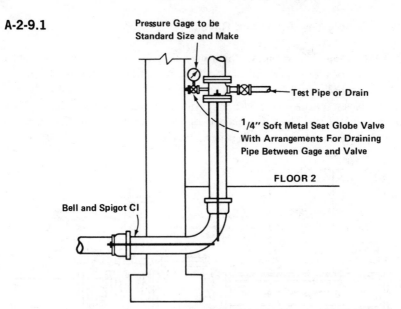

**Figure A-2-9.1   Test Pipe on Water Supply with Outside Control.
Also applicable to an interior riser.**

2-9.2 Gages.

2-9.2.1   A pressure gage shall be installed on the riser or feed main at or near each test pipe, with a connection not smaller than ¼ in. This gage connection shall be equipped with a shutoff valve and with provision for draining.

2-9.2.2   The required pressure gages shall be of approved type and shall have a maximum limit not less than twice the normal working pressure at the point where installed. They shall be installed to permit removal, and shall be located where they will not be subject to freezing.

3

SYSTEM COMPONENTS

3-1 Piping.

3-1.1 Piping Specifications.

3-1.1.1 Pipe or tube used in sprinkler systems shall be of the materials in Table 3-1.1.1 or in accordance with 3-1.1.2 through 3-1.1.6. The chemical properties, physical properties and dimensions of the materials listed in Table 3-1.1.1 shall be at least equivalent to the standard cited in the table. Pipe and tube used in sprinkler systems shall be designed to withstand a working pressure of not less than 175 psi (12.1 bars).

Table 3-1.1.1

Materials and Dimensions	Standard
Ferrous Piping (Welded and Seamless)	
Welded and Seamless Steel Pipe for Ordinary Uses, Spec. for Black and Hot-Dipped Zinc Coated (Galvanized)	ANSI/ASTM A120
†Spec. for Black and Hot-Dipped Zinc Coated (Galvanized) Welded and Seamless Steel Pipe for Fire Protection Use ...	ASTM A795
†Spec. for Welded and Seamless Steel Pipe	ANSI/ASTM A53
Wrought Steel Pipe	ANSI B36.10
Spec. for Elec.-Resistance Welded Steel Pipe	ASTM A135
Copper Tube (Drawn, Seamless)	
†Spec. for Seamless Copper Tube	ASTM B75
†Spec. for Seamless Copper Water Tube	ASTM B88
Spec for General Requirements for Wrought Seamless Copper and Copper-Alloy Tube	ASTM B251
Brazing Filler Metal (Classification BCuP-3 or BCuP-4)	AWS A 5.8
Solder Metal, 95-5 (Tin-Antimony-Grade 95TA)	ASTM B32

†Denotes pipe or tubing suitable for bending (*see 3-1.1.7*) according to ASTM standards.

In addition to specifying piping materials for use in sprinkler systems, this section also permits the use of pipe or tube of materials indicated in Table 3-1.1.1 made to standards other than those shown in the table provided the requirements of the other standards meet or exceed the cited standards.

3-1.1.2*   When welded and seamless steel pipe listed in Table 3-1.1.1 is used and joined by welding as referenced in 3-12.2 or by roll grooved pipe and couplings as referenced in 3-12.3, the minimum nominal wall thickness for pressures up to 300 psi (20.7 bars) shall be in accordance with Schedule 10 for sizes up to 5 in.; 0.134 in. (3.40 mm) for 6 in.; and 0.188 in. (4.78 mm) for 8- and 10-in. pipe; or as modified in 3-1.1.5 or as defined in 3-1.1.6.

When the thin wall pipe is joined by welding or roll grooved pipe and couplings, the groove forming process or welding should not result in a thinner wall thickness other than that due to the normal tolerances associated with the groove forming or welding process. The expression "minimum nominal wall thickness" recognizes that pipe standards permit plus or minus variations from the stated wall thickness.

3-1.1.3   When steel pipe listed in Table 3-1.1.1 is used and joined by threaded fittings referenced in 3-12.1 or by couplings used with pipe having cut grooves, the minimum wall thickness shall be in accordance with Schedule 30 (in sizes 8 in. and larger) or Schedule 40 (in sizes less than 8 in.) pipe for pressures up to 300 psi (20.7 bars).

When pipe is threaded or grooves are cut, material is lost and the use of thin wall pipe could result in too little material remaining between the inside diameter and the root diameter of the thread or groove.

A threaded assembly which has been investigated for suitability in automatic sprinkler installations and listed for this service is acceptable. See 3-12.1.2 and A-3-12.1.2.

3-1.1.4*   Copper tube as specified in the standards listed in Table 3-1.1.1, used in sprinkler systems, shall have a wall thickness of Type K, L or M.

Formal Interpretation

Question: Is use of copper tubing (Type K) with brazed joints acceptable for underground service from a city water main to a sprinkler system?

Answer: Yes — in accordance with 3-1.1.1 and 3-12.4. Copper tubing conforming with Table 3-1.1.1 and brazed joints conforming with 3-12.4 are acceptable.

Table A-3-1.1.2   Steel Pipe Dimensions

Nominal Pipe Size in.	Outside Diameter in.	(mm)	Schedule 10[1] Inside Diameter in.	(mm)	Wall Thickness in.	(mm)	Schedule 30 Inside Diameter in.	(mm)	Wall Thickness in.	(mm)	Schedule 40 Inside Diameter in.	(mm)	Wall Thickness in.	(mm)
1	1.315	(33.4)	1.097	(27.9)	0.109	(2.8)	—	—	—	—	1.049	(26.6)	0.133	(3.4)
1¼	1.660	(42.2)	1.442	(36.6)	0.109	(2.8)	—	—	—	—	1.380	(35.1)	0.140	(3.6)
1½	1.900	(48.3)	1.682	(42.7)	0.109	(2.8)	—	—	—	—	1.610	(40.9)	0.145	(3.7)
2	2.375	(60.3)	2.157	(54.8)	0.109	(2.8)	—	—	—	—	2.067	(52.5)	0.154	(3.9)
2½	2.875	(73.0)	2.635	(66.9)	0.120	(3.0)	—	—	—	—	2.469	(62.7)	0.203	(5.2)
3	3.500	(88.9)	3.260	(82.8)	0.120	(3.0)	—	—	—	—	3.068	(77.9)	0.216	(5.5)
3½	4.000	(101.6)	3.760	(95.5)	0.120	(3.0)	—	—	—	—	3.548	(90.1)	0.226	(5.7)
4	4.500	(114.3)	4.260	(108.2)	0.120	(3.0)	—	—	—	—	4.026	(102.3)	0.237	(6.0)
5	5.563	(141.3)	5.295	(134.5)	0.134	(3.4)	—	—	—	—	5.047	(128.2)	0.258	(6.6)
6	6.625	(168.3)	6.357	(161.5)	0.134[2]	(3.4)	—	—	—	—	6.065	(154.1)	0.280	(7.1)
8	8.625	(219.1)	8.249	(209.5)	0.188[2]	(4.8)	8.071	(205.0)	0.277	(7.0)	—	—	—	—
10	10.75	(273.1)	10.37	(263.4)	0.188[2]	(4.8)	10.14	(257.6)	0.307	(7.8)	—	—	—	—

NOTE 1: Schedule 10 defined to 5 in. (127 mm) nominal pipe size by ASTM A 135.
NOTE 2: Wall thickness specified in 3-1.1.2.

Table A-3-1.1.4 Copper Tube Dimensions

| Nominal Tube Size | Outside Diameter | | Type K | | | | Type L | | | | Type M | | | |
| | | | Inside Diameter | | Wall Thickness | | Inside Diameter | | Wall Thickness | | Inside Diameter | | Wall Thickness | |
in.	in.	(mm)	in.	(mm)	in.	(mm)	in.	(mm)	in.	(mm)	in.	(mm)	in.	(mm)
¾	0.875	(22.2)	0.745	(18.9)	0.065	(1.7)	0.785	(19.9)	0.045	(1.1)	0.811	(20.6)	0.032	(0.8)
1	1.125	(28.6)	0.995	(25.3)	0.065	(1.7)	1.025	(26.0)	0.050	(1.3)	1.055	(26.8)	0.035	(0.9)
1¼	1.375	(34.9)	1.245	(31.6)	0.065	(1.7)	1.265	(32.1)	0.055	(1.4)	1.291	(32.8)	0.042	(1.1)
1½	1.625	(41.3)	1.481	(37.6)	0.072	(1.8)	1.505	(38.2)	0.060	(1.5)	1.527	(38.8)	0.049	(1.2)
2	2.125	(54.0)	1.959	(49.8)	0.083	(2.1)	1.985	(50.4)	0.070	(1.8)	2.009	(51.0)	0.058	(1.5)
2½	2.625	(66.7)	2.435	(61.8)	0.095	(2.4)	2.465	(62.6)	0.080	(2.0)	2.495	(63.4)	0.065	(1.7)
3	3.125	(79.4)	2.907	(73.8)	0.109	(2.8)	2.945	(74.8)	0.090	(2.3)	2.981	(75.7)	0.072	(1.8)
3½	3.625	(92.1)	3.385	(86.0)	0.120	(3.0)	3.425	(87.0)	0.100	(2.5)	3.459	(87.9)	0.083	(2.1)
4	4.125	(104.8)	3.857	(98.0)	0.134	(3.4)	3.905	(99.2)	0.110	(2.8)	3.935	(99.9)	0.095	(2.4)
5	5.125	(130.2)	4.805	(122.0)	0.160	(4.1)	4.875	(123.8)	0.125	(3.2)	4.907	(124.6)	0.109	(2.8)
6	6.125	(155.6)	5.741	(145.8)	0.192	(4.9)	5.845	(148.5)	0.140	(3.6)	5.881	(149.4)	0.122	(3.1)
8	8.125	(206.4)	7.583	(192.6)	0.271	(6.9)	7.725	(196.2)	0.200	(5.1)	7.785	(197.7)	0.170	(4.3)
10	10.13	(257.3)	9.449	(240.0)	0.338	(8.6)	9.625	(244.5)	0.250	(6.4)	9.701	(246.4)	0.212	(5.4)

Since Type M tube is the thinnest walled, it has the largest inside diameter, which is an advantage in hydraulically designed systems. It is also the least costly. Therefore, Type M tube is most commonly used when bending is not required. See 3-1.1.7.

3-1.1.5   Other types of pipe or tube may be used, but only those investigated and listed for this service by a testing and inspection agency laboratory.

Formal Interpretation

Question 1: Is it the intent of the Committee that paragraphs 3-1.1.5 and 3-13.1.2 permit the use of piping products that differ from those specifically described in NFPA 13, if they have been investigated and listed for this service by the nationally recognized testing and inspection agency laboratory?

Answer: Yes. It permits use of different materials and different dimensions than those specifically described elsewhere.

Question 2: Is it the intent of the Committee that paragraphs 3-1.1.5 and 3-13.1.2 permit a manufacturer to pursue a testing program and obtain a listing of this product by a nationally recognized testing and inspection agency laboratory, even though that product may differ from those detailed by specifications or piping products described in NFPA 13?

Answer: Yes.

There is no prohibition against the use of any material, as long as it is listed for use in sprinkler systems. This section is intended to encourage development of either more efficient or cost-effective materials.

3-1.1.6   Whenever the word pipe is used in this standard it shall be understood to also mean tube.

For the purposes of this standard the word pipe refers to any conduit for transporting water.

3-1.1.7 Pipe Bending.   Bending of steel pipe (Schedule 40) and copper tube (Type K & L) may be accomplished when bends are made in conformance with good installation practices and show no kinks, ripples, distortions, reduction in diameter, or any noticeable deviations from round. The minimum radius of a bend shall be 6 pipe diameters for pipe sizes 2 in. and smaller, and 5 pipe diameters for pipe sizes 2½ in. and larger.

ASTM A120 and ASTM A135 pipe cannot be used for bending because the bending operation might cause the pipe to split along the seam whereas this does not occur with the seamless ASTM A53 or ASTM A795 pipe. Type M copper tube and Schedule 10 steel pipe cannot be used for bending because the bending results in a wall thickness too thin whereas with the heavier wall Type L & K copper tubing and Schedule 40 steel pipe there is adequate material left after bending. The specifications included in this section are intended to preclude any changes in the flow characteristics of the pipe or tube after it has been bent.

3-2* Definitions.   *(See Figure A-3-2.)*

Risers.   The vertical pipes in a sprinkler system.

All vertical pipes in a system are included in the definition of risers with the exception of the section of aboveground pipe directly connected to the water supply which is referred to as the system riser.

System Riser.   The abovegound supply pipe directly connected to the water supply.

Feed Mains.   Mains supplying risers or cross mains.

Cross Mains.   Pipes directly supplying the lines in which the sprinklers are placed.

Branch Lines.   Lines of pipe, from the point of attachment to the cross main (or similar connection) to the end sprinkler, in which the sprinklers are directly placed.

Formal Interpretation

Question: On a hydraulically calculated system, does a 1½-in. and 2 in. (pipe size) loop with sprinklers fed directly off the loop constitute a "branch line" as specified in Section 3-2 and 3-5.1?

Answer: Yes. A calculated system, as indicated in Figure 1, complies with the definition of a branch line in Section 3-2. The reference to branch lines in Section 3-5.1 is for a limitation to the number of sprinklers on a branch line for pipe schedule systems. Figure 1 is for a hydraulically calculated system and the requirements for number of sprinklers per branch line are superseded. See Section 7-1.1.2.

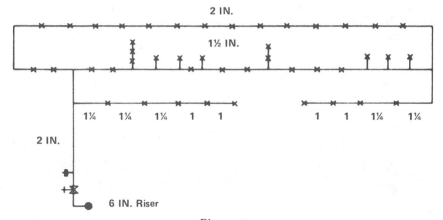

Figure 1

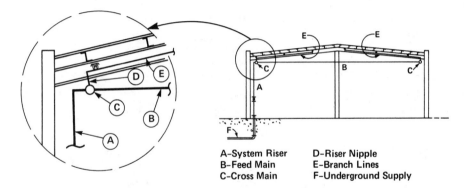

A–System Riser   D–Riser Nipple
B–Feed Main   E–Branch Lines
C–Cross Main   F–Underground Supply

Figure A-3-2   Building Elevation Showing Parts of Sprinkler Piping System.

3-3 Area Limitation.

3-3.1   The maximum floor area to be protected by sprinklers supplied by any one system riser or combined system riser on any one floor shall be as follows:

Light Hazard — 52,000 sq ft (4831 m²)

Ordinary Hazard — 52,000 sq ft (4831 m²)

Extra Hazard — 25,000 sq ft (2323 m²)

Storage — High piled storage (as defined in 4-1.3.9) and storage covered by other NFPA standards — 40,000 sq ft (3716 m²).

Exception: *When single systems serve both high piled storage or storage covered by other NFPA standards and ordinary hazard areas, the storage area coverage shall not exceed 40,000 sq ft (3716 m²) and the total area coverage shall not exceed 52,000 sq ft (4831 m²).*

A single system with an 8-in. supply protecting both ordinary hazard areas and solid piled, palletized, or racked storage in excess of 12 ft (3.7 m) high may have a coverage area of up to 52,000 sq ft (4831 m²), but not more than 40,000 sq ft (3716 m²) of that coverage area may be high piled storage. The piping for the ordinary hazard portions of the systems is sized in accordance with Section 3-6. Fewer sprinklers are allowed on each size pipe than are allowed for light hazard occupancies.

Formal Interpretation

Question: Would the 52,000 sq ft maximum area on one floor apply to several buildings being supplied by one riser assembly? Can the 52,000 sq ft be exceeded if the individual buildings are not in excess of 52,000 sq ft and are separated by a clear space of 20 feet or more?

Answer: No. The 52,000 sq ft area limitation and the 20 ft separating distance are not relevant to the problem. Each building should have its individual system riser.

The specified maximum coverage areas are the maximum floor area per system on any one floor. There is no limit on the number of floors, in that each floor constitutes a separate fire area. Openings are assumed to be protected as outlined in 4-4.8. Thus in a single-story 312,000-sq ft (28 986-m²) ordinary hazard occupancy at least six systems would be required. However, if the occupancy were six stories high and had 52,000 sq ft (4831 m²) on each floor, it could be protected by a single system. High-rise buildings in excess of 100 stories have been protected by a single system. A maximum of 40,000 sq ft (3716 m²) of solid piled, palletized, or rack storage may be protected by a single system. However, that system may also protect ordinary hazard areas provided the total coverage area does not exceed 52,000 sq ft (4831 m²). The standard formerly limited system size by the number of sprinklers on a system; for example, 400 sprinklers for ordinary hazard. The 400 sprinklers times 130 sq ft maximum spacing developed the 52,000 sq ft (4831 m²) limitation. These area limitations are not related to the hydraulics of the system, nor are they related to its operating characteristics. Rather the area limitations are judgmental factors as to the maximum area within a single fire division that it is felt should be protected by a single system (or that could be out of service at any one time).

This standard should be used for solid piled storage or rack storage heights of 12 ft (3.7 m) or less. For heights above 12 ft (3.7 m), storage standards such as the following should be used: NFPA 231, *General Storage*; NFPA 231C, *Rack Storage of Materials*; NFPA 231D, *Storage of Rubber Tires*; NFPA 231E, *Storage of Baled Cotton* and NFPA 231F, *Storage of Roll Paper*.

3-4* Pipe Schedules.

A-3-4 Long Runs of Pipe.   When the construction or conditions introduce unusually long runs of pipe or many angles, in risers or feed mains, an increase in pipe size over that called for in the schedules may be required to compensate for increased friction losses.

3-4.1   The pipe schedule sizing provisions shall not apply to hydraulically designed systems. Sprinkler systems having sprinklers with orifices other than ½ in. nominal shall be hydraulically designed or evaluated.

Sections 3-4, 3-5, 3-6 and 3-7 are applicable only to pipe schedule systems designed for use in conjunction with water supplies outlined in Table 2-2.1(A). Systems designed hydraulically in accordance with Chapter 7 permit the selection of pipe sizes appropriate to the characteristics of the water supply indicated by Table 2-2.1(B) and supersede the pipe schedule tables. The pipe schedule sizing is based on ½-inch orifice sprinklers. It must not be assumed that these schedules will prove adequate when sprinklers of other than ½-inch orifice are used.

3-4.2   The number of automatic sprinklers on a given pipe size on one floor shall not exceed the number given in Sections 3-5, 3-6, or 3-7 for a given occupancy.

Formal Interpretation

Question No. 1: Referring to the following figure (Figure 2), does Section 3-4.2 permit the piping supplying multiple floors to be sized for the maximum number of sprinklers on one floor or must it be sized for all sprinklers connected to it?

Answer: It may be sized to supply the maximum number of sprinklers on one floor. Referring to the figure, the pipe in question may be 1¼ in. for light hazard or ordinary hazard occupancies.

Question No. 2: Is the sizing of piping to supply sprinklers on any one floor restricted to system risers?

Answer: No.

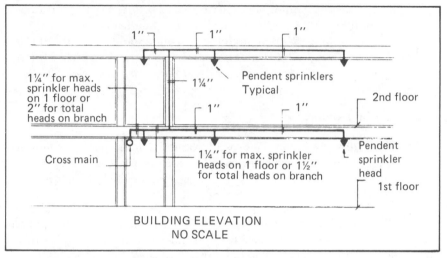

Figure 2

3-4.3 Size of Risers. Each system riser shall be sized to supply all sprinklers on the riser on any one floor as determined by the standard schedules of pipe sizes in Sections 3-5, 3-6, or 3-7.

3-4.4 Slatted Floors, Large Floor Openings, Mezzanines, and Large Platforms. Buildings having slatted floors, or large unprotected floor openings without approved stops, shall be treated as one area with reference to the pipes sizes, and the feed mains or risers shall be of the size required for the total number of sprinklers.

Fire will readily spread through grated or slatted floors. Therefore, areas with such divisions must be treated as a single fire area when sizing pipe. This is applicable to both pipe schedule and hydraulically designed systems.

3-5 Schedule for Light Hazard Occupancies.

3-5.1 Branch lines shall not exceed 8 sprinklers on either side of a cross main.

Exception: When occupancy is classified as light hazard and when more than 8 sprinklers on a branch line are necessary, lines may be increased to 9 sprinklers by making the two end lengths 1 in. and 1 1/4 in., respectively, and the sizes thereafter standard. Ten sprinklers may be placed on a branch line making the two end lengths 1 in. and 1 1/4 in., respectively, and feeding the tenth sprinkler by a 2 1/2 in. pipe.

The amount of pressure lost to friction in long runs of small diameter pipe is significant. To ensure that pressure is not excessively reduced, the number of sprinklers on a branch line is limited to 8 sprinklers in light hazard occupancies. The Exception allows up to 10 sprinklers on branch lines in light hazard occupancies providing pipe sizing is increased to reduce the friction loss.

3-5.2   Pipe sizes shall be in accordance with Table 3-5.2.

Table 3-5.2

Steel		Copper	
1 in. pipe .... 2 sprinklers		1 in. tube .... 2 sprinklers	
1¼ in. pipe .... 3 sprinklers		1¼ in. tube .... 3 sprinklers	
1½ in. pipe .... 5 sprinklers		1½ in. tube .... 5 sprinklers	
2 in. pipe .... 10 sprinklers		2 in. tube .... 12 sprinklers	
2½ in. pipe .... 30 sprinklers		2½ in. tube .... 40 sprinklers	
3 in. pipe .... 60 sprinklers		3 in. tube .... 65 sprinklers	
3½ in. pipe ....100 sprinklers		3½ in. tube ....115 sprinklers	
4 in. pipe .... See 3-3.1		4 in. tube .... See 3-3.1	

Exception: Each area requiring more than 100 sprinklers and without subdividing partitions (not necessarily fire walls) shall be supplied by feed mains or risers sized for ordinary hazard occupancies.

A system in a light hazard occupancy is limited to protection of 52,000 sq ft (4831 m^2). Piping is sized in accordance with 3-5.2, unless there are more than 100 sprinklers in an area without subdividing partitions. Piping supplying such sprinklers must be sized in accordance with Section 3-6 to reduce the friction loss.

3-5.3   When sprinklers are installed above and below ceilings [*see Figure 3-5-3(A), Figure 3-5.3(B) and Figure 3-5.3(C)*] and such sprinklers are supplied from a common set of branch lines, such branch lines shall not exceed 8 sprinklers above and 8 sprinklers below any ceiling on either side of the cross main. Pipe sizing, up to and including 2½ in., shall be as shown in Table 3-5.3.

Formal Interpretation

Question: Is the ceiling, as referenced in Section 3-5.3, required to be a rated assembly (i.e., one-hour, two-hours, etc.)?

Answer: No.

Since a fire either above or below a ceiling in a light hazard occupancy will normally be contained in that area by operating sprinklers, the standard permits in light hazard pipe schedule systems more sprinklers on a branch line than are permitted on a branch line protecting a single area and also permits more sprinklers to be supplied by a given size pipe.

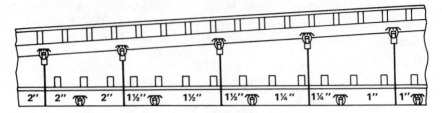

For SI Units: 1 in. = 25.4 mm.

Figure 3-5.3(A)   Arrangement of Branch Lines Supplying Sprinklers Above and Below a Ceiling.

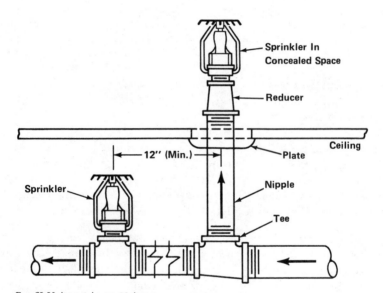

For SI Units: 1 in. = 25.4 mm.

Figure 3-5.3(B)   Sprinkler on Riser Nipple from Branch Line in Lower Fire Area.

Table 3-5.3   Number of Sprinklers Above and Below a Ceiling

Steel		Copper	
1 in.	2 sprinklers	1 in.	2 sprinklers
1¼ in.	4 sprinklers	1¼ in.	4 sprinklers
1½ in.	7 sprinklers	1½ in.	7 sprinklers
2 in.	15 sprinklers	2 in.	18 sprinklers
2½ in.	50 sprinklers	2½ in.	65 sprinklers

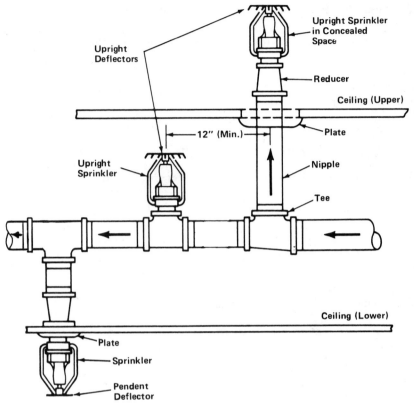

For SI Units: 1 in. = 25.4 mm.

Figure 3-5.3(C)   Arrangement of Branch Lines Supply Sprinklers Above and Below Ceilings.

3-5.3.1*   When the total number of sprinklers above and below a ceiling exceeds the number specified in Table 3-5.3 for 2½-in. pipe, the pipe supplying such sprinklers shall be increased to 3 in. and sized thereafter according to the schedule shown in Table 3-5.2 for the number of sprinklers above or below a ceiling, whichever is larger.

A-3-5.3.1   For example, a 2½-in. steel pipe, which is permitted to supply 30 sprinklers, may supply a total of 50 sprinklers where not more than 30 sprinklers are above or below a ceiling.

The probability of a fire involving the areas both above and below a ceiling in a light hazard occupancy and opening in excess of 50 sprinklers is extremely remote. To size piping for in excess of 50 sprinklers the pipe is increased to 3 in. to reduce friction loss.

3-6 Schedule for Ordinary Hazard Occupancies.

3-6.1   Branch lines shall not exceed 8 sprinklers on either side of a cross main.

Exception: When occupancy is classified as ordinary hazard and when more than 8 sprinklers on a branch line are necessary, lines may be increased to 9 sprinklers by making the two end lengths 1 in. and 1 ¼ in., respectively, and the sizes thereafter standard. Ten sprinklers may be placed on a branch line making the two end lengths 1 in. and 1 ¼ in., respectively, and feeding the tenth sprinkler by a 2 ½ -in. pipe.

The amount of pressure lost to friction in long runs of small diameter pipe is significant. To ensure that pressure is not excessively reduced, the number of sprinklers on a branch line is limited to 8 spinklers in ordinary hazard occupancies. The exception allows up to 10 sprinklers on branch lines in ordinary hazard occupancies providing pipe sizing is increased to reduce friction loss.

3-6.2   Pipe sizes shall be in accordance with Table 3-6.2(a).

The pipe sizes specified in Table 3-6.2(a) were originally developed taking into consideration "normal" water supplies and friction loss. The pipe sizes tend to be somewhat smaller in branch lines and larger in cross-mains, feed-mains, and system risers than with systems designed hydraulically as outlined in Chapter 7.

Table 3-6.2(a)

Steel		Copper	
1 in. pipe	2 sprinklers	1 in. tube	2 sprinklers
1¼ in. pipe	3 sprinklers	1¼ in. tube	3 sprinklers
1½ in. pipe	5 sprinklers	1½ in. tube	5 sprinklers
2 in. pipe	10 sprinklers	2 in. tube	12 sprinklers
2½ in. pipe	20 sprinklers	2½ in. tube	25 sprinklers
3 in. pipe	40 sprinklers	3 in. tube	45 sprinklers
3½ in. pipe	65 sprinklers	3½ in. tube	75 sprinklers
4 in. pipe ....100 sprinklers		4 in. tube ....115 sprinklers	
5 in. pipe ....160 sprinklers		5 in. tube ....180 sprinklers	
6 in. pipe ....275 sprinklers		6 in. tube ....300 sprinklers	
8 in. pipe .... See 3-3.1 and		8 in. tube .... See 3-3.1 and	
3-3.1 Exception		3-3.1 Exception	

Exception: When the distance between sprinklers on the branch line exceeds 12 ft (3.7 m), or the distance between the branch lines exceeds 12 ft (3.7 m), the number of sprinklers for a given pipe size shall be in accordance with Table 3-6.2(b).

This exception recognizes the increased friction loss when either distance between sprinklers on branch lines or the distance between branch lines exceeds 12 ft (3.7 m). It further recognizes the need for adjacent sprinklers to reinforce each other toward the edges of the patterns in order to provide a more uniform overall density. Increasing the pipe sizing as required therefore reduces the pressure loss due to friction making available greater flowing pressure with a resultant increased discharge and improved distribution.

Table 3-6.2(b)

Steel	Copper
2½ in. pipe .... 15 sprinklers	2½ in. tube .... 20 sprinklers
3 in. pipe .... 30 sprinklers	3 in. tube .... 35 sprinklers
3½ in. pipe .... 60 sprinklers	3½ in. tube .... 65 sprinklers

For other pipe and tube sizes, see Table 3-6.2(a).

3-6.3   When sprinklers are installed above and below ceilings and such sprinklers are supplied from a common set of branch lines, such branch lines shall not exceed 8 sprinklers above and 8 sprinklers below any ceiling on either side of the cross main. Pipe sizing up to and including 3 in. shall be as shown in Table 3-6.3 [*see Figures 3-5.3(B) and 3-5.3(C)*].

Table 3-6.3   Number of Sprinklers Above and Below a Ceiling

Steel	Copper
1 in. .... 2 sprinklers	1 in. .... 2 sprinklers
1¼ in. .... 4 sprinklers	1¼ in. .... 4 sprinklers
1½ in. .... 7 sprinklers	1½ in. .... 7 sprinklers
2 in. ....15 sprinklers	2 in. ....18 sprinklers
2½ in. ....30 sprinklers	2½ in. ....40 sprinklers
3 in. ....60 sprinklers	3 in. ....65 sprinklers

Since a fire either above or below a ceiling in an ordinary hazard occupancy will normally be contained in that area by operating sprinklers, the standard permits in ordinary hazard pipe schedule systems more sprinklers on a branch line than are permitted on a branch line protecting a single area and also permits more sprinklers to be supplied by a given pipe size.

3-6.3.1*   When the total number of sprinklers above and below a ceiling exceeds the number specified in Table 3-6.3 for 3-in. pipe, the pipe supplying such sprinklers shall be increased to 3½ in. and sized thereafter according to the schedule shown in Table 3-6.2(a) for the number of sprinklers above or below a ceiling, whichever is larger.

A-3-6.3.1   For example, a 3-in. steel pipe, which is permitted to supply 40 sprinklers, may supply a total of 60 sprinklers where not more than 40 sprinklers are above or below a ceiling.

The probability of a fire involving the area both above and below a ceiling in an ordinary hazard occupancy and opening in excess of 60 sprinklers is extremely remote. To size piping or tubing for systems with more than 60 sprinklers with steel pipe, and 65 sprinklers with copper tubing, the pipe or tubing size is increased to reduce friction loss.

3-7 Schedule for Extra Hazard Occupancies.

3-7.1   Branch lines shall not exceed 6 sprinklers on either side of cross main. The number of sprinklers for a given pipe size shall be in accordance with Table 3-7.1.

Table 3-7.1

Steel		Copper	
1 in. pipe	1 sprinkler	1 in. tube	1 sprinkler
1¼ in. pipe	2 sprinklers	1¼ in. tube	2 sprinklers
1½ in. pipe	5 sprinklers	1½ in. tube	5 sprinklers
2 in. pipe	8 sprinklers	2 in. tube	8 sprinklers
2½ in. pipe	15 sprinklers	2½ in. tube	20 sprinklers
3 in. pipe	27 sprinklers	3 in. tube	30 sprinklers
3½ in. pipe	40 sprinklers	3½ in. tube	45 sprinklers
4 in. pipe	55 sprinklers	4 in. tube	65 sprinklers
5 in. pipe	90 sprinklers	5 in. tube	100 sprinklers
6 in. pipe	150 sprinklers	6 in. tube	170 sprinklers
8 in. pipe	See 3-3.1	8 in. tube	See 3-3.1

The amounts of pressure lost to friction in long runs of small diameter pipe are significant. To ensure reasonable friction loss, the number of sprinklers on a branch line is limited to 6 sprinklers in extra hazard occupancies. Fewer sprinklers are allowed on each size pipe than are allowed for ordinary hazard occupancies.

3-7.2   Open sprinkler and deluge systems shall be hydraulically calculated according to applicable standards.

Exception: Open sprinklers for exposure protection. See Chapter 6.

Pipe schedule systems are designed with a margin of safety with recognition that only some of the sprinklers will operate in a fire. In open and deluge systems water discharges from all sprinklers including those outside the fire area. Therefore, the system must be specifically designed to provide the density of coverage required to extinguish or control the fire over the entire protected area.

3-8 Special Provisions Applicable to Piping.

3-8.1   For sprinklers in storage racks see NFPA 231C, *Standard for Rack Storage of Materials.*

Racked storage presents unique, rapidly developing high heat release shielded fires which are addressed in NFPA 231C, *Standard for Rack Storage of Materials.* That standard gives direction with regard to sprinkler location, sprinkler spacing, and water densities based on a large-scale testing program dealing with rack storage. Systems are installed in accordance with this standard except when specifically modified by NFPA 231C.

3-8.2* Provision for Flushing Systems.   All sprinkler systems shall be arranged for flushing. Readily removable fittings shall be provided at the end of all cross mains. All cross mains shall terminate in 1¼ in. or larger pipe. All branch lines on gridded systems shall be arranged to facilitate flushing. (*See NFPA 13A, Inspection, Testing and Maintenance of Sprinkler Systems.*)

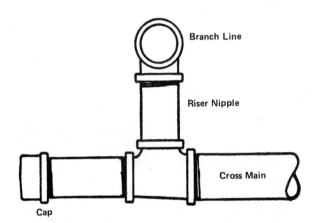

Figure A-3-8.2(a)   Screw-type Cap.

The readily removable fitting at the end of each crossmain is not limited with regard to maximum size. Branch lines in gridded systems must be installed in such a manner that the piping can be readily disconnected. Further guidance is provided in NFPA 13A, *Recommended Practice for the Care and Maintenance of Sprinkler Systems.* It is important to provide for flushing, particularly where

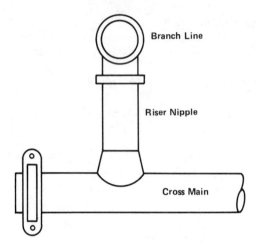

Figure A-3-8.2(b)   Groove-type Cap.

the system is supplied from a nonpotable water supply such as ponds, streams, or lakes or where the system has been or might alternate between wet- and dry-pipe operation (a practice which is to be discouraged). In either of these cases, foreign material, which can block the system or block individual sprinklers on the system, may be introduced, and it is necessary that provisions be provided to readily flush out and remove this material.

3-8.3 Stair Towers.   Stairs, towers or other construction with incomplete floors, if piped on independent risers, shall be treated as one area with reference to pipe sizes.

A fire in a stair or tower may tend to open a large percentage of the protecting sprinklers. An independent or separate riser supplying such an area must be sized to supply all the sprinklers in the area.

3-8.4 Return Bends.   When piping on wet systems is concealed, with sprinklers installed in pendent position below a ceiling, return bends shall be used when the water supply to the sprinkler system is from a raw water source, millpond, or from open top reservoirs. Return bends shall be connected to the tops of branch lines in order to avoid accumulation of sediment in the drop nipples. In new systems the return bend pipe and fittings shall be 1 in. in size. In revamping existing systems, where it is not necessary to retain sprinklers in the concealed space, ½-in. or ¾-in. close nipples inserted in the existing sprinkler fittings may be used with 1-in. pipe and fittings for the other portions of the return bend. When water supply is potable, return bends are not required on wet systems. (*See Figure 3-8.4.*)

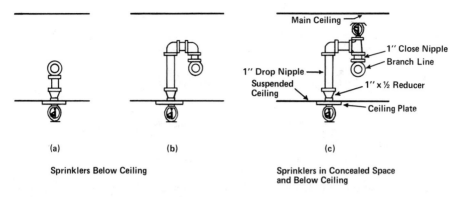

(a)

(b)

Sprinklers Below Ceiling

Main Ceiling

1" Close Nipple

Branch Line

1" Drop Nipple

Suspended
Ceiling

1" x ½ Reducer

Ceiling Plate

(c)

Sprinklers in Concealed Space
and Below Ceiling

Figure 3-8.4   Pendent Sprinklers at Suspended Ceiling.

Formal Interpretation

Question: In revamping an existing wet-pipe fire sprinkler system which has a potable water supply, and where it is not necessary to retain sprinklers in the concealed space above a noncombustible ceiling, would it be correct to use close nipples of the sprinkler thread size inserted in the existing sprinkler fittings with 1 in. pipe and fittings for the other portions of the drop?

Answer: Yes.

The purpose of return bends is to prevent collection of sedimentation in drop nipples which might be taken directly off the bottom of branch lines in wet-pipe sprinkler systems, particularly where the systems are supplied from water sources which might contain excessive sedimentation. Return bends are not necessary where a potable water supply is used. Return bends are also useful where centering of sprinklers in ceiling tile or exact positioning of sprinklers is desirable from an esthetic or architectural viewpoint. When standard sprinklers rather than dry pendent sprinklers are installed on dry systems, return bends are required in all cases due to the possibility of scale buildup. In such instances the sprinklers should be installed in accordance with 5-2.2. When it is necessary in revamping existing systems to provide protection under false ceilings and to retain existing sprinklers in the concealed space, the existing line fittings must be replaced with fittings providing 1 in. outlets. [See Figure 3-8.4(c).]

3-8.5 Dry Pipe Underground.   When necessary to place pipe which will be under air pressure underground, the pipe shall be protected against corrosion (*see 3-10.2*), or unprotected cast or ductile iron pipe may be used when joined with a gasketed joint listed for air service underground.

Corrosion forming on the exterior of cast iron or ductile iron pipe insulates that pipe from further corrosion. Such pipe is normally used to transport water which is less likely to leak than air. When it is to be subject to air pressure, it must have gasketed joints, and these joints must be specifically listed for use under air pressure.

3-8.6* One and One-Half-Inch Hose Connections.   One and one-half-inch (1½-in.) hose used for fire purposes only may be connected to wet sprinkler systems only, subject to the following restrictions:

(a) Hose stations supply pipes shall not be connected to any pipe smaller than 2½ in.

Exception: For hydraulically designed loops and grids the minimum size pipe between the hose stations supply pipe and the source may be 2 in.

(b) Pipe shall be minimum 1 in. for horizontal runs up to 20 ft (6.1 m), minimum 1¼ in. for the entire run for runs between 20 and 80 ft (6.1 and 24.4 m), and minimum 1½ in. for the entire run for runs greater than 80 ft (24.4 m).

Formal Interpretation

Question: When a horizontal run of pipe 50 ft long supplies a 1½ inch hose station, is the Committee's intent that:

(a) The first 20 ft of the run be 1 inch and the remainder 1¼ inch pipe?

(b) The entire 50 ft be 1¼ inch pipe?

Answer: The entire 50 ft should be 1¼ inch pipe.

(c) Piping shall be at least 1 in. for vertical runs.

(d) When the pressure at any hose station outlet exceeds 100 psi (6.9 bars), an approved device shall be installed at the outlet to reduce the pressure at the outlet to 100 psi (6.9 bars).

One and one-half inch hose connections are restricted to wet systems only. The maintenance problems due to loss of air pressure through the hose valves in other types of systems would outweigh their value. The pressure at the outlet is restricted to 100 psi (6.9 bars) due to the danger to the untrained operator in using a hose subject to high pressure.

A-3-8.6   One and one-half (1½) in. hose connections for use in storage occupancies and other locations where standpipe systems are not required are covered by this standard. When Class II standpipe systems are required see the appropriate provisions of NFPA 14, *Standard for the Installation of Standpipe and Hose Systems*, with respect to hose stations and water supply for hose connections from sprinkler systems.

One and one-half inch hose conections supplied from sprinkler systems have been successful in extinguishment and fire control for many years. It is not felt that the requirements of Class II standpipe systems in NFPA 14, *Standard for the Installation of Standpipe and Hose Systems*, should apply to hose connections attached to sprinkler systems inasmuch as this would impose water supply requirements on the sprinkler system in excess of those required for the sprinklers.

3-8.7*   Hose Connections for Fire Department Use.   In buildings of light or ordinary hazard occupancy, 2½-in. hose valves for fire department use may be attached to wet-pipe sprinkler system risers subject to the following restrictions:

(a) Sprinklers shall be under separate floor control valves.

(b) The minimum size of the riser shall be 4 in. unless hydraulic calculations indicate smaller size riser will satisfy sprinkler and hose stream demand.

(c) For completely sprinklered buildings, the water supply for sprinklers need not be added to standpipe demand as determined from NFPA 14, *Standard for the Installation of Standpipe and Hose Systems*.

Exception: When the sprinkler system demand, including hose stream allowance, indicated in Table 2-2.1(B) exceeds the requirements of NFPA 14, the values in Table 2-2.1(B) shall be used.

(d) For partially sprinklered buildings, the sprinkler demand, not including hose stream allowance, as indicated in Table 2-2.1(B) shall be added to the requirements given in NFPA 14.

(e) Each combined sprinkler and standpipe riser shall be equipped with a riser control valve to permit isolating a riser without interrupting the supply to other risers from the same source of supply.

(f) For fire department connections serving standpipe and sprinkler refer to Section 2-7.

For completely sprinklered buildings in which the risers supply both sprinklers and 2½-in. hose valves for fire department use the water supply must meet the requirements of NFPA 14 or the requirements of this standard, whichever are greater. However, when the occupancy is only partially sprinklered, the requirements of this standard, excluding hose stream allowance, and that of NFPA 14 must be combined, as the sprinkler system could be subjected to a fire starting outside the protected area. It should be recognized that, in many cases, 2½-in. hose valves will in fact be used by the fire department, and as such either on-site water supplies or pumping capacity for such hose valves should not be considered necessary. This additional hose stream usage within the building would normally be provided by the fire department pumper pumping into the fire department connection.

A-3-8.7   Combined automatic sprinkler and standpipe risers should not be interconnected by sprinkler system piping.

When risers are used to supply combination sprinkler and standpipe systems, guidance is needed to avoid confusion when systems are controlled by more than one valve in remotely located areas. When repairs or alterations are necessary this could lead to extensive water damage. No appreciable improvement in reliability is realized by cross connection.

3-9 System Test Pipes.

3-9.1 Wet Systems.

3-9.1.1*   A test pipe not less than 1 in. in diameter terminating in a smooth bore corrosion resistant orifice giving a flow equivalent to one sprinkler shall be provided for each system. The test connection valve shall be readily accessible. The discharge shall be to the outside, to a drain connection capable of accepting full flow under system pressure or to another location where water damage will not result.

The primary function of the wet system inspector's test is to test the operation of the water flow alarm device(s) at a flow equivalent to that of one operating sprinkler. The purpose of the remote branch line location is of course to provide the most severe testing requirement.

A-3-9.1.1   This test pipe should be in the upper story, and the connection should preferably be piped from the end of the most remote branch line. The discharge should be at a point where it can be readily observed. In locations where it is not practical to terminate the test pipe outside the building, the test pipe may terminate into a drain capable of accepting full flow under system pressure. (*See A-3-11.4.1.*) In this event, the test connection should be made using an approved sight test connection containing a smooth bore corrosion resistant orifice giving a flow equivalent to one sprinkler simulating the least flow from an individual sprinkler in the system. (*See Figures A-3-9.1.1 and A-3-9.1.2.*) The test valve should be located at an accessible point, and preferably not over 7 ft (2.1 m) above the floor. The control valve on the test connection should be located at a point not exposed to freezing.

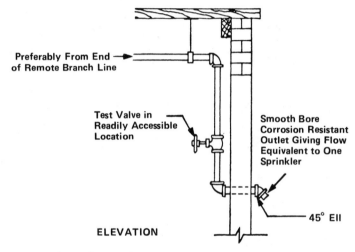

Preferably From End →
of Remote Branch Line

Test Valve in
Readily Accessible
Location

Smooth Bore
Corrosion Resistant
Outlet Giving Flow
Equivalent to One
Sprinkler

45° Ell

ELEVATION

For SI Units: 1 ft = 0.3048 m.

NOTE: Not less than 4 ft (1.2 m) of exposed test pipe in warm room beyond valve when pipe extends through wall to outside.

Figure A-3-9-1.1   System Test Pipe on Wet-Pipe System.

Ideally the inspector's test is located at the end of the most remote branch line in the upper story. Such a location is not, however, required. (*See 3-17.1.*)

3-9.1.2*   In multistory buildings where waterflow alarm devices are provided at each riser on each floor or where more than one alarm device is provided in one sprinkler system, a test pipe shall be provided for testing each alarm device.

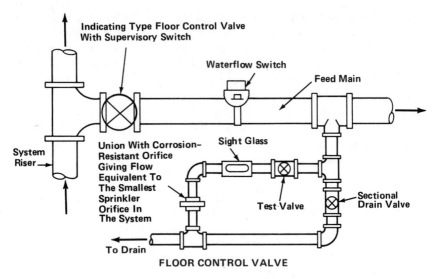

Figure A-3-9.1.2   Floor Control Valve.

The intent of this section is to require inspector's test connections located in a manner which will provide for the testing of all water flow alarm devices. It permits both an electrically operated alarm and a water motor alarm supplied from an alarm valve to be tested through a single test pipe.

3-9.2* Dry-Pipe Systems.   A test pipe not less than 1 in. in size terminating in a smooth bore corrosion resistant orifice to provide a flow equivalent to one sprinkler of a type installed on the particular system shall be installed on the end of the most distant sprinkler pipe in the upper story and be equipped with a readily accessible 1-in. shut-off valve and plug at least one of which shall be brass. In lieu of a plug, a nipple and cap may be used.

Exception: When a dry-pipe system is subdivided by the use of check valves according to 5-2.3.1, a separate system test valve shall be installed at the remote point of each section that is checked off.

The dry-pipe system inspector's test is needed to provide the capability to measure the approximate time from the opening of the most distant sprinkler in the system until water flows from that sprinkler. For that reason it must be located at the end of the farthest line in the top story of the protected occupancy. The valve must be sealed with a plug or nipple and cap when not in use to avoid leakage of air and to avoid accidental tripping of the dry-pipe valve.

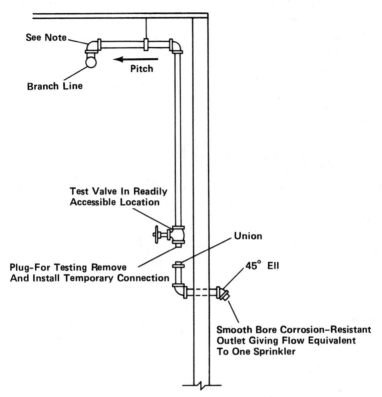

NOTE: To minimize condensation of water in the drop to the test connection, provide a nipple-up off of the branch line.

Figure A-3-9.2   System Test Pipe on Dry-Pipe System.

It is permissible to install the test valve on the drop pipe for the purpose of accessibility. When this is done the connection should be made to the top of the branch line to minimize condensation buildup and moisture should be drained periodically as is done with low point drains in areas subject to freezing. Manufacturer's instructions should be followed to avoid accidental operation of dry-pipe valves equipped with quick-opening devices.

3-9.3   A test pipe shall be used on a preaction system using supervisory air.

3-9.4   A test pipe is not required on a deluge system.

3-10* Protection of Piping.

A-3-10 Protection of Piping Against Damage Due to Impact.   Sprinkler piping should be located so as to minimize the possibility of damage due to impact by mobile material handling equipment and other vehicles. For example, risers adjacent to structural columns and out of vehicle travel routes are generally safe, as are feed mains and cross mains shielded by heavy structural members such as girders.

3-10.1 Protection of Piping Against Freezing.

3-10.1.1   When portions of systems subject to freezing and temperatures cannot be reliably maintained at or above 40°F (4°C) sprinklers shall be installed as a dry-pipe or pre-action system in such areas.

Exception: Small unheated areas may be protected by antifreeze systems. (See Section 5-5.)

Piping only areas subject to freezing as dry-pipe or pre-action systems, with the remainder of the installation piped as a wet system, is preferable to piping the entire system dry because of faster response, reduced corrosion, and increased reliability of a wet system.

3-10.1.2*   When supply pipes, risers, system risers or feed mains pass through open areas, cold rooms, passageways or other areas exposed to freezing, the pipe shall be protected against freezing by insulating coverings, frostproof casings or other reliable means capable of maintaining a minimum 40°F (4°C).

A-3-10.1.2   In areas subject to freezing climates, when piping extends through an exterior wall, as for fire department connections, system test pipes, or drains, a minimum of 4 ft (1.2 m) of pipe should be maintained between the wall and the section of piping containing water.

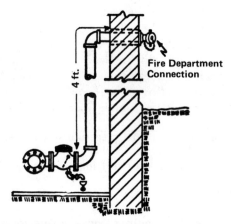

For SI Units: 1 ft = 0.3048 m.

Figure A-3-10.1.2   Minimum Clearance to Avoid Freezing.

The recommended 4 ft (1.2 m) of pipe to be maintained between the exterior wall and the piping containing water is for the purpose of providing a frost break.

3-10.2 Protection of Piping Against Corrosion.

3-10.2.1*   Where corrosive conditions are known to exist due to moisture or fumes from corrosive chemicals, or both, types of piping, fittings and hangers that resist corrosion shall be used or a protective coating shall be applied to all unprotected exposed surfaces of the sprinkler system to resist corrosion (*see 3-16.4*).

A-3-10.2.1   Types of locations where corrosive conditions may exist include bleacheries, dye-houses, metalplating processes, animal pens and certain chemical plants.

If corrosive conditions are not of great intensity and humidity is not abnormally high, good results can be obtained by a protective coating of red lead and varnish or by a good grade of commercial acid-resisting paint. The paint manufacturer's instructions should be followed in the preparation of the surface and in the method of application.

Where moisture conditions are severe but corrosive conditions are not of great intensity, copper tube or galvanized steel pipe, fittings and hangers may be suitable. The threaded ends of steel pipe should be painted.

In instances where the piping is not readily accessible and where the exposure to corrosive fumes is severe, either a protective coating of high quality may be employed or resort may be made to the use of some form of corrosion resisting material.

The referenced protective paints and coatings are not applied to the automatic sprinklers. Sprinklers must be provided with corrosion-resistant coatings applied by the manufacturer in accordance with 3-16.4.

3-10.2.2   Where water supplies are known to have unusual corrosive properties and threaded or cut grooved steel pipe is to be used, wall thickness shall be in accordance with Schedule 30 (in sizes 8 in. or larger) or Schedule 40 (in sizes less than 8 in.).

While corrosion protection for exposed surfaces is independent of wall thickness and acceptable pipe joining methods, the use of threaded thin wall pipe must be avoided where subject internally to water having unusual corrosive properties because of the lack of material between the inside diameter and the root diameter of the threads. If threaded pipe is used, it should be Schedule 30 or heavier for pipe sizes 8 in. or larger and Schedule 40 for pipe sizes less than 8 in. The use of welded or roll grooved Schedule 10 pipe would be satisfactory.

3-10.2.3*   Steel pipe in overhead feed mains running from one building to another, where exposed to the weather, shall be galvanized, or otherwise protected against corrosion.

A-3-10.2.3   It is important when protected steel pipe (galvanized, dipped and wrapped, coated, etc.) is used that particular care is taken to see that all exposed threads, wrench marks, or abrasions which have penetrated through the protection be repaired, sealed and/or properly coated.

3-10.2.4   When steel pipe is used underground as a connection from a system to sprinklers in a detached building, the pipe shall be protected against corrosion before being buried.

3-10.3* Protection of Piping Against Damage Where Subject to Earthquakes.

Earthquake design criteria have been included in this standard for fifty years. The performance of automatic sprinkler systems in earthquakes has been good. Following the San Fernando Earthquake in 1971 (6.6 Richter magnitude), the Pacific Fire Rating Bureau surveyed 973 sprinklered buildings and reported that "if a sprinklered building fared well, so did the sprinkler system."

It is not the intent of this standard to specify where earthquake design provisions must be used, but to provide design criteria for those systems where, according to the authority having jurisdiction, the system is subject to earthquakes.

A-3-10.3 Protection of Piping Against Damage Where Subject to Earthquakes.

3-10.3.1*   Sprinkler systems shall be protected to minimize or prevent pipe breakage where subject to earthquakes as follows:

(a) Piping shall be made flexible where necessary.

(b) Piping shall be tied to the structure for minimum relative movement, but allowing for expansion, and differential movement within and between structures.

A-3-10.3.1   Most earthquake-related strains are imparted to the sprinkler piping by the building. If the piping could be isolated from the building, earthquake strains could enter only through the riser. It is differential building movement between parts with different natural frequencies that damages sprinkler systems. Any method designed to protect sprinkler systems from earthquake-induced strains should combine flexibility and dampening as appropriate.

Flexibility is provided by requirements for flexible couplings and swing joints in 3-10.3.2 and 3-10.3.3, and clearances in 3-10.3.4. Dampening is provided by the requirements for limited bracing in 3-10.3.5.

3-10.3.2* Couplings.

A-3-10.3.2   Strains on sprinkler piping can be greatly lessened and, in many cases, damage prevented by increasing the flexibility between major parts of the sprinkler system. One part of the piping should never be held rigidly and another part allowed to move freely without provision for relieving the strain. Flexibility can be provided by use of listed flexible couplings, joining grooved end pipe at critical points and by allowing clearances at walls and floors.

Tank or pump risers should be treated the same as sprinkler risers for their portion within a building. The discharge pipe of tanks on buildings should have a control valve above the roof line so any pipe break within the building can be controlled.

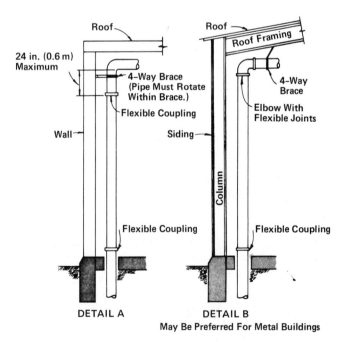

Figure A-3-10.3.2(a)   Riser Details.

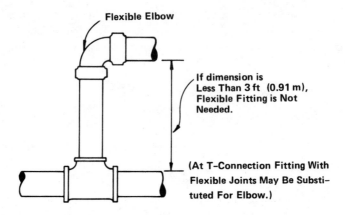

Figure A-3-10.3.2(b)   Detail at Short Riser.

Piping 3 in. or smaller in size is pliable enough so that flexible couplings are not usually necessary.

3-10.3.2.1   Listed flexible pipe couplings joining grooved end pipe shall be provided as flexure joints to allow individual sections of piping 3½ in. or larger to move differentially with the individual sections of the building to which it is attached. Couplings shall be arranged to coincide with structural separations within a building. They shall be installed:

(a) Within 24 in. (610 mm) of the top and bottom of all risers.

Exception No. 1:   In risers less than 3 ft (0.9 m) in length flexible couplings may be omitted.

Exception No. 2:   In risers 3 to 7 ft (0.9 to 2.1 m) in length, one flexible coupling is adequate.

(b) At the ceiling of each intermediate floor in multi-story buildings.

(c) At each side of concrete or masonry walls 2 to 3 ft (0.6 to 0.9 m) from wall surface.

(d)* On one side of building expansion joints.

A flexible or flexure joint is a pipe joint made using a listed flexible pipe coupling joining grooved end pipe. (*See Figure 3.1.*) "Rigid-type" mechanical couplings which do not permit movement at the grooved connection are not considered flexible couplings.

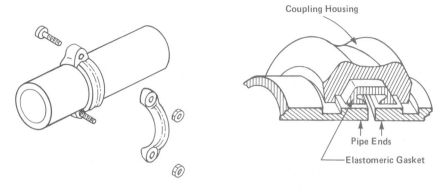

Figure 3.1.   Flexible pipe coupling.

A-3-10.3.2.1(d)   Sprinkler piping above the first floor may cross structural separations such as expansion joints if the piping is specifically designed with flexible connections at each crossing and able to accommodate the calculated differential motions during earthquakes, but not less than 4 in. (102 mm). In lieu of calculations, flexibility can be made at least twice the actual separations at right angles to the separation as well as parallel to it.

A building expansion joint is usually a bituminous fiber strip used to separate blocks or units of concrete to prevent cracking due to expansion as a result of temperature changes. In this case, the flexible coupling required on one side by 3-10.3.2(d) will suffice.

For seismic separation joints, considerably more flexibility is needed, particularly for piping above the first floor. Figure 3.2 shows a method of providing additional flexibility through the use of swing joints.

3-10.3.2.2   Listed plain-end couplings shall not be installed in sprinkler systems subject to earthquakes.

3-10.3.3 Fittings.   Additional fittings and devices with flexible joints shall be installed where necessary.

3-10.3.3.1   Fittings with flexible joints shall be installed at the top of drops to hose lines regardless of piping size.

3-10.3.3.2   Drops to sprinklers in racks shall be equipped with swing joints assembled with flexible fittings between the rack and the overhead sprinkler system.

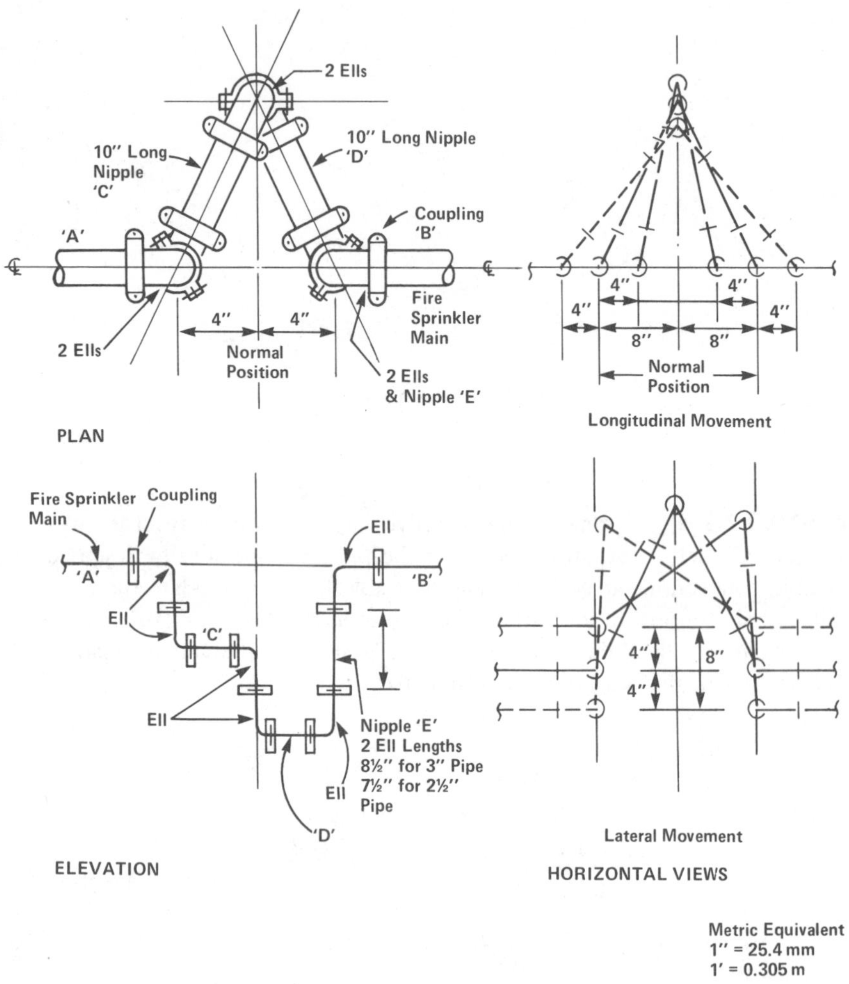

PLAN

2 Ells

10" Long Nipple 'D'

10" Long Nipple 'C'

Coupling 'B'

'A'

Fire Sprinkler Main

2 Ells

4"   4"
Normal Position

2 Ells & Nipple 'E'

Longitudinal Movement

4"   4"
4"   4"
8"   8"
Normal Position

ELEVATION

Fire Sprinkler Coupling Main

Ell

'A'   'B'

Ell

'C'

Ell

Nipple 'E'
2 Ell Lengths
8½" for 3" Pipe
7½" for 2½" Pipe

Ell

'D'

Lateral Movement

4"   8"
4"

HORIZONTAL VIEWS

Metric Equivalent
1" = 25.4 mm
1' = 0.305 m

Figure 3.2.   Seismic flexible joint.

Exception: Flexible fittings are not required in the swing joints on drops 3 in. or less in size.

3-10.3.4* Clearance. Clearance shall be provided around all piping extending through walls, floors, platforms and foundations, including drains, fire department connections and other auxiliary piping.

(a) Minimum clearance on all sides shall be not less than 1 in. (25 mm) for pipes 1 in. through 3½ in. and 2 in. (51 mm) for pipe sizes 4 in. and larger.

Exception No. 1:   When clearance is provided by a pipe sleeve, a nominal diameter 2 in. (50 mm) larger than the nominal diameter of the pipe is acceptable for pipe sizes 1 in. through 3 ½ in. and the clearance provided by a pipe sleeve of nominal diameter 4 in. larger than the nominal diameter of the pipe is acceptable for pipe sizes 4 in. and larger.

Exception No. 2:   No clearance is necessary for piping passing through gypsum board or equally frangible construction, which is not required to have a fire-resistance rating.

Exception No. 3:   When piping enters a building through a basement wall and ground water conditions make providing clearance a problem, the end of the pipe may be attached firmly to the wall, with provisions to allow flexing to take place outside the building. The pipe shall be connected to the riser with fittings with flexible joints.

(b) When required the clearance shall be filled with a flexible material such as mastic.

A-3-10.3.4   While clearances are necessary around the sprinkler piping to prevent breakage due to building movement, suitable provision should also be made to prevent passage of water, smoke or fire.

Drains, fire department connections, and other auxiliary piping connected to risers should not be cemented into walls or floors; similarly, pipes which pass horizontally through walls or foundations should not be cemented solidly or strains will accumulate at such points.

When risers or lengths of pipe extend through suspended ceilings, they should not be fastened to the ceiling framing members.

Although Appendix Section B-3-1 recommends thimbles or sleeves for sprinkler piping passing through floors of concrete or waterproof construcion to prevent floor leakage, it exempts areas subject to earthquakes. Adequate clearance must be provided to avoid damage to the sprinkler system in the event of an earthquake. Pipe sleeves are a means to that end but not the exclusive means.

3-10.3.5* Sway Bracing of Piping Where Subject to Earthquakes.

It is the intent to brace the main sprinkler piping so that it will withstand a horizontal force equal to 50 percent of the weight of the water-filled piping (Horizontal Acceleration A^H = 0.5 g). It is felt that if the piping is designed to withstand this force without breaking or permanently deforming, the system will be reasonably safe from earthquake forces.

Lateral bracing must be spaced such that the piping will not fail as a beam under the uniform load of its expected horizontal "weight." A maximum spacing of 40 ft (12.2 m) is specified.

Branch lines are not required to be braced laterally except where movement could damage other equipment, because the smaller piping used for branches is capable of considerable shaking without damage. Furthermore, bracing of mains will prevent excessive movement.

It is also the intent to provide longitudinal bracing for each run of main piping. One brace per run should be sufficient so long as the brace is sized to accommodate the weight of the piping and the piping is held in a straight line by intermediate hangers and lateral braces. Where "doglegs" or turns are made in the mains, an additional longitudinal brace should be provided for each run of feed or cross main piping.

Lateral braces and longitudinal braces are both "two-way braces" in that they prevent piping from shaking back and forth in one direction. "Four-way bracing" requires the simultaneous action of lateral and longitudinal braces. Unlike the requirements for flexible couplings, braces are required for feed and cross mains regardless of their size. As indicated in 3-10.3.1, all braces must be fastened directly to the building structure.

A-3-10.3.5 Sway Bracing.

Location of Bracing.   [*See Figure A-3-10.3.5(a).*]

Two-way braces are either longitudinal or lateral depending on their orientation with the axis of the piping. [*See Figures A-3-10.3.5(a), (b), (c) and (d)*]. The simplest form of two-way brace is a piece of steel pipe or angle. Because the brace must act in both compression and tension, it is necessary to size the brace to prevent buckling.

An important aspect of sway bracing is its location.

In Building 1, the relatively heavy main will pull on the branch lines when shaking occurs. If the branch lines are held rigidly to the roof or floor above, the fittings can fracture due to the induced stresses.

Bracing should be on the main as indicated at Location "B." With shaking in the direction of the arrows, the light branch lines will be held at the fittings. When a branch line can pound against a piece of equipment, such as a space heater or a structural member, a lateral brace should be installed on the branch line to help prevent rupture.

A four-way brace is indicated at Location "A" [*also see Figure A-3-10.3.5(g)*]. This keeps the riser and main lined up and also prevents the main from shifting.

In Building 1, the branch lines are flexible in a direction parallel to the main, regardless of building movement. The heavy main cannot shift under the roof or floor, and it also steadies the branch lines.

While the main is braced, the flexible couplings on the riser allow the sprinkler system to move with the floor or roof above, relative to the floor below.

Figures A-3-10.3.5(b), (c) and (d) show typical locations of sway bracing for pipe schedule, gridded, and looped sprinkler systems.

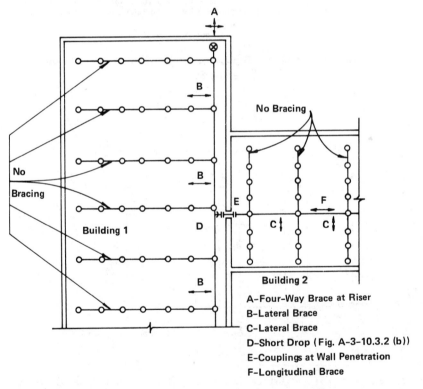

Figure A-3-10.3.5(a)   Earthquake Protection for Sprinkler Piping.

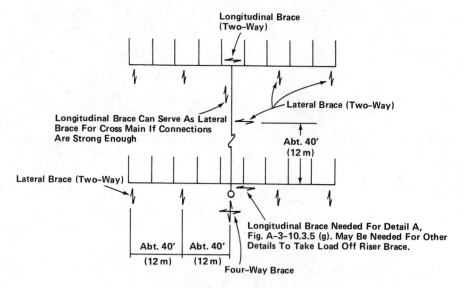

Figure A-3-10.3.5 (b)   Typical Location of Bracing on a Pipe
Schedule System.

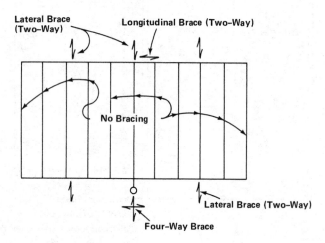

Figure A-3-10.3.5 (c)   Typical Location of Bracing on a Gridded System.

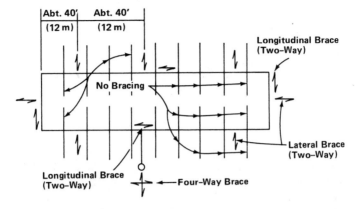

Figure A-3-10.3.5(d)   Typical Location of Bracing on a Looped System.

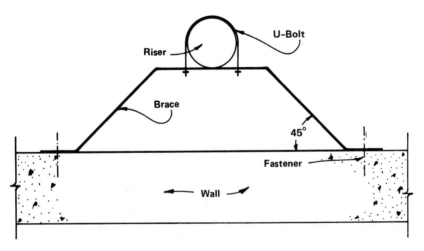

Figure A-3-10.3.5(e)   Detail of Four-Way Brace at Riser.

Formal Interpretation

Question No. 1: Is it the intent of Sections 3-10.3.5.10 and 3-10.3.5.2 of NFPA 13 to require lateral earthquake bracing for 2 in. and smaller diameter cross mains?

Answer: Yes.

Question No. 2: If the response to question one is yes, is it the intent of Section 3-10.3.5.4 to permit hanger rods less than 6 in. long to act as lateral earthquake bracing when toggle hangers are used to support cross mains suspended below metal lathe and plaster ceilings?

Answer: No. Section 3-10.3.1(b) requires that the piping be tied to the building structure.

Question No. 3: Is it the intent of Section 3-10.3.4 of NFPA 13 to require a nominal clearance of 2 in. for pipes larger than 3½ in.?

Answer: It is the intent of Section 3-10.3.4 to require a minimum 2 in. clearance for pipes larger than 3½ in. except that a nominal 2 in. clearance is acceptable when pipe sleeves are utilized.

Question No. 4: Is it the intent of Section 3-10.3.5 of NFPA 13 to require a four-way earthquake brace only at the top of risers in a multi-story building?

Answer: A four-way brace is required at the top of the riser in all buildings subject to earthquake damage. Four-way bracing is not required for intermediate floors in multi-story buildings.

Question No. 5: Is it the intent of Section 3-10.3.5 of NFPA 13 to require a four-way earthquake brace above the roof line when the sprinkler riser is used to supply 2½ in. hose outlets on each floor and above the roof?

Answer: No. The four-way brace just below the roof slab is adequate.

Table A-3-10.3.5   Typical Brace Sizes

Item	$l/r = 200$	Item	l/r 200
Angles		Flats	
1½ X 1½ X ¼ in.	4 ft 10 in.	1½ X ¼ in.	1 ft 2 in.
2 X 2 X ¼ in.	6 ft 6 in.	2 X ¼ in.	1 ft 2 in.
2½ X 2 X ¼ in.	7 ft 0 in.	2 X ⅜ in.	1 ft 9 in.
2½ X 2½ X ¼ in.	8 ft 2 in.	Pipe	
3 X 2½ X ¼ in.	8 ft 10 in.	1 in.	7 ft 0 in.
3 X 3 X ¼ in.	9 ft 10 in.	1¼ in.	9 ft 0 in.
Rods		1½ in.	10 ft 4 in.
¾ in.	3 ft 1 in.	2 in.	13 ft 1 in.
⅞ in.	3 ft 7 in.		

For SI Units: 1 in. = 25.4 mm; 1 ft = 0.3048 m.

Pipe sizes are ANSI B36.10 Schedule 40.

The ratio of length of brace to radius of gyration of brace cross-section is kept as close to 200 as practical in order to provide dampening for system vibrations. The value of 200 is chosen as the traditional limit for determining the maximum allowable fiber-stress for secondary compression members. Above 200 there would be a concern for buckling of the brace.

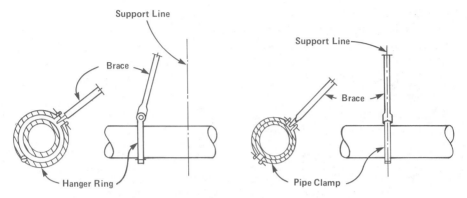

Figure 3.3.   Details of connection to brace pipe. Arrangement on left is not acceptable because hanger ring is not snug and brace is not perpendicular to pipe. Arrangement on right is acceptable.

3-10.3.5.1* Sway bracing shall be designed to withstand a force in tension or compression equivalent to not less than half the weight of water-filled piping.

A-3-10.3.5.1 Details of Bracing.

In the design of sway braces, the slenderness ratio l/r should approximately be 200 where "l" is the distance between the center lines of support and "r" is the least radius of gyration, both in inches (mm). Suggested lengths of shapes used for sway bracing are shown in Table A-3-10.3.5.

Sway bracing should be tight and concentric. Details should be laid out in advance so that suitable fittings are available on the job site. All parts and fittings of a brace should lie in a straight line to avoid eccentric loading on fittings and fasteners. [See Figure A-3-10.3.5.1(a).]

Connection to the pipe can be made with a pipe clamp or U-bolt. One bolt of the pipe clamp can pass through a flattened end of pipe or one leg of an angle. (The other leg and filet of the angle can be cut away.) Pipe rings should be avoided because they result in a loose fit. Once the pipe is able to vibrate within a loose fitting, the bolts in the ring assembly can be fractured.

The brace should not be connected to a tab welded to the pipe.

The brace can be attached to the structural system directly through a leg of an angle or a flattened portion of pipe. Where dimensions are tight or some play must be allowed, a special fitting can be used [*see Figure A-3-10.3.5.1(b)*]. This threads on an end of pipe. Rotation of the flat around the bolt allows play in the angle of the brace without sacrificing snugness.

Some adjustment can be provided in a pipe brace by use of a left-hand/right-hand coupling. However, this adds to the cost of earthquake protection, and care should be taken that sufficient thread is engaged or else the threads will shear.

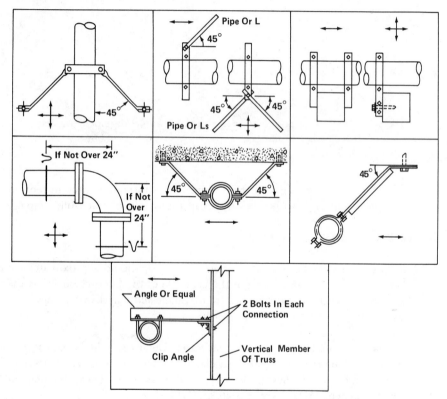

Figure A-3-10.3.5.1(a)   Acceptable Types of Sway Bracing.

Figure A-3-10.3.5.1(b)   Special Fitting.

In wood joisted buildings, U-hangers with legs bent out 10 degrees from the vertical satisfy the purpose of two-way braces. When piping is hung from C-clamps, positive bracing is needed because the clamp can slip under seismic shaking.

Where "C-type" clamps are used on flanges of steel beams, the type equipped with a retaining strap is often used to avoid the possibility of the clamps sliding off the flange. (*See Figure A-3-15.1.*)

3-10.3.5.2   Feed and cross mains shall be braced with a two-way longitudinal sway brace.

3-10.3.5.3*   Tops of risers shall be secured against drifting in any direction, utilizing a four-way sway brace.

A-3-10.3.5.3   A four-way brace is usually provided at the riser. It may provide longitudinal and lateral bracing for the adjacent mains. [*See Figures A-3-10.3.5(a) and A-3-10.3.2(a).*]

For risers, a U-bolt and flat bent at 45 degree angles can be used [*see Figure A-3-10.3.5(e)*].

The four-way brace at the top of the riser keeps the riser and main lined up and prevents the main from shifting. For multistory buildings, the intent is to require the four-way brace only at the highest floor. Where a combined sprinkler/standpipe riser extends through the ceiling to a roof outlet, the four-way brace located below the roof is sufficient.

3-10.3.5.4   Lateral sway bracing spaced at a maximum of 40 ft (12 m) centers shall be provided for feed and cross mains.

Exception: Sway bracing may be omitted when hanger rods less than 6 in. (152 mm) long are used.

Formal Interpretation

Question: If horizontal runs having flexible couplings for purposes other than the requirement for earthquake protection are provided, can the sway bracing indicated in 3-10.3.5.8 be omitted when the hanger rods are less than 6 in. long?

Answer: Yes.

Short hanger rods are expected to limit the lateral movement of mains, eliminating the need for lateral bracing.

3-10.3.5.5   Bracing shall be attached directly to feed and cross mains.

The intent is to disallow bracing of the feed and cross mains through the branch lines, and to emphasize that branch lines do not require bracing.

3-10.3.5.6   A length of pipe shall not be braced to sections of the building which will move differentially.

3-10.3.5.7   The last length of pipe at the end of a feed or cross main shall be provided with a lateral brace. Lateral braces may also act as longitudinal braces if they are within 24 in. (610 mm) of the center line of the piping braced longitudinally.

3-10.3.5.8   When additional flexible couplings are used in horizontal piping for purposes other than the requirements for earthquake protection (usually for ease of installation), a sway brace shall be provided within 24 in. (610 mm) of each such coupling.

Formal Interpretation

Question: Depicted below are two cases of sway bracing fire main piping. The first case shows the recommended bracing as per NFPA 13. The second case shows the same main cut into 10 ft 6 in. lengths so that it may be installed into the building ceiling. Please clarify if the method shown on Case No. 2 is correct per Section 3-10.3.5.8 in the present NFPA 13 pamphlet.

Answer: Neither of the examples is correct. Assuming that all joints are made with flexible couplings, a sway brace is required within 24 in. of all 3 couplings in Case 1 and all 5 couplings in Case 2. If mechanical grooved fittings of the rigid type were used, additional bracing beyond the normal requirements for earthquake protection would not be required.

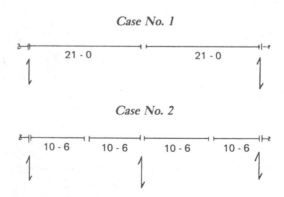

Case No. 1

21 - 0   21 - 0

Case No. 2

10 - 6   10 - 6   10 - 6   10 - 6

The intent is to prevent excessive movement of the piping, possibly resulting in "bellows" or accordion effects. Additional bracing is not required where "rigid" type mechanical couplings are used.

3-10.3.5.9   Where "U" hook hangers are used on branch lines, the pipe shall be secured to the end hanger by a wrap-around-type "U" hook. (*See Figure A-3-15.1.*)

In earthquake-prone areas, the wrap-around U hook at the end of the branch lines is intended to prevent the branches from whipping and possibly bouncing out of the hangers.

3-10.3.5.10   U-type hangers used to support a system will satisfy most of the requirements for sway bracing except that, in general, the longitudinal brace in Figure A-3-10.3.5(b) shall also be required for 2½ in. and larger diameter piping. U-type hangers used as lateral braces shall have legs bent out 10 degrees from the vertical.

Formal Interpretation

Question No. 1: What is the maximum length at which a U hook will satisfy sway bracing requirements as outlined in Section 3-10.3.5.10 of NFPA 13?

Answer: There is no maximum length restriction.

Question No. 2: Is it the intent of Section 3-10.3.5.8 of NFPA 13 to require sway bracing at each flexible coupling when using 21 in. or 24 in. lengths of pipe? Please refer to the attached sketch (Figure 3).

Answer: Yes.

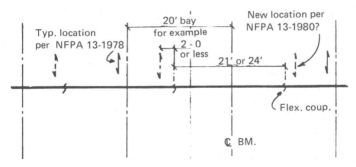

Figure 3

The ability of a brace to withstand the force expected to be applied to it in the event of an earthquake should be determined by adding the weight of the piping which would contribute to its load.

a. For the loads on lateral braces on cross mains, add one-half the weight of branch line piping within the zone of the influence of the brace to the weight of the portion of the cross main within the zone or influence of the brace. (See Figure 3.4, examples 1, 3, 6, and 7.)

b. For the loads on longitudinal braces on cross mains, consider only one-half the weight of the cross main and feed mains within the zone of influence of the brace (See examples 2, 4, 5, and 8.)

c. For the 4-way brace at the riser, add the longitudinal and lateral loads within the zone of influence of the brace. (See examples 2, 3, 4, 5, 6, 7, and 8.)

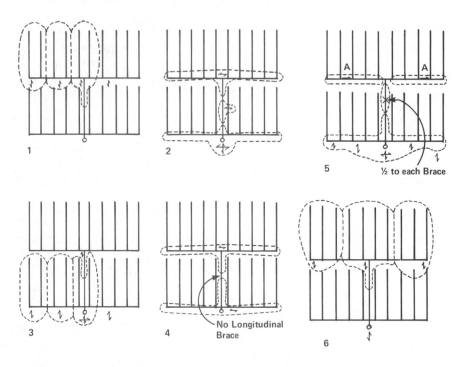

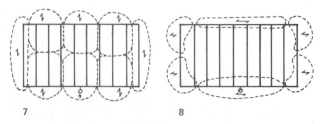

Figure 3.4.   Examples of load distribution to bracing.

d. Care should be taken that the total expected load within the zone of influence of a brace does not exceed the strength of the brace.

e. If the load is excessive for the braces, add additional braces to reduce the zone of influence for each brace.

f. Care should be taken that connection details are proper such that the expected loads do not exceed the capacities of fasteners.

3-11 Drainage.

3-11.1 Pitching of Piping for Drainage.

3-11.1.1* All sprinkler pipe and fittings shall be so installed that the system may be drained.

A-3-11.1.1 All piping should be arranged where practicable to drain to the main drain valve.

Ideally the entire system is pitched to drain at the main drain. Where this cannot be done, auxiliary drainage facilities as required in this section are sized and arranged with consideration of the type of system and the volume of piping trapped.

3-11.1.2 On wet-pipe systems, sprinkler pipes may be installed level. Trapped piping shall be drained in accordance with 3-11.3.

In pipe schedule systems the reduction in pipe sizes from the supply to the branch lines has the effect of a slight pitch when the piping is installed level, because of the slope of the reducing fittings which results from reduction in pipe size.

3-11.1.3 On those portions of pre-action systems subject to freezing and on dry-pipe systems, sprinkler pipe on branch lines shall be pitched at least ½-in. in 10 ft (4 mm/m) and the pipe of cross and feed mains shall be given a pitch of not less than ¼-in. in 10 ft (2 mm/m). A pitch of ¾-in. to 1-in. (19 mm to 25 mm) shall be provided for short branch lines and ½-in. in 10 ft (4 mm/m) for cross and feed mains in refrigerated areas and in buildings of light construction which may settle under heavy loads.

Because of the danger of freeze-ups all dry-pipe system horizontal piping is pitched. The amount of pitch required varies with the potential for freeze-up.

3-11.2 System, Sectional or Main Drain Connections and Drain Valves. *(See Figures 3-11.2 and A-3-9.1.2.)*

Formal Interpretation

Question: Would the pressure gage arrangement shown in Figure 4 be considered an acceptable alternate to that shown in Figure 3-11.2 and Figure A-2-9.1?

Answer: No. The pressure gage location, as illustrated, will not give a true residual reading. It will indicate an excessive pressure drop.

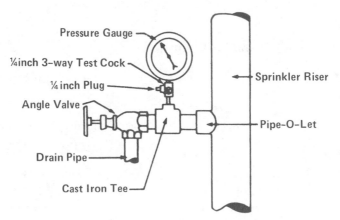

Drain Connection for Sprinkler Riser

Figure 4

3-11.2.1   Provisions shall be made to properly drain all parts of the system.

The intent of this section is to provide the facilities needed to drain any portion of a system in an efficient manner without undue risk of water damage while sizing and arranging the drainage equipment to meet this need in a cost-effective manner.

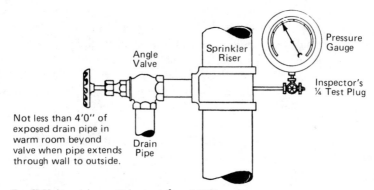

For SI Units: 1 in. = 25.4 mm; 1 ft = 0.3048 m.

Figure 3-11.2   Drain Connection for Sprinkler Riser.

3-11.2.2   On all risers 4 in. or larger, 2-in. drain pipes and valves shall be provided.

3-11.2.3   On risers 2½ in. to 3½ in. inclusive, drain pipes and valves not smaller than 1¼ in. shall be provided.

3-11.2.4   On smaller risers, drain pipe and valves not smaller than ¾ in. shall be provided.

3-11.2.5   Each interior sectional control valve shall be provided with a sectional drain valve sized as above so as to drain that portion of the system controlled by the sectional valve. These drains shall discharge either outside or to a drain connection.

Piping controlled by sectional valves represents a significant segment of the sprinkler system. The drain valves provided with such sectional control valves serve the same function for that part of the system as the main drain does for the entire system. These valves must be sized in accordance with 3-11.2.2, 3-11.2.3 and 3-11.2.4 and must be piped to discharge in a safe location.

3-11.2.6   The test valves required by 2-9.1 may be used as main drain valves.

3-11.3 Auxiliary Drains.

3-11.3.1   Auxiliary drains shall be provided when a change in piping direction prevents drainage of sections of branch lines or mains through the main drain valve.

Drainage facilities are desirable for trapped piping. The type of facility, its size and its arrangement are dependent on the type of system and the volume of trapped piping.

3-11.3.2 Auxiliary Drains for Wet-Pipe Systems.

3-11.3.2.1   When capacity of trapped sections of pipes is 5 gal (18.9 L) or less, the auxiliary drain shall consist of a nipple and cap or brass plug not less than ¾ in. in size.

Exception: Auxiliary drains are not required for piping which can be drained be removing a single sprinkler.

It is rarely necesary to drain a trapped section of piping in a wet system. Therefore, a nipple and cap or a plug is adequate when the volume of the trapped section is 5 gal (18.9 L) or less. A nipple and cap are preferable to a plug because of the comparative ease with which they can be removed. A plug must be brass to provide dissimilar metals making it easier to remove the plug. The Exception recognizes that the small volume of water trapped in supplying one sprinkler can be drained as readily through the sprinkler connection as through a plug. It would be unrealistic to require a drain where pendent sprinklers installed below a ceiling are supplied directly from branch lines concealed above the ceiling.

3-11.3.2.2   When capacity of isolated trapped sections of pipe is more than 5 gal (18.9 L), the auxiliary drain shall consist of a valve not smaller than ¾ in. size and plug, at least one of which shall be brass. In lieu of a plug, a nipple and cap may be used.

A valved drain is required when the trapped section in a wet system exceeds 5 gal (18.9 L) so that the trapped piping may be emptied in a reasonable length of time without water damage.

3-11.3.2.3   Tie-in drains are not required on wet-pipe systems.

Tie-in drains are cross-connections of the ends of trapped branch lines which are piped to a single drain valve. They are not required on wet systems because such piping is very seldom drained.

3-11.3.3 Auxiliary Drains for Dry-Pipe Systems.

B-3-11.3.3   The time required to remove water from a trapped section of the system is important. In extreme cases the time required to drain the system may allow water to freeze.

3-11.3.3.1   When capacity of trapped sections of pipe is 5 gal (18.9 L) or less, the auxiliary drain shall consist of a valve not smaller than ¾ in. and plug, at least one of which shall be brass. In lieu of a plug, a nipple and cap may be used.

Exception: Auxiliary drains are not required for a drop nipple when installed in accordance with 5-2.2.

Formal Interpretation

Question: Is the 5 gal limit intended to apply to each individual trapped section of pipe?

Answer: Yes.

Because dry-pipe systems are subject to freezing, trapped areas of 5 gal (18.9 L) or less must be provided with drain valves to remove water which has entered the system either due to dry-pipe valve tripping or condensation of moisture from the pressurized air in the system. An auxiliary drain is not required when pendent sprinklers in a heated area are supplied by piping which is also in a heated area.

3-11.3.3.2   When capacity of isolated trapped sections of pipe is more than 5 gal (18.9 L), the auxiliary drain shall consist of two 1-in. valves, and one 2-in. by 12-in. (305-mm) condensate nipple or equivalent. (*See Figure 3-11.3.3.*)

(Also see commentary under 3-11.3.2.3.) The condensate nipple illustrated in Figure 3-11.3.3 or its equivalent is required for each section of trapped piping in a dry-pipe system with more than 5 gal (18.9 L) capacity. The condensate nipple provides the capability to collect and remove moisture from the system while minimizing the potential for excessive loss of air pressure and possible unplanned operation of a dry-pipe valve. The upper valve is normally open, allowing moisture to enter the chamber with the lower valve closed and sealed to avoid leaks. To drain the chamber, the upper valve is closed to temporarily isolate it from the system and the lower valve is opened to remove the moisture.

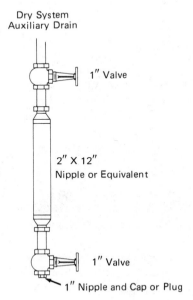

Dry System
Auxiliary Drain

1″ Valve

2″ X 12″
Nipple or Equivalent

1″ Valve

1″ Nipple and Cap or Plug

For SI Units: 1 in. = 25.4 mm; 1 ft = 0.3048 m.

Figure 3-11.3.3

3-11.3.3.3* Tie-in drains shall be provided for multiple adjacent trapped branch lines and shall be a minimum of 1-in. Tie-in drain lines shall be pitched a minimum of ½ in. in 10 ft (4 mm/m).

A-3-11.3.3.3 The size of tie-in drain lines should be increased as necessary to provide efficient removal of trapped water. Consideration should be given to the volume of water trapped, elevation head available and the time required to discharge water. Water or condensation or both that remains after initial drain down should not be permitted to collect since freezing may cause failure of drain/system piping.

When two or more adjacent branch lines are trapped in a dry-pipe or a pre-action system the ends of the lines must be piped together and run to a low point drain of at least 1 in. (25.4 mm) in size equipped with a drain valve and nipple and cap or plug to facilitate removal of moisture from the system.

It is extremely important that tie-in drains be properly pitched and that condensate be removed from these tie-in drains prior to freezing weather. Typically, tie-in drains are smaller sized piping connected to a rather large volume of piping and, with these smaller drains, freeze-ups are possible.

3-11.3.4 Auxiliary Drains for Pre-Action Systems.

3-11.3.4.1 When trapped sections of pipe are in areas subject to freezing, auxiliary drains shall conform to 3-11.3.3.

3-11.3.4.2 When trapped sections of pipe are in areas not subject to freezing, auxiliary drains shall consist of a valve not smaller than ¾ in. and plug, at least one of which shall be brass. In lieu of a plug, a nipple and cap may be used.

Exception No. 1: Auxiliary drains are not required for piping to a single sprinkler.

Exception No. 2: When the capacity of the trapped section of piping is 5 gal (18.9 L) or less, auxiliary drains shall consist of a ¾ -in. valve and plug, at least one of which shall be brass. In lieu of a plug, a nipple and cap may be used.

A drain valve is required for trapped sections of pre-action system piping of 5 gal (18.9 L) or less which are not subject to freezing due to the need to remove water which could interfere with the functioning of the low pressure air maintained on most pre-action systems for supervision.

3-11.4 Discharge of Drain Valves.

3-11.4.1* Direct interconnections shall not be made between sewers and sprinkler drains of systems supplied by public water. The drain discharge shall be in conformity with any health or water department regulations.

A-3-11.4.1 When possible, the main sprinkler riser drain should discharge outside the building at a point free from the possibility of causing water damage. When not possible to discharge outside the building wall, the drain should be piped to a sump which in turn should discharge by gravity or be pumped to a waste water drain or sewer. The main sprinkler riser drain connection should be of a size to carry off water from the fully open drain valve while it is discharging under normal water system pressures. When this is not possible, a supplementary drain of equal size should be provided for test purposes with free discharge, located at or above grade.

The prohibition regarding connection of sprinkler drains and sewers is intended to prevent any harmful element entering the sprinkler system by way of the drain connection and then entering the public water system by way of the sprinkler system. The valve between the sprinkler system and the drain connection and the check valve between the sprinkler system itself and the public water supply provide some protection, but this additional regulation is considered necessary to provide better assurance.

3-11.4.2 When drain pipes are buried underground, approved corrosive-resistant pipe shall be used.

3-11.4.3 Drain pipes shall not terminate in blind spaces under the building.

The drain discharge must be piped in a manner that allows the operator to ascertain that the drain is not obstructed and that it is not causing water damage.

3-11.4.4 Drain pipes when exposed shall be fitted with a turned down elbow.

The downturn elbow discourages the use of the drain piping as a refuse receptacle and minimizes the possibility of damage to property or wetting of passers-by.

3-11.4.5* Drain pipes shall be arranged as not to expose any part of the sprinkler system to freezing conditions.

A-3-11.4.5   When exterior ambient temperatures are subject to freezing, 32°F (0°C) or less, at least 4 ft (1.2 m) of pipe should be installed beyond the valve, in a warm room.

The recommended 4 ft (1.2 m) of pipe beyond the valve provides a frost break which protects against water in the system freezing due to exterior cold acting on the pipe. Additionally, a good maintenance program is required to properly remove all moisture from trapped unheated areas in dry-pipe and pre-action systems. See NFPA 13A, *Recommended Practice for the Care and Maintenance of Sprinkler Systems*, Section 4-8.2.

3-12 Joining of Pipe and Fittings.

3-12.1 Threaded Pipe and Fittings.

3-12.1.1   All threaded fittings and pipe shall have threads cut to ANSI/ASME Standard B1.20.1. Care shall be taken that the pipe does not extend into the fitting sufficiently to reduce the waterway.

Poor workmanship can result in threads that allow pipe protrusions to partially obstruct fitting openings. If such joints are permitted they will seriously restrict the flow in the system and greatly increase the pressure lost to friction and thereby impair the operation of the system.

3-12.1.2*   Steel pipe with wall thicknesses less than Schedule 30 (in sizes 8 in. and larger) or Schedule 40 (in sizes less than 8 in.) shall not be joined by threaded fittings, unless a threaded assembly has been investigated for suitability in automatic sprinkler installations and listed for this service.

This section relates back to 3-1.1.3 and in effect permits an exception allowing for new technology to develop an innovative threaded assembly for use with thinner walled pipe and to submit it to the testing laboratories for listing.

A-3-12.1.2   Some steel piping material having lesser wall thickness than specified in 3-12.1.2 has been listed for use in sprinkler systems when joined with threaded connections. The service life of such products may be significantly less than that of Schedule 40 steel pipe and it should be determined if this service life will be sufficient for the application intended.

All such threads should be checked by the installer using working ring gages conforming to the Basic Dimensions of Ring Gages for USA (American) Standard Taper Pipe Threads, NPT as per ANSI/ASME B1.20.1, Table 8.

Threading of listed thin wall pipe requires careful workmanship and good quality control. See 3-10.2.3 and 3-15.1.11 Exception for additional information.

The anticipated service life of this pipe in relation to Schedule 40 steel pipe will vary with pipe size. Refer to the product listings for specific information.

3-12.1.3 Joint compound or tape shall be applied to the threads of the pipe and not in the fitting.

Joint compound or tape applied to a fitting rather than to the threads of the pipe forms a ridge inside the pipe when the joint is made, reducing the inside diameter of the pipe and thereby adversely affecting the flow rate. The operation of systems other than wet systems involves a high velocity flow when the valve trips. The impact of the water tends to break off chunks of joint compound and carry them in a mass to the opened sprinklers where they tend to obstruct the orifices.

3-12.2* Welded Piping.

3-12.2.1 Welding methods which comply with all of the requirements of AWS D10.9, *Standard for Building Service Piping*, Level AR-3, are acceptable means of joining fire protection piping.

Formal Interpretation

Question 1: Do welding methods and weld inspection conforming to the requirements of ANSI B31.1-1978 meet the intent of Section 3-12-2.1?

Question 2: Does qualification of welding procedures, welders, and welding operators to ANSI B31.1-1977, *Code for Power Piping*, and the requirements of the ASME *Boiler and Pressure Vessel Code* referenced therein meet the intent of Section 3-12-2.11?

Answer: Yes. The standard describes the minimum acceptable welding methods of procedure. Other standards requiring a higher level of weld quality, test procedures and welder qualification meet the intent of NFPA 13.

The AWS Standard D10.9 Level Ar-3 reflects the minimum welding requirements needed for the pressures found in sprinkler systems. As the above interpretation indicates, methods meeting higher standards are also acceptable.

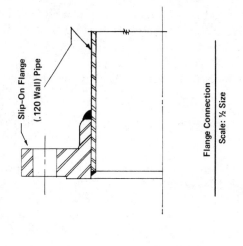

Slip-On Flange
(.120 Wall) Pipe

Flange Connection
Scale: ½ Size

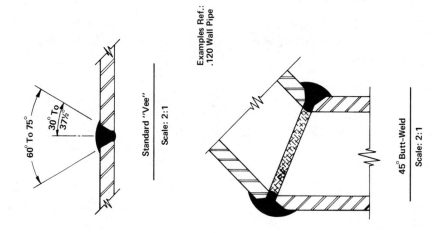

60° To 75°

30° To 37½°

Standard "Vee"
Scale: 2:1

Examples Ref.:
.120 Wall Pipe

45° Butt-Weld
Scale: 2:1

Figure A-3-12.2(a)   Butt-Weld Joints and Flange Connection.

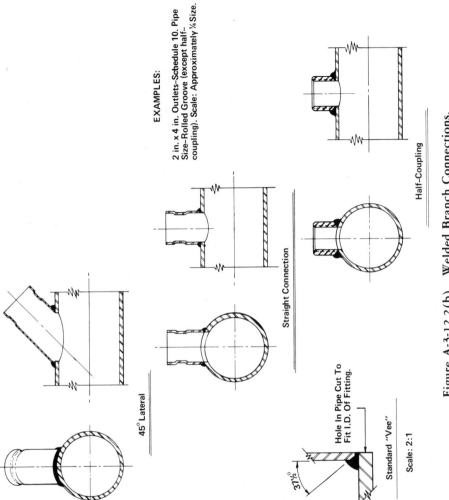

EXAMPLES:

2 in. x 4 in. Outlets–Schedule 10. Pipe Size–Rolled Groove (except half–coupling). Scale: Approximately ¼ Size.

45° Lateral

Straight Connection

Half–Coupling

Hole In Pipe Cut To Fit I.D. Of Fitting.

37½°

Standard "Vee"

Scale: 2:1

Figure A-3-12.2(b)   Welded Branch Connections.

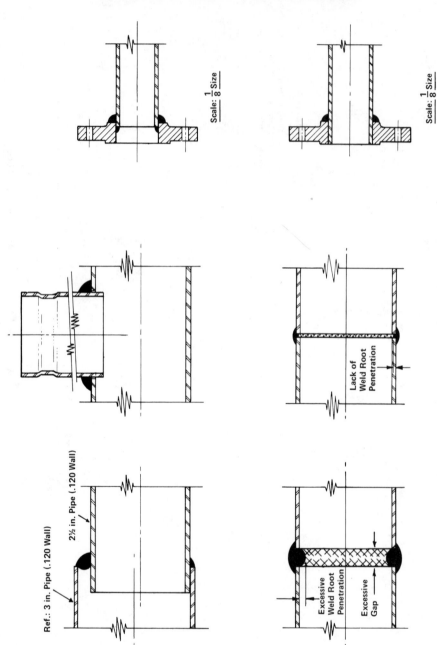

Figure A-3-12.2(c)   Unacceptable Weld Joints.

3-12.2.2* Welding sections of sprinkler piping in place inside the building shall not be permitted. Sections of branchlines, cross mains or risers may be shop welded.

Formal Interpretation

Question: Can repair welding be done to an existing sprinkler system while in place?

Answer: No. Section 3-12.2 does not address itself to welding inside a building.

The subcommittee interprets shop welding as meaning either on the sprinkler contractor's own premises or in a construction shack on the plant site sufficiently remote from the place of installation so as not to present any exposure hazard. On-site welding inside the building being protected is not permitted.

Exception: Welding sections of sprinkler piping in place inside new buildings under construction may be permitted only when the construction is noncombustible and no combustible contents are present and when the welding process is performed in accordance with NFPA 51B, tandard for Fire Prevention in Use of Cutting and Welding Processes.

Welding in place in existing occupancies is not permitted. It is permitted only in new buildings when its use does not introduce a potential fire hazard.

A-3-12.2.2 As used in this standard "Shop" in the term "Shop Welded" means either:

(a) At the sprinkler contractor's or fabricator's premise.

(b) An approved welding area at the building site.

The definition of shop is very broad and encompasses almost any location other than in places where the welding cannot be safely performed.

3-12.2.3 Welding procedures, welders and welding machine operators shall be qualified as required by 3-12.2.11.

3-12.2.4 Welded fittings and welded formations manufactured, fabricated, or joined in conformance with a qualified welding procedure as set forth herein are an acceptable product under this standard, provided that materials and wall thickness are compatible with other sections of this standard.

Welded fittings and formations conforming with this section 3-12.2 do not require laboratory listing or specific acceptance elsewhere in this standard.

3-12.2.5   No welding shall be performed if there is impingement of rain, snow, sleet or high wind on the weld area of the pipe product.

Welding must not be performed under conditions introducing a hazard to the mechanic or which would compromise the welder's ability to produce a satisfactory product.

3-12.2.6   When welded outlets are formed:

(a) Holes in piping shall be cut to full inside diameter of fitting or shaped, contoured nipple.

(b) Discs shall be retrieved.

(c) Openings in piping shall be smooth.

(d) All slag and other welding residue shall be removed.

(e) Fittings or shaped, contoured nipples of any length shall not penetrate beyond the internal diameter of the piping.

When welded outlets are formed the reasons for the requirements are as follows:
a. The area of flow must not be restricted.
b. Retrieval of discs ensures that holes have been cut.
c. Rough edges cause turbulence with resultant loss in system pressure.
d. Welding residue could obstruct opened sprinklers.
e. Penetration beyond the internal diameter of the pipe would restrict flow and cause turbulence.

3-12.2.7   When reducing a pipe size in the run of a main, cross main, or branch, a suitable reducing fitting designed for that purpose shall be used.

Fabricated reducing fittings will not have characteristics comparable to fitting specifically designed for the purpose.

3-12.2.8   Torch cutting and welding shall not be permitted as means of modifying or repairing sprinkler systems.

Torch cutting and welding are restricted to new installations and then only when in compliance with the Exception to 3-12.2.2. This restriction bears in mind possible introduction of ignition sources while protection is impaired.

3-12.2.9   When welding is planned, contractor shall specify the section to be shop welded on drawings and the type of fittings or formations to be used.

3-12.2.10   Sections of shop welded piping shall be joined by means of flanged or flexible gasketed joints or other approved fittings.

Exception: See 3-12.2.2.

When sections of piping are shop welded, the sections are usually joined to the piping in place by means of flanged or flexible gasketed joints because of the ease of assembly.

3-12.2.11 Qualifications.

3-12.2.11.1   A welding procedure shall be prepared and qualified before any welding is done. Qualification of the welding procedure to be used and the performance of welders and welding operators is required and shall comply with the requirements of American Welding Society Standard AWS D10.9, Level AR-3.

3-12.2.11.2   Each contractor or fabricator shall be responsible for all welding installed by him. Each contractor or fabricator shall have an established written quality assurance procedure related to control of the requirements of 3-12.2.6, available to the authority having jurisdiction.

The contractor or fabricator is directly responsible for qualifying procedures and welders in accordance with this standard as well as for the quality of the welds performed by his employees.

3-12.2.11.3   Each contractor or fabricator shall be responsible for qualifying any welding procedure that he intends to have used by personnel of his organization.

3-12.2.11.4   Each contractor or fabricator shall be responsible for qualifying all of the welders and welding machine operators employed by him in compliance with the requirements of AWS D10.9, Level AR-3.

3-12.2.12 Qualifications Records.   The contractor or fabricator shall maintain certified records, which are available to the authority having jurisdiction, of the procedures used and the welders or welding machine operators employed by him. Records shall show the date and the results of procedure and performance qualifications.

3-12.3 Groove Joining Methods.

3-12.3.1   Pipe joined with mechanical grooved fittings shall be joined by a listed combination of fittings, gaskets and grooves. When grooves are cut or rolled on the pipe they shall be dimensionally compatible with the fitting.

Exception: Steel pipe with wall thicknesses less than Schedule 30 (in sizes 8 in. and larger) or Schedule 40 (in sizes less than 8 in.) shall not be joined by fittings used with pipe having cut grooves.

Formal Interpretation

Question: Is it the intent of Section 3-12.3.1 to include rolled grooves in Schedule 40 pipe?

Answer: Yes. Grooves may be rolled on Schedule 40 pipe provided the instructions of the manufacturer of the groove rolling equipment are followed.

Unlike threads, grooves are not fabricated to a standard. Therefore, the listing of a fitting entails the combination of the fitting, the gasket, and the groove. See Figure 3.1.

3-12.3.2   Mechanical grooved couplings including gaskets used on dry-pipe systems shall be listed for dry-pipe service.

3-12.4* Brazed and Soldered Joints.   Joints for the connection of copper tube shall be brazed.

Exception No. 1:   Solder joints may be permitted for wet-pipe systems in Light Hazard Occupancies where the temperature clasfication of the installed sprinklers is Ordinary or Intermediate.

Exception No. 2:   Solder joints may be permitted for wet-pipe systems in Ordinary Hazard-Group 1 Occupancies where the piping is concealed.

A-3-12.4   The fire hazard of the brazing process should be suitably safeguarded.

Section 3-12.4 restricts the use of soldered joints to circumstances under which the system is water filled and the exposure to heat will not be adequate to compromise the integrity of the joint.

3-12.5 Other Types.   Other types of joints shall be made or installed in accordance with the requirements of the listing for this service.

This section encourages innovative technology by accepting listed joints not specifically addressed in the standard.

3-12.6 End Treatment.   After cutting, pipe ends shall have burrs and fins removed.

Piping must be reamed in order not to reduce the inside diameter of the pipe and in order to remove any rough edges from the end of the pipe. The removal of rough edges is particularly important when the fitting utilized has internal gaskets.

3-12.6.1   When using listed fittings, the pipe and its end treatment shall be in accordance with the manufacturer's installation instruction and the listing.

The end treatment required will vary with the fitting utilized. For example, for some but not all mechanical type fittings, it is required that the varnish be removed from the exterior wall of the pipe entering the joint.

3-13 Fittings.

3-13.1 Type of Fittings.

3-13.1.1   Fittings used in sprinkler systems shall be of the materials listed in Table 3-13.1.1 or in accordance with 3-13.1.2. The chemical properties, physical properties and dimensions of the materials listed in Table 3-13.1.1 shall be at least equivalent to the standards cited in the table. Fittings used in sprinkler systems shall be designed to withstand the working pressures involved, but not less than 175 psi (12.1 bars) cold water [125 psi (8.6 bars) saturated steam] pressure.

Table 3-13.1.1

Material and Dimensions	Standard
Cast Iron	
Cast Iron Threaded Fittings, Class 125 and 250	ANSI B16.4
Cast Iron Pipe Flanges and Flanged Fittings	ANSI B16.1
Malleable Iron	
Malleable Iron Threaded Fittings, Class 150 and 300	ANSI B16.3
Steel	
Factory-made Wrought Steel Buttweld Fittings	ANSI B16.9
Buttwelding Ends for Pipe, Valves, Flanges and Fittings	ANSI B16.25
Spec. for Piping Fittings of Wrought Carbon Steel and Alloy Steel for Moderate and Elevated Temperatures	ASTM A234
Steel Pipe Flanges and Flanged Fittings	ANSI B16.5
Forged Steel Fittings, Socket Welded and Threaded	ANSI B16.11
Copper	
Wrought Copper and Bronze Solder-Joint Pressure Fittings	ANSI B16.22
Cast Bronze Solder-Joint Pressure Fittings	ANSI B16.18

Fittings of the types and materials indicated in Table 3-13.1.1 must be manufactured to the standards indicated or to standards meeting or exceeding the indicated standards. Any fittings for use in a sprinkler system must be designed for a working pressure of at least 175 psi (12.1 bars). See 3-13.1.3 regarding requirements where pressure exceeds 175 psi (12.1 bars).

3-13.1.2* Other types of fittings may be used, but only those investigated and listed for this service.

Exception: Welded fittings or formations as permitted in 3-12.2.

A-3-13.1.2 Rubber gasketed pipe fittings and couplings should not be installed where ambient temperatures can be expected to exceed 150°F (66°C) unless listed for this service. If the manufacturer further limits a given gasket compound, those recommendations should be followed.

Other types of fittings are acceptable if they are specifically listed for sprinkler system use, with the exception of welded formations which must be fabricated in compliance with 3-12.2 but do not require listing. This section again encourages innovative technology and the development of more cost-effective or efficient fittings or joining methods.

3-13.1.2.1* When unique characteristics of a fitting, such as a tendency to rotate, require support in addition to that required in Section 3-15, restraint shall be provided in accordance with its listing.

A-3-13.1.2.1 Unless properly restrained, gravitational forces on unsupported nonvertical branches can cause the pipe to rotate out of position.

There have been instances of unsupported branch outlets rotating out of position. Any fitting having the potential for such rotation must have as part of its listing appropriate measures to ensure against that circumstance.

3-13.1.3 Fittings used in sprinkler systems shall be extra heavy pattern where pressures exceed 175 psi (12.1 bars).

Exception No. 1: Standard weight pattern cast-iron fittings 2 in. size and smaller may be used where pressures do not exceed 300 psi (20.7 bars).

Exception No. 2: Standard weight pattern malleable iron fittings 6 in. size and smaller may be used where pressures do not exceed 300 psi (20.7 bars).

Exception No. 3: Fittings may be used for system pressures up to the limits specified in listings by a testing laboratory.

Formal Interpretation

Question: Is the interpretation correct that the use of standard weight cast-iron fittings is permissible in sizes up to and including 2 in. where normal pressure in the piping system exceeds 175 psi? In sizes 2½ in. and larger, cast-iron fittings shall be of the extra heavy pattern where normal pressure exceeds 175 psi?

Answer: Yes.

This section allows the use of cast-iron fittings 2 in. size and smaller and malleable iron fittings 6 in. size and smaller to be used up to 300 psi (30.7 bars), because the rupture strength of those fittings allows an adequate factor of safety.

3-13.1.4 Where water pressures are 175 to 300 psi (12.1 to 20.7 bars), the ANSI standards permit the use of standard wall pipe and extra heavy valves. Until pressure ratings for valves are standardized, the manufacturers' ratings shall be observed.

Standard wall pipe may be used for working pressures up to 300 psi (30.7 bars). However, extra heavy valves must be utilized for working pressures in excess of 175 psi (12.1 bars). Flanges must be in compliance with 3-13.1.3.

3-13.1.5* When risers are 3 in. in size or larger, a flanged joint or mechanical coupling shall be used at the riser at each floor.

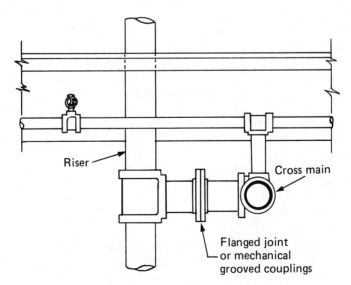

Figure A-3-13.1.5   One Arrangement of Flanged Joint at Sprinkler Riser.

The flanged joint or mechanical coupling at the riser provides a facility to isolate a section of the system, so that the remainder of the system may be returned to service while repairs or modifications proceed.

3-13.2* Couplings and Unions.   Screwed unions shall not be used on pipe larger than 2 in. Couplings and unions of other than screwed type shall be of types approved specifically for use in sprinkler systems. Unions, screwed or mechanical couplings, or flanges may be used to facilitate installation.

Screwed unions larger than 2 in. in size present a maintenance problem due to their tendency to develop leaks. There are a variety of acceptable fittings available for use in connecting portions of systems.

A-3-13.2   Approved flexible connections are permissible and encouraged for sprinkler installations in racks to reduce possibility of physical damage. When flexible tubing is used it should be located so that it will be protected against mechanical injury.

3-13.3 Reducers and Bushings.   A one-piece reducing fitting shall be used wherever a change is made in the size of the pipe.

Exception: Hexagonal or face bushings may be used in reducing the size of openings of fittings when standard fittings of the required size are not available.

Bushings may be used only when reducing fittings are unavailable. They have poorer flow characteristics than do reducing fittings and have a greater tendency to leak.

Formal Interpretation

Question: Is it the intent of 3-13.3 to exclude the use of bushings when the required reducing fitting is manufactured?

Answer: Yes. The intent is to exclude the use of bushings when the one-piece reducing fitting is available on the market at the time the system is fabricated.

3-14 Valves.

3-14.1 Types of Valves to Be Used.

3-14.1.1   All valves on connections to water supplies and in supply pipes to sprinklers shall be listed indicating valves, unless a nonindicating valve, such as an underground gate valve with approved roadway box complete with T-wrench, is accepted by the authority having jurisdiction.

Such valves shall not close in less than 5 seconds when operated at maximum possible speed from the fully open position. This is to avoid damage to piping by water hammer.

The following may not incorporate indicating devices as part of the valve, but the valve assembly described shall qualify as an indicating valve:

(a) A listed underground gate valve equipped with a listed indicator post,

(b) A listed water control valve assembly which is normally open and requires constant energy application to close and keep closed,

(c) A listed water control valve assembly which has a reliable position indication connected to a remote supervisory station.

> All valves controlling the flow of water to sprinklers must be listed and must incorporate a method of readily determining that the valve is open. This method of indicating position may be part of the valve itself, such as a rising stem, or a feature of the assembly as described in the section.
>
> The one exception is the use of roadway boxes where the valve must be installed in a circumstance where nothing can extend above the surface and the authority having jurisdiction accepts them in lieu of valve pits.

3-14.1.2   Drain valves and test valves shall be of approved type of 175 psi (12.1 bars) cold water [125 psi (8.6 bars) saturated steam] pressure rating.

3-14.1.3   Check valves shall be listed and shall be installed in a vertical or horizontal position in accordance with their listing.

> Check valves as used in this section means check valves in the sprinkler system waterway. It is not intended to require listing of check valves not affecting system reliability, such as those used in the cross-connection of drains. Not all valves are listed for both horizontal and vertical use. Therefore, care must be taken to ensure that valves are only installed in accordance with their listing.

3-14.2* Valves Controlling Sprinkler Systems.

A-3-14.2 Valves Controlling Water Supplies.

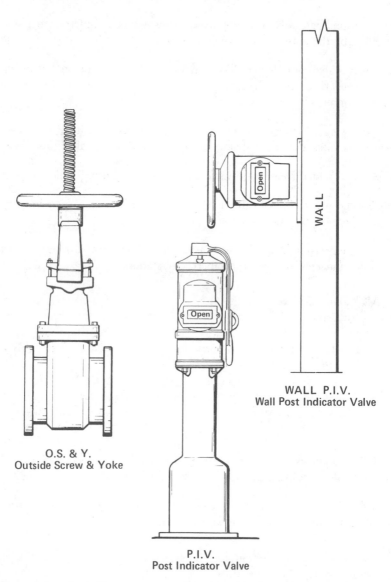

WALL P.I.V.
Wall Post Indicator Valve

O.S. & Y.
Outside Screw & Yoke

P.I.V.
Post Indicator Valve

Figure 3.5.   Sprinkler system water control valves (Insurance Services Office).

Formal Interpretation

Question: For a high-rise building with two risers supplying both sprinklers and 2½ in. hose outlets with a connection from each of the risers to a spinkler loop main on each floor, where each connection from the riser to the loop contains an indicating type pressure regulating control valve, waterflow switch, and O.S. & Y. control valve - is the valve arrangement indicated below acceptable?

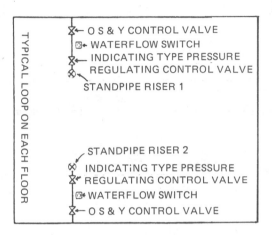

Figure 5

Answer: Yes. The Committee answer is with respect to the valve arrangement only and excludes all other aspects of the system.

3-14.2.1* Each system shall be provided with a listed indicating valve so located as to control all sources of water supply except fire department connections.

A-3-14.2.1 A water supply connection should not extend into or through a building unless such connection is under the control of an outside listed indicating valve or an inside listed indicating valve located near outside wall of the building.

All valves controlling water supplies for sprinkler system or portions thereof including floor control valves should be accessible to authorized persons during emergencies. Permanent ladders, clamped treads on risers, chain-operated hand wheels, or other accepted means should be provided when necessary.

Outside control valves are suggested in the following order of preference:

(a) Listed indicating valves at each connection into the building at least 40 ft (12.2 m) from buildings if space permits.

(b) Control valves installed in a cut-off stair tower or valve room accessible from outside.

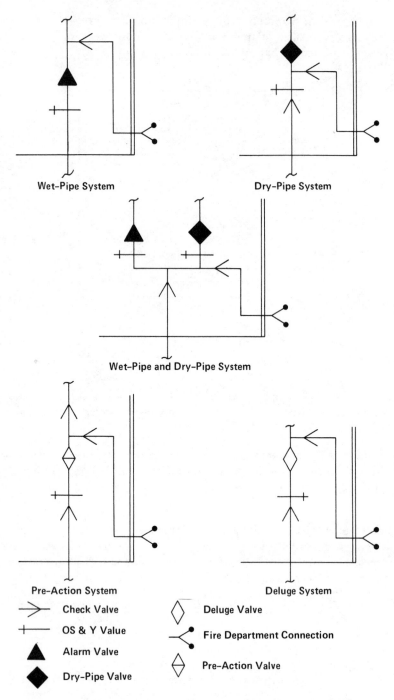

Figure A-3-14.2   Examples of Acceptable Valve Arrangements.

(c) Valves located in risers with indicating posts arranged for outside operation.

(d) Key operated valves in each connection into the building.

Each system must be provided with a valve which isolates that system from all automatic sources of supply. It is preferred, but not always possible, for the fire department connection to be on the system side of the valve so that, with the valve shut, the fire department can supply water to the system through the fire department connection. See 2-7.3.

Formal Interpretation

Question: According to 3-14.2.1 and 3-14.2.2, would listed indicating valve be required at "A" in Figure 6?

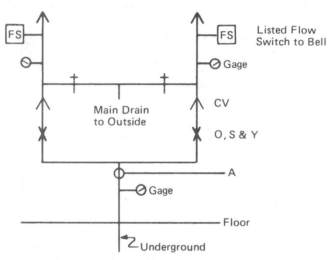

Figure 6   Sectional Detail of Automatic Sprinkler Risers

Answer: A valve is not required at point "A" for the sketch provided. A more comprehensive answer would require consideration of the configuration of the outside underground piping and the control valves on it in addition to the location of the fire department connection. The pressure gage indicated below point "A" is not required.

3-14.2.2   At least one listed indicating valve shall be installed in each source of water supply except fire department connections.

3-14.2.3   Valves on connections to water supplies, sectional control valves and other valves in supply pipes to sprinklers shall be supervised open by one of the following methods:

(a) Central station, proprietary or remote station signaling service,

(b) Local signaling service which will cause the sounding of an audible signal at a constantly attended point,

(c) Locking valves open,

(d) Sealing of valves and approved weekly recorded inspection when valves are located within fenced enclosures under the control of the owner.

Exception: Underground gate valves with roadway boxes need not be supervised.

The methods of supervision are listed in their order of preference. Closed valves represent 30 percent of all sprinkler system failures. The importance of establishing procedures to ensure that control valves are in the fully opened position cannot be overemphasized. See NFPA 13A, *Recommended Practice for the Care and Maintenance of Sprinkler Systems.*

It should be noted that either central station, proprietary, or remote station alarm service are far superior methods of supervision than the other choices listed. These particular methods have not, however, been mandated due to the cost of such a system, particularly in existing large, spread out plants.

3-14.2.4   When there is more than one source of water supply, a check valve shall be installed in each connection.

Exception: When cushion tanks are used with automatic fire pumps, no check valve is required in the cushion tank connection.

Each source of supply must have a check valve to isolate the supplies from each other. In a fire situation, water flows from the highest pressure source. For example, consider a system supplied by a gravity tank, a city connection and a fire department connection with the gravity tank being the highest pressure source. Initial flow would be from the gravity tank. If its pressure dropped below the city pressure, the tank's check valve would close and supply would be from the city. When the fire department connected onto the system, the check valves in both automatic supplies would close and supply would be through the fire department pumps via the fire department connection.

A cushion tank, when installed, is a part of the fire pump supply and as such is subject to the pump's check valve.

3-14.2.5*   A check valve shall be installed in each water supply connection if there is a fire department connection on the system.

A-3-14.2.5   Pits for underground valves, except those located at the base of a tank riser, are described in NFPA 24, *Standard for Installation of Private Fire Service Mains and Their Appurtenances.* For pits protecting valves located at the base of a tank riser, refer to NFPA 22, *Standard for Water Tanks for Private Fire Protection.*

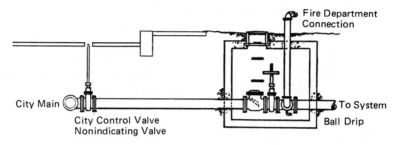

Figure A-3-14.2.5   Pit for Gate Valve, Check Valve and Fire Department Connection.

If a check valve were not installed in an automatic supply, pressure developed via the fire department connections would be dissipated into the supply. All sources, whether automatic or non-automatic, must have check valves, so that pressure from the highest pressure source is concentrated in the system.

3-14.2.6*   When a single wet-pipe sprinkler system is equipped with a fire department connection the alarm valve is considered a check valve and an additional check valve shall not be required.

A-3-14.2.6   When a system having only one dry-pipe valve is supplied with city water and fire department connection it will be satisfactory to install the main check valve in water supply connection immediately inside of the building; in case there is no outside control the system indicating valve should be placed at the wall flange ahead of all fittings.

Formal Interpretation

Question: Is an approved indicating valve required on the system side of an alarm valve, in addition to the approved indicating or indicator post valve on the supply side of the alarm valve?

Answer: No. It was never the intent of the Committee to require a control valve on the system side of the alarm valve, only on the supply side. Hence, only one control valve per sprinkler system.

An alarm valve is a check valve which has been designed to provide facilities to activate water flow alarms. Another check valve would be redundant.

3-14.2.7 In a city connection serving as one source of supply the city valve in the connection may serve as one of the required valves. A listed indicating valve or an indicator post valve shall be installed on the system side of the check valve. (*See Figure A-3-14.2.5.*)

Exception: When a wet-pipe sprinkler system is equipped with an (alarm) check valve, a gate valve is not required on the system side of the (alarm) check valve.

The intent of requiring gate valves on both sides of a check valve is to provide facilities for isolating the check valve for maintenance or repair while permitting the system to remain in service. If the system has only one automatic supply, a gate valve on the system side of the check valve is not required.

It should be recognized that the alarm valve is a specialized device and that control valves are not required on the system side of alarm check valves.

3-14.3* Identification of Valves. When there is more than one control valve, permanently marked identification signs indicating the portion of the system controlled by each valve shall be provided.

Embossed plastic tape, pencil, ink, crayon, etc., shall not be considered permanent markings. The sign shall be secured with noncorrosive wire, chain, or other means.

A-3-14.3 All control, drain, and test valves should be provided with identification signs.

Signs identifying control valves must be of a type and installed in a manner that will result in their remaining permanently affixed and being clearly legible.

Figure 3.6. Typical sprinkler control valve identification sign, Style "A."

3-15 Hangers.

3-15.1* General.   Type of hangers and installation methods shall be in accordance with the requirements of Section 3-15.

Exception: Hangers and installation methods certified by a registered professional engineer for the following:

(a) Designed to support five times the weight of the water-filled pipe plus 250 lb (114 kg) at each point of piping support.

(b) These points of support are enough to support the sprinkler system.

(c) Ferrous materials are used for hanger components.

Detailed calculations shall be submitted, when required by the reviewing authority, showing stresses developed both in hangers and piping and safety factors allowed.

The Exception is granted where hangers and installation methods are certified by a registered professional engineer in accordance with (a), (b) and (c).

A-3-15.1   Branch line hangers under metal decking may be attached by drilling or punching vertical members and using through bolts. The distance from the bottom of the bolt hole to the bottom of the vertical member should be not less than ⅜ in. (9.5 mm).

3-15.1.1   Hangers and their components shall be ferrous.

Exception: Nonferrous components which have been proven by fire tests to be adequate for the hazard application and are listed for this purpose and are in compliance with the other requirements of this section.

Formal Interpretation

Question: Does Subsection 3-15.1 Exception mean that pipe supports need not be designed to accommodate a loading equal to five times the weight of water filled pipe plus 250 lb at each support when the following are accomplished?

1. A detailed piping stress and pipe support analysis which substantiates stresses, loads and safety factors is performed. All calculations are retained as part of the permanent project records.

Type of pipe analysis performed:

(a) 2½ in. and larger piping — Verified Computer Analysis

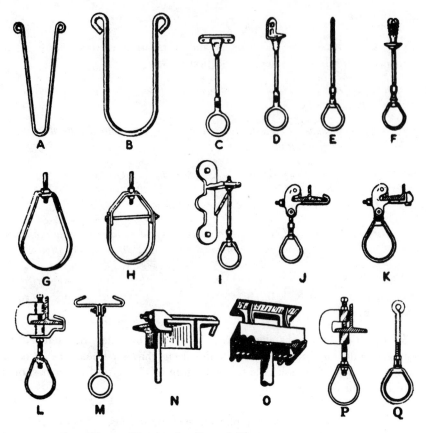

A — U-type Hanger for Branch Lines.
B — U-type Hanger for Cross Mains and Feed Mains.
C — Adjustable Clip for Branch Lines.
D — Side Beam Adjustable Hanger.
E — Adjustable Coach Screw Clip for Branch Lines.
F — Adjustable Swivel Ring Hanger with Expansion Shield.
G — Adjustable Flat Iron Hanger.
H — Adjustable Clevis Hanger.
I — Cantilever Bracket.
J — "Universal" I-beam Clamp.
K — "Universal" Channel Clamp.
L — C-type Clamp with Retaining Strap.
M — Cener I-beam Clamp for Branch Lines.
N — Top Beam Clamp.
O — "CL-Universal" Concrete Insert.
P — C-type Clamp without Retaining Strap.
Q — Eye Rod and Ring Hanger.
R — Wrap-around U Hook.

Figure A-3-15.1   Common Types of Acceptable Hangers.

(b) 2 in. and smaller piping — Computer Analysis or in accordance with a company approved design standard.

2. Ferrous materials are utilized for piping and piping support components.

3. Types of pipe supports and installations are in accordance with industry standards (MSS SP-58 and SP-69) and the requirements of Section 3-15 of NFPA 13.

Answer: No. The standard mandates all pipe supports must be designed to support 5 times the weight of water filled pipe plus 250 lb at each point of piping support. This is the same requirement used by laboratories when reviewing hanger supports for listing purposes.

3-15.1.2 The components of hanger assemblies which directly attach to the pipe or to the building structure shall be listed.

Exception: *Mild steel hangers formed from rods need not be listed.*

Hanger assembly components must be listed. The Exception is granted for mild steel rod hangers because data has been submitted substantiating that in the sizes required in this section the rods will support, at maximum permitted spacing, 5 times the weight of water filled Schedule 40 pipe plus 250 lb (114 kg).

Formal Interpretation

Question: If we comply with Section 3-15, do we have any further burden of proof other than "support the added load of water filled pipe plus a minimum load of 250 lb applied at the point of hanging"?

Answer: No. The Committee feels there has been a misinterpretation of the requirements of Section 3-15.1.4, which were quoted in part, as this applies only to the structure to which the hanger is attached. If the installation otherwise meets the requirement of Section 3-15, the requirements of 3-15.1 Exception as to certification by a registered professional engineer do not apply.

Formal Interpretation

Question: Does the method of hanging sprinkler piping as shown in Figure 7 comply with Section 3-15? It is capable of supporting 5 times the weight of water filled piping plus 250 lb and has been certified by a registered professional engineer.

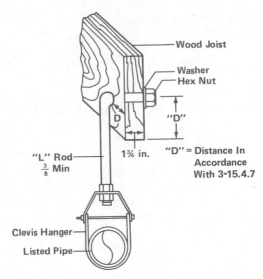

Figure 7

Answer: Yes. Certification by a registered professional engineer also includes compliance with other requirements of Section 3-15.1 Exception, namely: "(b) These points of support are enough to support the sprinkler system. (c) Ferrous materials are used for hanger components."

"Detailed calculations shall be submitted, when required by the reviewing authority, showing stresses developed both in hangers and piping and safety factors allowed."

3-15.1.3* Sprinkler piping or hangers shall not be used to support non-system components.

A-3-15.1.3 The rules covering the hanging of sprinkler piping take into consideration the weight of water filled pipe plus a safety factor. No allowance has been made for the hanging of non-system components from sprinkler piping.

System piping and hangers may be used to support such system components as tubing or wiring for deluge or pre-action system detectors.

3-15.1.4 Sprinkler piping shall be substantially supported from the building structure which must support the added load of the water-filled pipe plus a minimum of 250 lb (114 kg) applied at the point of hanging.

Formal Interpretation

Question: Is it the intent of Section 3-15.1.4 to require all hangers (main, crossmain and branch) to support the added load of the water filled pipe plus a minimum of 250 lb applied at the point of hanging?

Answer: Yes.

3-15.1.5   Sprinkler piping shall be supported independently of the ceiling sheathing.

Exception: Toggle hangers shall be used only for the support of pipe 1 ½ in. or smaller in size under ceilings of hollow tile or metal lath and plaster.

Toggle hangers are not permitted to be used with gypsum wallboard or other less substantial types of ceiling materials.

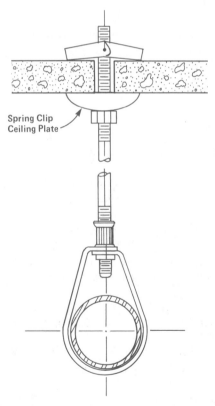

Figure 3.7.   Toggle hanger for pipe 1 ½ in. or smaller.

3-15.1.6   When sprinkler piping is installed below ductwork, piping shall be substantially supported from the building structure or from the steel angles supporting the ductwork provided the angles conform to Table 3-15.1.7.

If the steel angles supporting the duct work have dimensions less than those required by Table 3-15.1.7, the pipe must be supported by a hanger connected to the building structure or the duct hangers must be increased in size to satisfy the requirements of the table.

Sprinkler piping must be supported from the building structure with the exception of 1½ in. and smaller piping which may be supported under ceilings of hollow tile or metal lath and plaster with toggle hangers.

3-15.1.7*   For trapeze hangers, the minimum size of steel angle or Schedule 40 pipe span between purlins or joists shall be as shown in Table 3-15.1.7, all angles to be used with longer leg vertical. Any other sizes or shapes giving equal or greater section modulus will be acceptable. The trapeze member shall be secured to prevent slippage.

A-3-15.1.7   Table 3-15.1.7 assumes that the load is located at the midpoint of the span of the trapeze member. If the load is applied at other than the midpoint, for the purpose of sizing the trapeze member, an equivalent length of trapeze may be used, derived from the formula

$$L = \frac{4ab}{a + b}$$ where "L" is the equivalent length, "a" is the distance from one support to the load and "b" is the distance from the other support to the load.

Table 3-15.1.7 provides the Schedule 40 pipe sizes and the angle iron sizes required for trapeze bars supporting various spans and pipe sizes. Other sizes or shapes such as unistrut can be used provided they have equal or greater strength.

3-15.1.8   The size of hanger rods and fasteners required to support the steel angle iron or pipe indicated in Table 3-15.1.7 shall comply with 3-15.4.

3-15.1.9   Eye rods and ring hangers shall be secured with necessary lock washers to prevent lateral motion at the point of support.

3-15.1.10   Holes through concrete beams may also be considered as a substitute for hangers for the support of pipes.

Holes through building materials other than concrete may not be considered as a substitute for hangers.

Table 3-15.1.7 Trapeze Members – One foot, six inches to ten foot spans.

Span of Trapeze Bars / Pipe Size	2½" or Less	3"	3½"	4"	5"	6"	8"	10"
1' 6"	1½" x 1½" x 3/16" 1" Pipe	1½" x 1½" x 3/16" 1" Pipe	1½" x 1½" x 3/16" 1" Pipe	2" x 1½" x 3/16" 1" Pipe	2" x 1½" x 3/16" 1¼" Pipe	2½" x 1½" x 3/16" 1¼" Pipe	3" x 2" x 3/16" 1½" Pipe	3" x 2" x ¼" 2" Pipe
2' 0"	1½" x 1½" x 3/16" 1" Pipe	2" x 1½" x 3/16" 1" Pipe	2" x 1½" x 3/16" 1" Pipe	2" x 1½" x 3/16" 1¼" Pipe	2½" x 1½" x 3/16" 1¼" Pipe	2½" x 1½" x 3/16" 1½" Pipe	3" x 2" x 3/16" 2" Pipe	3" x 2" x ¼" 2" Pipe
2' 6"	2" x 1½" x 3/16" 1" Pipe	2" x 1½" x 3/16" 1" Pipe	2" x 1½" x 3/16" 1¼" Pipe	2½" x 1½" x 3/16" 1¼" Pipe	2½" x 1½" x 3/16" 1½" Pipe	3" x 2" x 3/16" 2" Pipe	3" x 2" x ¼" 2" Pipe	3" x 2" x ¼" 2" Pipe
3' 0"	2" x 1½" x 3/16" 1" Pipe	2" x 1½" x 3/16" 1¼" Pipe	2½" x 1½" x 3/16" 1¼" Pipe	2½" x 1½" x 3/16" 1¼" Pipe	3" x 2" x 3/16" 1½" Pipe	3" x 2" x 3/16" 2" Pipe	3½" x 2½" x ¼" 2½" Pipe	3½" x 2½" x 3/16" 2½" Pipe
4' 0"	2½" x 1½" x 3/16" 1¼" Pipe	2½" x 1½" x 3/16" 1¼" Pipe	2½" x 1½" x 3/16" 1½" Pipe	3" x 2" x 3/16" 1½" Pipe	3" x 2" x 3/16" 2" Pipe	3" x 2" x ¼" 2" Pipe	3½" x 2½" x 3/16" 2½" Pipe	4" x 3" x 3/16" 3½" Pipe
5' 0"	2½" x 1½" x 3/16" 1¼" Pipe	2½" x 1½" x 3/16" 1½" Pipe	3" x 2" x 3/16" 1½" Pipe	3" x 2" x 3/16" 2" Pipe	3" x 2" x ¼" 2" Pipe	3½" x 2½" x 3/16" 2½" Pipe	4" x 3" x 3/16" 2½" Pipe	5" x 3½" x 3/16" 4" Pipe
6' 0"	2½" x 1½" x 3/16" 1½" Pipe	3" x 2" x 3/16" 2" Pipe	3" x 2" x 3/16" 2" Pipe	3" x 2" x ¼" 2" Pipe	3½" x 2½" x 3/16" 2½" Pipe	4" x 3" x 3/16" 2½" Pipe	4" x 3" x 3/16" 3" Pipe	5" x 3½" x 3/16" 4" Pipe
7' 0"	3" x 2" x 3/16" 2" Pipe	3" x 2" x 3/16" 2" Pipe	3" x 2" x ¼" 2" Pipe	3" x 2" x ¼" 2½" Pipe	3½" x 2½" x 3/16" 2½" Pipe	4" x 3" x 3/16" 3" Pipe	5" x 3½" x 3/16" 3" Pipe	6" x 4" x ¼" 4" Pipe
8' 0"	3" x 2" x 3/16" 2" Pipe	3" x 2" x ¼" 2½" Pipe	3" x 2" x ¼" 2½" Pipe	3½" x 2½" x 3/16" 2½" Pipe	3½" x 2½" x 3/16" 3" Pipe	4" x 3" x 3/16" 3" Pipe	5" x 3½" x 3/16" 3½" Pipe	6" x 4" x ¼" 4" Pipe
9' 0"	3" x 2" x 3/16" 2" Pipe	3" x 2" x ¼" 2½" Pipe	3½" x 2½" x 3/16" 2½" Pipe	3½" x 2½" x 3/16" 3" Pipe	3½" x 2½" x 3/16" 3½" Pipe	4" x 3" x 3/16" 3½" Pipe	5" x 3½" x 3/16" 4" Pipe	6" x 4" x ⅜" 5" Pipe
10' 0"	3" x 2" x ¼" 2½" Pipe	3" x 2" x ¼" 2½" Pipe	3½" x 2½" x 3/16" 2½" Pipe	3½" x 2½" x 3/16" 3" Pipe	4" x 3" x 3/16" 3½" Pipe	5" x 3½" x 3/16" 3½" Pipe	6" x 4" x ¼" 4" Pipe	6" x 4" x ⅜" 5" Pipe

For SI units: 1 in. = 25.4 mm; 1 ft = 0.3048 m.

3-15.1.11* Maximum Distance Between Hangers.   With steel pipe or copper tube as specified in 3-1.1.1, the maximum distance between hangers shall not exceed 12 ft (3.7 m) for 1- and 1¼-in. sizes nor 15 ft (4.6 m) for sizes 1½-in. and larger except as provided in 3-15.6. (*See Figure A-3-15.1.11.*)

Exception: Threaded lightweight steel pipe shall have a maximum distance between hangers not exceeding 12 ft (3.6 m) for pipe sizes 3 in. or less.

With the exception of where cross mains receive additional support from hangers in the branch lines as specified in 3-15.6, hanger spacing cannot exceed 15 ft (4.6 m) for 1½ and larger pipe nor 12 ft (3.7 m) for 1¼ and smaller pipe including ¾ in. copper pipe in hydraulically designed systems. In instances where pipe or tubing does not have the beam strength anticipated by these spacing specifications, additional hangers will probably be necessary.

Because of the lack of material between the inside diameter and the root diameter of the thread, hangers are required to be spaced closer for threaded thin wall pipe.

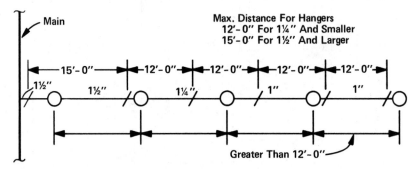

For SI Units: 1 in. = 25.4 mm; 1 ft = 0.3048 m.

Figure A-3-15.1.11   Distance Between Hangers.

3-15.1.12   When sprinkler piping is installed in storage racks as defined in NFPA 231C, piping shall be substantially supported from the storage rack structure or building in accordance with all applicable provisions of Section 3-15.

NFPA 231C, *Standard for Rack Storage of Materials*, stipulates requirements for such areas as sprinkler location, sprinkler spacing and water supplies. However, everything not modified by NFPA 231C is installed in accordance with the requirements of this standard.

3-15.2 Hangers in Concrete.

3-15.2.1 Listed inserts set in concrete may be installed for the support of hangers. Wood plugs shall not be used.

Inserts listed for use in concrete have been tested for such use. Wood plugs used in concrete do not have the required retention to be used for the support of piping.

3-15.2.2 Listed expansion shields for supporting pipes under concrete construction may be used in a horizontal position in the sides of beams. In concrete having gravel or crushed stone aggregate, expansion shields may be used in the vertical position to support pipes 4 in. or less in diameter.

Expansion shields may be installed either horizontally or vertically. Shields installed horizontally are not subject to a direct downward pull and are therefore preferable. Pipe 4 in. and smaller may be supported vertically by expansion shields at the spacing indicated in 3-15.1.11. For larger piping, see 3-15.2.3.

3-15.2.3 For the support of pipes 5 in. and larger, expansion shields if used in the vertical position shall alternate with hangers connected directly to the structural members such as trusses and girders, or to the sides of concrete beams. In the absence of convenient structural members, pipes 5 in. and larger may be supported entirely by expansion shieids in the vertical position, but spaced not over 10 ft (3 m) apart.

Due to stress on expansion shields installed in the vertical position for pipe 5 in. or larger, shields must either be alternated with hangers providing other means of support or their maximum spacing reduced from 15 ft (4.6 m) to 10 ft (3.0 m) apart.

3-15.2.4 Expansion shields shall not be used in ceilings of gypsum or similar soft material. In cinder concrete, expansion shields shall not be used except on branch lines where they shall alternate with through bolts or hangers attached to beams.

The effectiveness of expansion shields depends on the gripping power of the material with which they are used. They will pull out of soft materials such as gypsum.

3-15.2.5 When expansion shields are used in the vertical position, the holes shall be drilled to provide uniform contact with the shield over its entire circumference. Depth of the hole shall be not less than specified for the type of shield used.

The listings of expansion shields, as with any other hanger, assume that they are properly installed.

3-15.2.6   Holes for expansion shields in the side of concrete beams shall be above the center line of the beam or above the bottom reinforcement steel rods.

Shields must be located for maximum strength. If a reinforcing rod is contacted when drilling for a shield, a new hole away from the reinforcing rod must be made to the required depth.

3-15.3 Powder Driven Studs and Welding Studs.

3-15.3.1*   Powder driven studs, welding studs, and the tools used for installing these devices shall be listed by a testing laboratory and installed within the limits of pipe size, installation position, and construction material into which they are installed as expressed in individual listings or approvals.

A-3-15.3.1   Powder driven studs should not be used in steel less than $\frac{3}{16}$ in. total thickness.

The limitations placed on powder actuated tools and stud welders vary with the individual tool. Limitations of each device and the components of its installation unit are based on its individual listing.

3-15.3.2   The ability of concrete to hold the studs varies widely according to type of aggregate and quality of concrete, and it shall be established in each case by testing concrete on the job to determine that the studs will hold a minimum load of 750 lb (341 kg) for 2-in. or smaller pipe, 1000 lb (454 kg) for 2½-, 3-, or 3½-in. pipe, and 1200 lb (545 kg) for 4- or 5-in. pipe.

The concrete in the individual occupancy must be tested to ascertain that powder driven studs will have not less than the indicated holding power.

3-15.3.3   When increaser couplings are used, they shall be attached directly to the powder driven stud or welding stud.

Listings commonly permit studs with a smaller diameter than required by Table 3-15.4.1. Rods in compliance with the table must be connected directly to the studs with increaser couplings.

3-15.3.4   Welded studs or other hanger parts shall not be attached by welding to steel less than US Standard, 12 gage.

Steel less than U.S. Standard, 12 gage lacks the necessary strength for welding.

3-15.4 Rods and "U" Hooks.

3-15.4.1   Hanger rod size shall be the same as that approved for use with the hanger assembly and the size of rods shall not be less than that given in Table 3-15.4.1.

Exception: Rods of smaller diameter may be used when the hanger assembly has been tested and listed by a testing laboratory and installed within the limits of pipe sizes expressed in individual listings or approvals. For rolled threads, the rod size shall not be less than the root diameter of the thread.

Table 3-15.4.1 indicates the smallest size rods that may be used as part of a hanger assembly. The Exception allows for smaller diameters if improved technology results in such rods securing listing.

Table 3-15.4.1

Pipe Size	Dia. of Rod		Pipe Size	Dia. of Rod	
	in.	mm		in.	mm
Up to and including 4 in.	⅜	9.5	5, 6 and 8 in.	½	12.7
			10 and 12 in.	⅝	15.9

3-15.4.2 "U" Hooks.   The size of the rod material of "U" hooks shall be not less than that given in Table 3-15.4.2.

Table 3-15.4.2

Pipe Size	Hook Material Diameter	
	in.	mm
Up to 2 in.	5⁄16	7.9
2½ in. to 6 in.	⅜	9.5
8 in.	½	12.7

3-15.4.3   The size of the rod material for eye rods shall not be less than specified in Table 3-15.4.3.

Table 3-15.4.3

Pipe Size	Diameter of Rod			
	With Bent Eye		With Welded Eye	
	in.	mm	in.	mm
Up to 4 in.	⅜	9.5	⅜	9.5
5-6 in.	½	12.7	½	12.7
8 in.	¾	19.1	½	12.7

3-15.4.4   Threaded sections of rods shall not be formed or bent.

Bending on the threaded section of hanger rods could result in cracks at the thread root and, ultimately, failure.

3-15.4.5 Screws.   For ceiling flanges and "U" hooks, screw dimensions shall be not less than those given in Table 3-15.4.5.

Exception: When the thickness of planking and thickness of flange does not permit the use of screws 2 in. (51 mm) long, screws of 1 ¾ in. (44 mm) long may be permitted with hangers spaced not over 10 ft (3 m) apart. When the thickness of beams or joists does not permit the use of screws 2 ½ in. (64 mm) long, screws 2 in. (51 mm) long may be permitted with hangers spaced not over 10 ft (3 m) apart.

The screw dimensions in Table 3-15.4.5 are intended for use with hangers spaced to the maximums permitted by 3-15.1.11. The Exception recognizes that it is not always possible to use the screws specified in the table and allows for use of shorter screws with closer hanger spacing.

Table 3-15.4.5

Pipe Size	2 Screw Flanges
Up to 2 in.	Wood Screw No. 18 x 1½ in.

Pipe Size	3 Screw Flanges
Up to 2 in.	Wood Screw No. 18 x 1½ in.
2½ in., 3 in., 3½ in.	Lag Screw ⅜ in. x 2 in.
4 in., 5 in., 6 in.	Lag Screw ½ in. x 2 in.
8 in.	Lag Screw ⅝ in. x 2 in.

Pipe Size	4 Screw Flanges
Up to 2 in.	Wood Screw No. 18 x 1½ in.
2½ in., 3 in., 3½ in.	Lag Screw ⅜ in. x 1½ in.
4 in., 5 in., 6 in.	Lag Screw ½ in. x 2 in.
8 in.	Lag Screw ⅝ in. x 2 in.

Pipe Size	"U" Hooks
Up to 2 in.	Drive Screw No. 16 x 2 in.
2½ in., 3 in., 3½ in.	Lag Screw ⅜ in. x 2½ in.
4 in., 5 in., 6 in.	Lag Screw ½ in. x 3 in.
8 in.	Lag Screw ⅝ in. x 3 in.

For SI Units: 1 in. = 25.4 mm.

3-15.4.6   The size bolt or lag (coach) screw used with an eye rod or flange on the side of the beam shall not be less than specified in Table 3-15.4.6.

Exception: When the thickness of beams or joists does not permit the use of screws 2½ in. (64 mm), screws 2 in. (51 mm) may be permitted with hangers spaced not over 10 ft (3 m) apart.

Table 3-15.4.6

Size of Pipe	Size of Bolt or Lag Screw		Length of Lag Screw Used with Wood Beams	
	in.	mm	in.	mm
Up to and including 2 in.	⅜	9.5	2½	64
2½ to 6 in. (inclusive)	½	12.7	3	76
8 in.	⅝	15.9	3	76

Drive screws, which as the name indicates are intended to be installed with a hammer, do not have the holding power of wood screws. They can only be installed horizontally to support 2-in. and smaller pipe only.

3-15-4.7   Drive screws shall be used only in a horizontal position as in the side of a beam. Wood screws shall not be driven. Nails are not acceptable for fastening hangers.

3-15.4.8   Screws in the side of a timber or joist shall be not less than 2½ in. (64 mm) from the lower edge when supporting branch lines, and not less than 3 in. (76 mm) when supporting main lines. This shall not apply to 2-in. (51-mm) or thicker nailing strips resting on top of steel beams.

Screws must be located to ensure that the weight of the water filled pipe is adequately supported in accordance with 3-15.1.2.

3-15.4.9   The minimum thickness of plank and the minimum width of lower face of beams or joists in which lag screw rods are used shall be as given in Table 3-15.4.9.

Table 3-15.4.9

Pipe Size	Nominal Plank Thickness		Nominal Width of Beam Face	
	in.	mm	in.	mm
Up to 2 in.	3	76	2	51
2½ in. to 3½ in.	4	102	2	51
4 in. and 5 in.	4	102	3	76
6 in.	4	102	4	102

Lag screw rods shall not be used for support of pipes larger than 6 in. All holes for lag screw rods shall be pre-drilled ⅛ in. (3.2 mm) less in diameter than the root diameter of the lag screw thread.

The thickness required in Table 3-15.4.9 is needed for compliance with 3-15.1.2. Predrilling holes ⅛ in. (3.2 mm) smaller than the root diameter of the lag screw thread combines ease of installation with assurance of the needed gripping of the lag into the wood.

3-15.5 Location of Hangers on Branch Lines.   This subsection applies to the support of steel pipe or copper tube as specified in 3-1.1.1, subject to the provisions of 3-15.1.10.

3-15.5.1   On branch lines, there shall be not less than one hanger for each length of pipe.

Exception: Hangers may be located as provided in 3-15.5.2 to 3-15.5.6 inclusive.

Because of unique characteristics of their fittings, some manufacturers recommend hanger locations exceeding the requirements of this section. Under such circumstances the manufacturer's instructions should be followed.

3-15.5.2   The distance between the hanger and centerline of upright sprinkler shall be no less than 3 in. (76 mm).

3-15.5.3*   The unsupported length between the end sprinkler and the last hanger shall be not more than 36 in. (914 mm) for 1-in. pipe, or 48 in. (1219 mm) for 1¼-in. pipe. When these limits are exceeded, the pipe shall be extended beyond the end sprinkler and supported by an additional hanger.

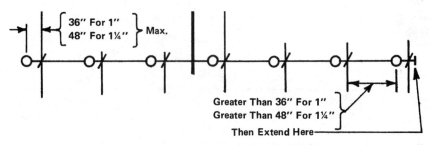

For SI Units: 1 in. = 25.4 mm; 1 ft = 0.3048 m.

Figure A-3-15.5.3   Distance Sprinkler to Hanger.

As Figure A-3-15.5.3 illustrates, excessive overhang is overcome by extending the branch line to the next structural member. The 36-in. (914-mm) overhang limitation is also applicable to ¾-in. copper pipe.

3-15.5.4* When sprinklers are less than 6 ft (1.8 m) apart, hangers may be spaced up to, but not exceeding, 12 ft (3.7 m). *(See Figure A-3-15.5.4.)*

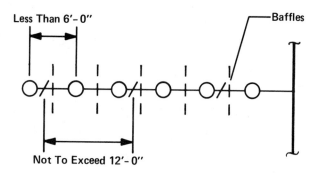

For SI Units: 1 in. = 25.4 mm; 1 ft = 0.3048 m.

Figure A-3-15.5.4 Distance Between Hangers. *(See 3-15.5.4.)*

Unless unique characteristics of the fitting utilized require support at each fitting, spacing hangers up to 12 ft (3.7 m) apart will adequately support the branch line piping when sprinklers are less than 6 ft (1.8 m) apart. In judging omission of hangers at each sprinkler under this section, consideration should be given to the flexibility of the joining methods being used at the individual sprinklers.

3-15.5.5 Starter lengths less than 6 ft (1.8 m) do not require a hanger, except on the end line of a side-feed system, or where an intermediate cross main hanger has been omitted.

3-15.5.6* Hangers are not required on 1-in. arms not over 12 in. (305 mm) long for copper tube, nor 24 in. (610 mm) long for steel pipe from branch lines or cross mains. *(See 3-13.1.2.1.)*

This section must be used in combination with 3-13.1.2.1. Fittings having a tendency to rotate unless restrained will either incorporate a mechanism to make rotation impossible or will as part of their listing have a requirement for a hanger on any arm regardless of length.

A-3-15.5.6   To take care of the thrust in a steeply pitched roof branch line, a clamp should be installed on the pipe just above the lowest hanger.

3-15.6 Location of Hangers on Cross Mains.   This subsection applies to the support of steel pipe only as specified in 3-1.1.1, subject to the provisions of 3-15.1.10. Intermediate hangers shall not be omitted for copper tube.

3-15.6.1*   On cross mains, there shall be at least one hanger between each two branch lines.

Exception No. 1:   In bays having two branch lines, the intermediate hanger may be omitted provided that a hanger attached to a purlin is installed on each branch line located as near to the cross main as the location of the purlin permits. [See Figure A-3-15.6.1(A).] Remaining branch line hangers shall be installed in accordance with 3-15.5.

Exception No. 2:   In bays having three or more branch lines, either side or centerfeed, one (only) intermediate hanger may be omitted provided that a hanger attached to a purlin is installed on each branch line located as near to the cross main as the location of the purlin permits. [See Figures A-3-15.6.1(B) and A-3-15.6.1(C).] Remaining branch line hangers shall be installed in accordance with 3-15.5.

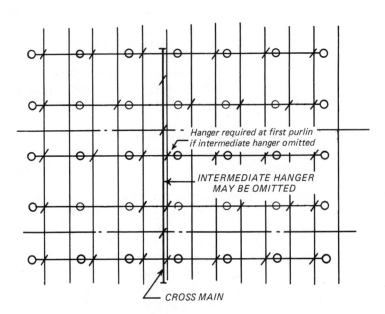

Figure A-3-15.6.1 (A)   Hangers on Cross Main.

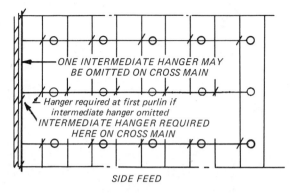

Figure A-3-15.6.1(B)   Hanger Omission on Side Feed System.

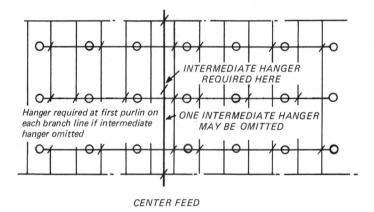

Figure A-3-15.6.1(C)   Hangers on Cross Main — Center Feed System.

Copper tube cross mains must have at least one hanger between each two branch lines. With steel pipe cross mains, the designer can either install a hanger between each two branch lines or install hangers on each branch line as near as possible to the cross main and delete one intermediate cross main hanger in each bay.

3-15.6.2   At the end of the cross main, intermediate trapeze hangers shall be installed unless the cross main is extended to the next framing member with an ordinary hanger installed at this point, in which event an intermediate hanger may be omitted in accordance with 3-15.6.1, Exceptions No. 1 and 2.

The option to delete the intermediate cross main hanger in accordance with the Exceptions to 3-15.6.1 is applicable to the last piece of cross main only if the main is extended to the next framing member and a hanger is also installed at that point.

3-15.7 Support of Risers.

3-15.7.1   Risers shall be supported by attachments directly to the riser or by hangers located on the horizontal connections close to the riser.

Risers must be adequately supported to avoid excess strain on fittings and joints.

3-15.7.2   In multistory buildings, riser supports shall be provided at the lowest level, at each alternate level above, above and below offsets, and at the top of the riser. Supports above the lowest level shall also restrain the pipe to prevent movement by an upward thrust when flexible fittings are used. Where risers are supported from the ground, the ground support constitutes the first level or riser support. Where risers are offset or do not rise from the ground the first ceiling level above the offset constitutes the first level of riser support.

Restraints are required at alternate levels and at the points most subject to stress.

3-15.7.3   Sprinkler and tank risers in vertical shafts, or in buildings with ceilings over 25 ft (7.6 m) high, shall have at least one support for each riser pipe section.

3-15.7.4   Clamps supporting pipe by means of set screws shall not be used.

To adequately support a riser, the full surface of a clamp must bear against the surface of the pipe.

B-3 Systems Components.

B-3-1 Sleeves for Pipe Risers.   *(See Figure B-3-1.)*

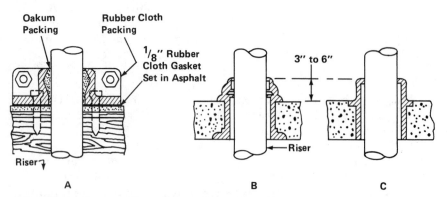

A–For Wood or Concrete Floors; B and C–For Concrete Floors.

Figure B-3-1   Watertight Riser Sleeves.

B-3-1.1 Sprinkler piping passing through floors of concrete or waterproof construction should have properly designed substantial thimbles or sleeves projecting 3 to 6 in. (76 to 152 mm) above the floor to prevent possible floor leakage, except in areas subject to earthquakes. (*See A-3-10.3.*) The space between the pipe and sleeve should be caulked with oakum or equivalent material. If floors are of cinder concrete, thimbles or sleeves should extend all the way through to protect the piping against corrosion.

B-3-1.2 Ordinary floors through which pipes pass should be made reasonably tight around the risers, except in areas subject to earthquakes. (*See A-3-10.3.*)

3-16 Sprinklers.

3-16.1 Types of Sprinklers. Some of the commonly used sprinklers are as follows:

(a) *Upright Sprinklers.* Sprinklers designed to be installed in such a way that the water spray is directed upwards against the deflector.

(b) *Pendent Sprinklers.* Sprinklers designed to be installed in such a way that the water stream is directed downward against the deflector.

Figure 3.8. An approved and listed sprinkler showing upright (left) and pendent (right) models of the same issue. Sprinklers shown are Reliable Model G.

(c) *Sidewall Sprinklers.* Sprinklers having special deflectors which are designed to discharge most of the water away from the nearby wall in a pattern resembling one quarter of a sphere, with a small portion of the discharge directed at the wall behind the sprinkler.

Figure 3.9.   Star sidewall sprinkler.

(d) *Extended Coverage Sidewall Sprinklers.*   Sprinklers with special extended, directional, discharge patterns.

Figure 3.10.   Extended coverage Grinnell Duraspeed Issue C sidewall sprinkler.

(e) *Open Sprinklers.*   Sprinklers from which the actuating elements have been removed.

Figure 3.11.   Grinnell open sprinkler.

(f) *Corrosion-Resistant Sprinklers.*   Sprinklers with special coatings or platings to be used in an atmosphere which would corrode an uncoated sprinkler.

Corrosion resistant sprinklers are usually standard sprinklers with a corrosion resistant coating, such as wax or lead.

(g) *Nozzles.*   Devices for use in applications requiring special discharge patterns, directional spray, fine spray, or other unusual discharge characteristics.

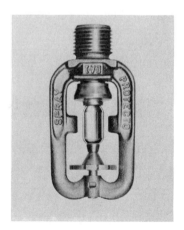

Figure 3.12.   Grinnell automatic Protectospray nozzle.

(h) *Dry Pendent Sprinklers.*   Sprinklers for use in a pendent position in a dry-pipe system or a wet-pipe system with the seal in a heated area.

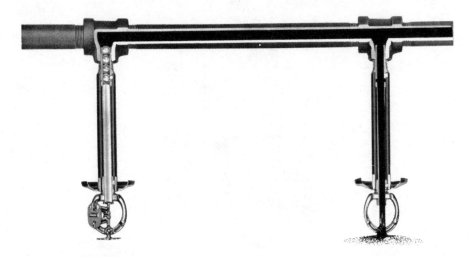

Figure 3.13.   Representative dry pendent sprinkler. When the solder holding the fusible link together melts, the levers and link parts are ejected away from the sprinkler. The inner tube, which also serves as an orifice, drops to a predetermined position, allowing the sealing elements to pass through the tube and away from the sprinkler. This, in turn, allows water to flow through the tube and strike the deflector, which distributes it in a spray pattern comparable to a ½-in. standard sprinkler. Shown is Reliable's Model C Dry Pendent.

(i) *Dry Upright Sprinklers.*   Sprinklers which are designed to be installed in an upright position, on a wet-pipe system, to extend into an unheated area with a seal in a heated area.

The dry upright sprinkler is similar to the dry pendent except that the sprinkler uses an upright deflector.

(j) *Ornamental Sprinklers.*   Sprinklers which have been painted or plated by the manufacturer.

(k) *Flush Sprinklers.*   Sprinklers in which all or part of the body, including the shank thread, is mounted above the lower plane of the ceiling.

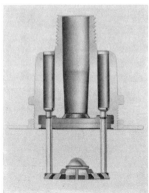

Figure 3.14.   Reliable Model A flush type automatic sprinkler before operation (left) and after operation (right).

(1) *Recessed Sprinklers.*   Sprinklers in which all or part of the body, other than the shank thread, is mounted within a recessed housing.

Figure 3.15.   A Central Model A recessed sprinkler.

(m) *Concealed Sprinklers.*   Recessed sprinklers with cover plates.

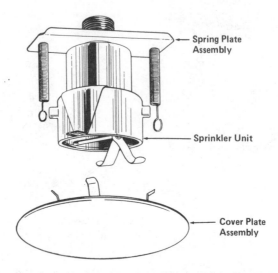

Figure 3.16.   Stargard Model G concealed ceiling sprinkler. Cover plate drops away when heat is applied to bottom side of plate.

(n) *Old Style Sprinklers.*   Sprinklers which direct only from 40 to 60 percent of the total water initially in a downward direction and which are designed to be installed with the deflector either upright or pendent.

(o) *Residential Sprinklers.*   Sprinklers which have been specifically listed for use in residential occupancies.

Figure 3.17.   Grinnell Model F954 residential sprinkler.

(p) *Intermediate Level Sprinklers.*   Sprinklers equipped with integral shields to protect their operating elements from the discharge of sprinklers installed at higher elevations.

Figure 3.18.   Star Model LD Intermediate Level Sprinkler.

(q) *Special Sprinklers.*   Sprinklers which have been tested and listed as prescribed in 4-1.1.3.

Special sprinklers are listed for installations with protection areas or distances between sprinklers different from those specified in this standard. Such listings are intended to be based on tests specifically designed to evaluate the sprinkler's performance under the special listing criteria.

(r) *Quick Response Sprinklers.*   A type of special sprinkler.

(s) *Large-Drop Sprinkler.*   A listed large-drop sprinkler is characterized by a K factor between 11.0 and 11.5, and proven ability to meet prescribed penetration,

Figure 3.19.   Viking Model A Large Drop Sprinkler.

cooling and distribution criteria prescribed in the large-drop sprinkler examination requirements. The deflector/discharge characteristics of the large-drop sprinkler generate large drops of such size and velocity as to enable effective penetration of the high-velocity fire plume.

The upright, the pendent, and the sidewall are the basic sprinklers. The remaining sprinklers defined in this section, with the exception of the nozzles and the old style sprinklers, are variations of the basic sprinklers modified to address specific needs. The primary difference between the old style sprinkler, which is designed for installation in either the upright or pendent position, and the current upright and pendent sprinklers is the design of the deflector, which for the old style sprinkler directed approximately 40 percent of the water up against the ceiling with the remainder of the water flowing to the fire. Nozzles are special application devices which have unique characteristics to meet specific needs.

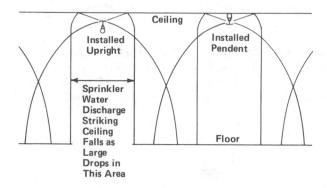

Figure 3.20. Principal distribution pattern of water from old-style sprinklers (previous to 1953).

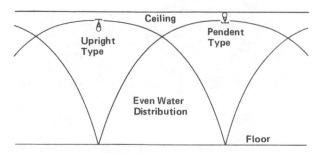

Figure 3.21. Principal distribution pattern of water from standard sprinklers (in use since 1953).

3-16.2 Use of Sprinklers.

3-16.2.1* Only listed sprinklers shall be used and shall be installed in accordance with their listing.

Exception: When construction features or other special situations require unusual water distribution, listed sprinklers may be installed in other positions than anticipated by their listing to achieve specific results.

Formal Interpretation

Question: Does 3-16.2.1 provide for the listing of sprinklers designed specifically for smaller protection areas or distances between sprinklers than specified elsewhere in NFPA 13?

Answer: No.

A-3-16.2.1(a) Upright sprinklers should be installed with the frame parallel to the branch line pipe to reduce to minimum the obstruction of the discharge pattern.

A-3-16.2.1(b) Large orifice sprinklers should not be used with pipe schedule systems unless their use is acceptable to the authority having jurisdiction and supported by hydraulic calculations.

Any limitations placed on a sprinkler, such as the hazard classifications where it may be used or limitations placed on its temperature rating, are included as part of its listing.

3-16.2.2 Sprinklers shall not be altered in any respect or have any type of ornamentation or coating applied after shipment from the place of manufacture.

3-16.2.3 Sprinklers shall not be used for system working pressures exceeding 175 psi (12.1 bars).

Exception: Higher design pressures may be used when sprinklers are listed for those pressures.

Currently all sprinklers are designed for a maximum working pressure of 175 psi (12.1 bars). The Exception allows for the future listing of higher pressure designs. See the Exception to 1-8.1.3.

3-16.2.4 Old style sprinklers shall not be used in a new installation.

Exception No. 1: For installation under piers and wharves where construction features require an upward discharge to wet the underside of decks and structural members supporting the decks, a sprinkler that projects water upward to wet the overhead shall be used. This can be accomplished by using standard pendent sprinklers installed in an upright position or by the use of old style sprinklers. See NFPA 87, Standard for the Construction and Protection of Piers and Wharves.

Exception No. 2:   Old style sprinklers shall be installed in fur storage vaults. See 4-4.17.3. Also see NFPA 81, Standard on Fur Storage and Fumigation and Cleaning.

Exception No. 3:   Listed old style sprinklers may be used when construction features or other special situations require unique water distribution.

The discharge characteristics of old style sprinklers are inferior to those of the upright and pendent sprinklers which provide a more even distribution of water over a greater area with a more consistent droplet size.

3-16.2.5   Sidewall sprinklers shall be installed only in light hazard occupancies.

Exception: Sidewall sprinklers specifically listed for use in ordinary hazard occupancies.

The discharge characteristics of sidewall sprinklers are inferior to those of upright or pendent sprinklers. Accordingly, they are confined to use in light hazard occupancies, unless the sprinkler has been specifically listed for use in ordinary hazard occupancies.

3-16.2.6*   Extended coverage sidewall sprinklers shall be installed only in accordance with their listing.

A-3-16.2.6   Multiple extended coverage sidewall sprinklers may be used in a single room or space not exceeding 9 ft (2.7 m) in height unless otherwise indicated in the listing. Where such sprinklers are placed back to back, a baffle should be employed so that the operation of one sprinkler will not prevent the other from operating.

The characteristics of standard sidewall sprinklers are such that the rules covering their installation found in Section 4-5 of this standard are applicable regardless of manufacturer. However, the characteristics of extended coverage sidewall sprinklers vary substantially as to area of coverage and required operating pressure. Each model of extended coverage sprinkler must be installed in accordance with the requirements of its specific listing.

3-16.2.7   Open sprinklers may be used to protect special hazards, for protection against exposures, or in other special locations.

Open sprinklers referenced in this section are standard sprinklers with the operating elements removed. For information on open window or cornice sprinklers designed for outside use for protection against exposure fires, see Chapter 6 of this standard.

3-16.2.8   When nonmetallic ceiling plates (escutcheons) are used they shall be listed.

3-16.2.9   Residential sprinklers may be used in dwelling units located in any occupancy provided they are installed in conformance with their listing and the positioning requirements of NFPA 13D, *Standard for the Installation of Sprinkler Systems in One- and Two-Family Dwellings and Mobile Homes*. One-half inch or larger orifice residential sprinklers may be used in dry-pipe systems when the design area is in compliance with Chapter 2.

The special water supply requirements for residential sprinklers on wet-pipe systems are not applicable when residential sprinklers are supplied by a dry-pipe system. (See 2-2.1.2.8 Exception and 7-4.4.) As with standard sprinklers, orifice sizes less than ½ in. are not permitted on dry-pipe systems. *[See 3-16.5.2(a) Exception.]*
Since residential sprinklers are tested for listing using a residential fire scenario they are permitted to be used in residential portions of all occupancies. In other portions of such occupancies, sprinklers listed for general usage and installed in accordance with the requirements of this standard are required.

Formal Interpretation

Question: Is NFPA 13D appropriate for use in multiple (three or more) attached dwellings under any condition?

Answer: No. NFPA 13D is appropriate for use only in one- and two-family dwellings and mobile homes. Buildings which contain more than two dwelling units shall be protected in accordance with NFPA 13. Section 3-16.2.9 of NFPA 13 permits residential sprinklers to be used in residential portions of other buildings provided all other requirements of NFPA 13, including water supplies, are satisfied.

Note: Building codes may contain requirements such as 2-hour fire separations which would permit adjacent dwellings to be considered unattached.

3-16.3 Replacement of Sprinklers.

3-16.3.1   When sprinklers are replaced, the replacement sprinkler shall be of the same type, orifice, and temperature rating unless conditions require a different type sprinkler be installed. The replacement sprinkler shall then be of a type, orifice, and temperature rating to suit the new conditions.

In replacing sprinklers, in addition to matching the type (upright, pendent, etc.), the orifice size and the temperature rating, care must be taken to ensure that the replacement sprinkler is installed in accordance with its listing as required by 3-16.2.1. It would be very easy, for example, to replace a sidewall sprinkler in an ordinary hazard occupancy with a sprinkler that is listed for light hazard use only.

3-16.3.2   Old style sprinklers may be replaced with old style sprinklers, or with the appropriate pendent or upright sprinkler.

3-16.3.3   Old style sprinklers shall not be used to replace pendent or upright sprinklers.

Because of their better discharge characteristics, greater coverage area per sprinkler is permitted for upright and pendent sprinklers than was permitted for the old style sprinklers.

3-16.3.4   Extreme care shall be exercised when replacing horizontal sidewall and extended coverage sidewall sprinklers to assure the correct replacement sprinkler is installed.

3-16.3.5   Sprinklers which have been painted or coated, except by the manufacturer, shall be replaced and shall not be cleaned by use of chemicals, abrasives, or other means. (*See 3-16.9.2.*)

3-16.4 Corrosion-Resistant, Wax-Coated or Similar Sprinklers.

Sprinklers with special coatings or platings are specifically listed for use in atmospheres that would corrode an uncoated sprinkler. Attempts to protect sprinklers by parties other than the manufacturer would result in ineffective protection of the sprinkler and/or serious impairment of its operation. The listing of sprinklers with protective or special coatings requires that the coatings be applied at the manufacturer's facility. Field application of these platings and coatings is not allowed.

3-16.4.1* Listed corrosion-resistant or special coated sprinklers shall be installed in locations where chemicals, moisture or other corrosive vapors exist sufficient to cause corrosion of such devices.

A-3-16.4.1 Examples of such locations are paper mills, packing houses, tanneries, alkali plants, organic fertilizer plants, foundries, forge shops, fumigation, pickle and vinegar works, stables, storage battery rooms, electroplating rooms, galvanizing rooms, steam rooms of all descriptions, including moist vapor dry kilns, salt storage rooms, locomotive sheds or houses, driveways, areas exposed to outside weather such as piers and wharves exposed to salt air, areas under sidewalks, around bleaching equipment in flour mills, all portions of cold storage buildings where a direct ammonia expansion system is used, and portions of any plant where corrosive vapors prevail.

3-16.4.2 Care shall be taken in the handling and installation of wax-coated or similar sprinklers to avoid damaging the coating.

3-16.4.3 Corrosion-resistant coatings shall be applied only by the manufacturer of the sprinkler.

Exception: Any damage to the protective coating occurring at the time of installation shall be repaired at once using only the coating of the manufacturer of the sprinkler in the approved manner so that none of the sprinkler will be exposed after installation has been completed.

3-16.5* Sprinkler Discharge Characteristics and Identification.

A-3-16.5 The following Table A-3-16.5 shows the nominal discharge capacities of approved sprinklers having a nominal ½-in. (12.7-mm) orifice at various pressures up to 100 psi (6.9 bars).

Table A-3-16.5

Pressure at Sprinkler lb per sq in.	Discharge gal per min	Pressure at Sprinkler lb per sq in.	Discharge gal per min
10	18	35	34
15	22	50	41
20	25	75	50
25	28	100	58

For SI Units: 1 gpm = 3.785 L/min; 1 psi = 0.0689 bar.

3-16.5.1 Table 3-16.5 shows the K factor, relative discharge and identification for sprinklers having different orifice sizes.

Exception: Special listed sprinklers may have pipe threads different from those shown in Table 3-16.5.

Table 3-16.5   Sprinkler Discharge Characteristics Identification

Nominal Orifice Size (in.)[1]	Orifice Type	"K"[2] Factor	Percent of Nominal ½ In. Discharge	Thread Type	Pintle	Nominal Orifice Size Marked On Frame
¼	Small	1.3-1.5	25	½ in. NPT	Yes	Yes
⁵⁄₁₆	Small	1.8-2.0	33.3	½ in. NPT	Yes	Yes
⅜	Small	2.6-2.9	50	½ in. NPT	Yes	Yes
⁷⁄₁₆	Small	4.0-4.4	75	½ in. NPT	Yes	Yes
½	Standard	5.3-5.8	100	½ in. NPT	No	No
¹⁷⁄₃₂	Large	7.4-8.2	140	¾ in. NPT	No	No
				or ½ in. NPT	Yes	Yes

[1]See A-3-16.5.2

[2]"K" factor is the constant in the formula $Q = K\sqrt{P}$

Where   Q=Flow in gpm
   P=Pressure in psi

For SI Units: $Qm = Km\sqrt{Pm}$

Where   Qm=Flow in L/min
   Pm=Pressure in bars
   Km=14 K

The orifice sizes are indicated as nominal in Table 3-16.5 because the various sprinklers differ slightly plus or minus from the norm. The exact orifice size of a sprinkler is indicated by its "K" factor. The "K" factor is used in determining the sprinkler's flow rate, which is of major consequence in hydraulically designed systems. See Chapter 7. In most instances sprinklers with ½-in. orifices and ½-in. pipe threads or sprinklers with ¹⁷⁄₃₂-in. orifice with ¾-in. pipe threads are installed. Sprinklers of other than those combinations are indicated by a pintle, which is a metal rod extending beyond the sprinkler's deflector.

3-16.5.2* For light hazard occupancies not requiring as much water as is discharged by a nominal ½ in. (12.7 mm) orifice sprinkler, sprinklers having a smaller orifice may be used subject to the following restrictions:

(a) Small orifice sprinklers shall not be used on dry-pipe, pre-action or combined dry-pipe and pre-action systems.

Exception: Outside sprinklers for protection from exposure fires. See Chapter 6.

On systems that normally are filled with air rather than water, there is a tendency to form more internal scale, therefore, small orifice sprinklers might become clogged.

(b) An approved strainer shall be provided in the riser or feed main which supplies sprinklers having orifices smaller than ⅜ in. (9.5 mm).

A-3-16.5.2 Small orifice sprinklers should not be used as a substitute for standard ½ in. and large orifice sprinklers to take advantage of available high water pressure.

Small orifice sprinklers are restricted to light hazard occupancies, because they are not as efficient in fire control as standard and large orifice sprinklers. They are restricted to wet systems due to the danger of the orifice being obstructed by scale and foreign material carried at the head of the high velocity water flowing in other types of systems.

Figure 3.22. Viking small orifice sprinkler.

3-16.5.3 For locations or conditions requiring more water than is discharged by a nominal ½ in. (12.7 mm) orifice sprinkler, a sprinkler having a larger orifice may be used. Special sprinklers having an orifice larger than ¹⁷⁄₃₂ in. (13.5 mm) may be installed in accordance with 4-1.1.3.

3-16.5.4 Sprinklers having orifice sizes exceeding ½ in. (12.7 mm) and having ½ in. NPT shall not be installed in new sprinkler systems.

Large orifice sprinklers with ½-in. national pipe threads are intended solely for use where conditions merit replacing existing ½-in. orifice sprinklers with sprinklers having ¹⁷⁄₃₂-in. orifices. Some sprinklers having larger orifices have been developed & listed with special conditions of use and are limited by those conditions.

Figure 3.23.   Viking large orifice sprinkler. This model is fitted with
¾-in. threads.

3-16.6* Temperature Ratings, Classifications and Color Coding.

A-3-16.6   Information regarding the highest temperature that may be encountered in any location in a particular installation may be obtained by use of a thermometer that will register the highest temperature encountered, which should be hung for several days in the questionable location with the plant in operation.

When an occupancy hazard normally may be expected to produce a fast-developing fire or a rapid rate of heat release, the use of sprinklers of high temperature classification, as a means of limiting the total number of sprinklers which might open in a fire, is recommended. Since the number of sprinklers which might be expected to open will be reduced where the water pressure effective in first operating sprinklers is at least 75 psi (5.2 bars) without the disadvantage of a potential increase in fire damage, this alternative should be given first consideration.

NOTE: Fire tests have shown that the number of sprinklers which might be expected to open, particularly under conditions where fast-developing fires may be expected, can be limited by the use of sprinklers of High Temperature Classification. This may be of advantage in reducing the number of sprinklers which would otherwise open outside the area directly involved in a fire and decrease the overall water demand. However, some increase in fire damage and fire temperatures may be expected when sprinklers of Intermediate or High Temperature Classification are used.

Some occupancies employ high temperature fumigation processes requiring consideration in the selection of sprinkler temperature ratings.

3-16.6.1 The standard temperature ratings of automatic sprinklers are shown in Table 3-16.6.1. Automatic sprinklers shall have their frame arms colored in accordance with the color code designated in Table 3-16.6.1 with the following exceptions:

Exception No. 1: *The color identification for coated sprinklers may be a dot on the top of the deflector, the color of the coating material or colored frame arms.*

Exception No. 2: *Color identification is not required for plated sprinklers, flush, recessed and concealed sprinklers or similar decorative types.*

Sprinklers are color coded in accordance with 3-16.6.1 to provide a ready means of establishing the temperature classifications of the sprinklers' operating elements. Table 3-16.6.1 indicates the range of temperatures for sprinklers in each classification and the maximum ceiling temperatures for which each classification may be installed.

Table 3-16.6.1   Temperature Ratings, Classification and Color Codings

Max. Ceiling Temp. °F	°C	Temperature Rating °F	°C	Temperature Classification	Color Code	Glass Bulb Colors
100	38	135 to 170	57 to 77	Ordinary	Uncolored	Orange or Red
150	66	175 to 225	79 to 107	Intermediate	White	Yellow or Green
225	107	250 to 300	121 to 149	High	Blue	Blue
300	149	325 to 375	163 to 191	Extra High	Red	Purple
375	191	400 to 475	204 to 246	Very Extra High	Green	Black
475	246	500 to 575	260 to 302	Ultra High	Orange	Black
625	329	650	343	Ultra High	Orange	Black

3-16.6.2 Ordinary temperature rated sprinklers shall be used throughout buildings.

Exception No. 1: *Where maximum ceiling temperatures exceed 100°F (38°C), sprinklers with temperature ratings in accordance with the maximum ceiling temperatures of Table 3-16.6.1 shall be used.*

Exception No. 2: *Intermediate and high temperature sprinklers may be used throughout Ordinary and Extra Hazard Occupancies.*

Exception No. 3: *Sprinklers of intermediate and high temperature ratings shall be installed in specific locations as required by 3-16.6.3.*

Exception No. 4: When permitted by other NFPA standards.

Higher temperature classification sprinklers are preferable for some types of fires where ordinary temperature classification sprinklers would tend to operate beyond the fire area and thereby reduce the water discharge density available over the fire. In some high heat release fires with high thermal updraft discharge, water from sprinklers is carried back toward the ceiling as steam and when it condenses on ordinary temperature sprinklers will cause them to operate. This is the phenomenon considered to be responsible for sprinklers operating beyond the fire area.

3-16.6.3 The following practices shall be observed when installing high temperature sprinklers, unless maximum expected temperatures are otherwise determined or unless high temperature sprinklers are used throughout.

(a) Sprinklers near unit heaters. Where steam pressure is not more than 15 psi (1 bar), sprinklers in the heater zone shall be high and sprinklers in the danger zone intermediate temperature classification.

(b) Sprinklers located within 12 in. (505 mm) to one side or 30 in. (762 mm) above an uncovered steam main, heating coil or radiator shall be intermediate temperature classification.

(c) Sprinklers within 7 ft (2.1 m) of a low pressure blow-off valve which discharges free in a large room shall be high temperature classification.

(d) Sprinklers under glass or plastic skylights exposed to the direct rays of the sun shall be intermediate temperature classification.

(e) Sprinklers in an unventilated concealed space under an uninsulated roof, or in an unventilated attic, shall be of intermediate temperature classification.

(f) Sprinklers in unventilated show windows having high-powered electric lights near the ceiling shall be intermediate temperature classification.

(g) Where a locomotive enters a building, sprinklers shall be located not nearer than 5 ft (1.5 m) from the center line of the track.

(h) For sprinklers protecting commercial-type cooking equipment and ventilation systems, temperature classifications of intermediate, high or extra high shall be provided as determined by use of a temperature measuring device (*see 4-4.18.2*).

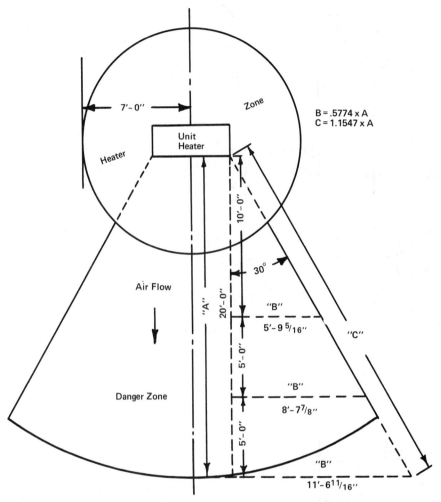

For SI Units: 1 in. = 25.4 mm; 1 ft = 0.3048 m.

Figure 3-16.6.3(a)   Heater and Danger Zones at Unit Heaters.

3-16.6.4   In case of change of occupancy involving temperature change, the sprinklers shall be changed accordingly.

3-16.7* Stock of Spare Sprinklers.

A-3-16.7   For equipment aboard vessels or in isolated locations, a greater number of sprinklers should be provided to permit equipment to be put back into service promptly after a fire. When a great number of sprinklers are likely to be opened by a flash fire, a greater number of sprinklers should be provided.

Table 3-16.6.3(A)   Distance of Sprinklers from Heat Sources

Type of Heat Condition	Ordinary Degree Rating	Intermediate Degree Rating	High Degree Rating
1. Heating Ducts (a) Above	More than 2 ft 6 in.	2 ft 6 in. or less	—
(b) Side and Below	More than 1 ft 0 in.	1 ft 0 in. or less	—
(c) Diffuser Downward Discharge Horizontal Discharge	Any distance except as shown under Intermediate	*Downward:* Cylinder with 1 ft 0 in. radius from edge, extending 1 ft 0 in. below and 2 ft 6 in. above *Horizontal:* Semi-cylinder with 2 ft 6 in. radius in direction of flow, extending 1 ft 0 in. below and 2 ft 6 in. above	—
2. Unit Heater (a) Horizontal Discharge	—	*Discharge Side:* 7 ft 0 in. to 20 ft 0 in. radius pie-shaped cylinder [see Figure 3-16.6.3(a)] extending 7 ft 0 in. above and 2 ft 0 in. below heater; also 7 ft 0 in. radius cylinder more than 7 ft 0 in. above unit heater	7 ft 0 in. radius cylinder extending 7 ft 0 in. above and 2 ft 0 in. below unit heater
(b) Vertical Downward Discharge [Note: For sprinklers below unit heater, see Figure 3-16.6.3(a).]	—	7 ft 0 in. radius cylinder extending upward from an elevation 7 ft 0 in. above unit heater	7 ft 0 in. radius cylinder extending from the top of the unit heater to an elevation 7 ft 0 in. above unit heater
3. Stream Mains (Uncovered) (a) Above	More than 2 ft 6 in.	2 ft 6 in. or less	—
(b) Side and Below	More than 1 ft 0 in.	1 ft 0 in. or less	—
(c) Blow-off Valve	More than 7 ft 0 in.	—	7 ft 0 in. or less

For SI Units: 1 in. = 25.4 mm; 1 ft = 0.03048 m.

Table 3-16.6.3(B)   Ratings of Sprinklers in Specified Locations

Location	Ordinary Degree Rating	Intermediate Degree Rating	High Degree Rating
Skylights	—	Glass or plastic	—
Attics	Ventilated	Unventilated	—
Peaked Roof: Metal or thin boards; concealed or not concealed; insulated or uninsulated	Ventilated	Unventilated	—
Flat Roof: Metal, not concealed; insulated or uninsulated	Ventilated or unventilated	Note: For uninsulated roof, climate and occupancy may require Intermediate sprinklers. Check on job.	—
Flat Roof: Metal; concealed: insulated or uninsulated	Ventilated	Unventilated	—
Show Windows	Ventilated	Unventilated	—

Note: A check of job condition by means of thermometers may be necessary.

3-16.7.1   There shall be maintained on the premises a supply of spare sprinklers (never less than six) so that any sprinklers that have operated or been damaged in any way may promptly be replaced. These sprinklers shall correspond as to types and temperature ratings with the sprinklers in the property. The sprinklers shall be kept in a cabinet located where the temperature to which they are subjected will at no time exceed 100°F (38°C).

3-16.7.2   A special sprinkler wrench shall also be provided and kept in the cabinet, to be used in the removal and installation of sprinklers.

3-16.7.3   The stock of spare sprinklers shall be as follows:

For equipments not over 300 sprinklers, not less than 6 sprinklers.

For equipments 300 to 1,000 sprinklers, not less than 12 sprinklers.

For equipments above 1,000 sprinklers, not less than 24 sprinklers.

Stock of spare sprinklers shall include all types and ratings installed.

The stock of spare sprinklers required is a minimum. Spare sprinklers of all types and ratings installed must be available. For an occupancy with a variety of types and ratings of sprinklers installed, the stock of spare sprinklers should be increased above the minimum.

3-16.8* Guards and Shields.   Sprinklers which are so located as to be subject to mechanical injury (in either the upright or the pendent position) shall be protected with approved guards.

A-3-16.8   Sprinklers under open gratings should be provided with shields. Shields over automatic sprinklers should not be less, in least dimension, than four times the distance between the shield and fusible element, except special sprinklers incorporating a built-in shield need not comply with this recommendation if approved for the particular application.

Any sprinkler suffering damage that could affect its efficiency, such as a bent deflector, must be replaced.

3-16.9 Painting and Ornamental Finishes.

3-16.9.1*   When the sprinkler piping is given any kind of coating, such as whitewash or paint, care shall be exercised to see that no automatic sprinklers are coated.

A-3-16.9.1   When painting sprinkler piping or painting in areas near sprinklers, the sprinklers may be protected by covering with a bag which should be removed immediately after the painting has been finished.

3-16.9.2* Sprinklers shall not be painted and any sprinklers which have been painted shall be replaced with new listed sprinklers of the same characteristics.

Exception: Factory applied coatings to sprinkler frames for identifying sprinklers of different temperature ratings in accordance with 3-16.6.1.

A-3-16.9.2 Painting of sprinklers may retard the thermal response of the heat responsive element, may interfere with the free movement of parts, and may render the sprinkler inoperative. Moreover, painting may invite the application of subsequent coatings, thus increasing the possibility of a malfunction of the sprinkler.

Painting is the primary, but not the only, example of a problem known as loading in which a buildup on the sprinkler would delay or prevent proper response. Any sprinkler subject to loading that cannot be readily dusted or blown away must be replaced.

3-16.9.3 Ornamental finishes shall not be applied to sprinklers by anyone other than the sprinkler manufacturer and only sprinklers listed with such finishes shall be used.

Any coatings for purpose of identification or ornamentation must be installed by the manufacturer in accordance with the sprinkler's listing.

3-17 Sprinkler Alarms.

3-17.1 Definition. A local alarm unit is an assembly of apparatus approved for the service and so constructed and installed that any flow of water from a sprinkler system equal to or greater than that from a single automatic sprinkler of the smallest orifice size installed on the system will result in an audible alarm on the premises within 5 minutes after such flow begins. For remote sprinkler water flow alarm transmission see 3-17.6.1.

3-17.2* Where Required. Local waterflow alarms shall be provided on all sprinkler systems having more than 20 sprinklers.

A-3-17.2 Central station, auxiliary, remote station, or proprietary protective signaling systems are a highly desirable supplement to local alarms, especially from a safety to life standpoint. (*See Section 8-5.*)

Identification Signs. Approved identification signs should be provided for outside alarm devices. The sign should be located near the device in a conspicuous position and should be worded as follows:

SPRINKLER FIRE ALARM — WHEN BELL RINGS CALL FIRE DEPART-MENT OR POLICE. (*See Figure A-3-17.2.*)

Figure A-3-17.2   Identification Sign.

This standard requires facilities to sound an audible water flow alarm on the premises for all systems having more than 20 sprinklers. It recommends, but does not require, locally sounding alarms for fewer than 20 sprinklers, nor does it require supplemental alarm systems. It requires alarm supervision of valves controlling sprinkler systems with nonfire protection connections (*see 5-6.1.6*) and recommends, but does not require, such supervision in other instances (*see 3-14.2.3*). It should be noted that the sprinkler alarms are as indicated water flow alarms and are not considered to be building evacuation alarms. In some applications they may, however, be utilized as such. The purpose for which the alarm is intended should be considered in the overall design of both actuating mechanism and audible alarm devices.

3-17.3 Water Flow Detecting Devices.

3-17.3.1 Wet-Pipe Systems.   The alarm apparatus for a wet-pipe system shall consist of listed alarm check valve or other listed waterflow detecting alarm device with the necessary attachments required to give an alarm.

On wet-pipe systems, alarm check valves may be prefered to other listed water flow detecting alarm devices, particularly where fluctuating pressure water supplies are provided. Fluctuating pressure can cause false alarms when other types of water flow detecting devices are used. On larger systems an alarm check valve makes it possible to introduce an excess pressure higher than the supply on the system side of the alarm check valve. Water flow is then detected by a drop in pressure on the system side of the alarm check. In other instances, the alarm check valve serves to prevent

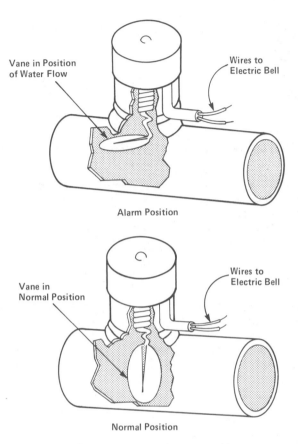

Figure 3.24.   Water flow detector.

water flow alarms on risers not involved in an actual water flow condition when fire pumps start or similar large fluctuations occur.

Formal Interpretation

Question: Would the use of a listed water flow switch and bell as the alarm apparatus meet the intent of 3-17.3.1?

Answer: Yes.

3-17.3.2 Dry-Pipe Systems.   The alarm apparatus for a dry-pipe system shall consist of listed alarm attachments to the dry-pipe valve. When a dry-pipe valve is located on the system side of an alarm valve, the actuating device of the alarms for the dry-pipe valve may be connected to the alarms on the wet-pipe system.

In some instances, small dry-pipe systems may be necessary in areas of otherwise heated buildings, such as outside canopies, loading docks, unheated penthouses, etc. In these cases, it may be more economical to install an auxiliary dry-pipe system supplied from the building's wet system. Under these circumstances, the water flow alarm from the dry-pipe system may be connected into the wet-pipe system alarm sounding or annunciating device.

3-17.3.3* Pre-Action and Deluge Systems.   In addition to the waterflow alarms required for systems having more than 20 sprinklers, all deluge and preaction systems shall be provided with listed alarm attachments actuated by the detection system.

A-3-17.3.3   A mechanical alarm (water motor gong) may also be required.

Inasmuch as pre-action and deluge valves are not actuated by the operation of an automatic sprinkler, a separate detection system is required to actuate the valves. This detection system gives an indication of fire prior to flow of water in the system. It may also be utilized to initiate other fire-related activities.

3-17.3.4*   Waterflow alarm indicators (paddle type) shall not be installed in dry-pipe, pre-action or deluge systems.

A-3-17.3.4   The surge of water when valve trips may seriously damage the device.

In addition to damaging the device itself, the high velocity flow could totally disengage the paddle and carry it downstream, until it lodged in the piping, obstructing it.

3-17.4 Attachments — General.

3-17.4.1*   An alarm unit shall include a listed mechanical alarm, horn or siren, or an approved weatherproof electric gong, bell, horn or siren.

A-3-17.4.1   Audible alarms are normally located on the outside of the building. Listed electric gongs, bells, horns, or sirens inside the building or a combination inside and outside are sometimes advisable.

The required audible local alarm may be either mechanically or electrically operated. The standard does not stipulate whether its location be indoors or outdoors, as this must be determined by individual circumstances. The location and number of audible alarms will normally be dictated by the purpose of the alarms (water flow or evacuation). They will also be dictated by normal operations within the protected premises and location at which the alarm would be expected to be heard.

3-17.4.2* Outdoor mechanical or electrically operated bells shall be of weatherproof and guarded type.

A-3-17.4.2 All alarm apparatus should be so located and installed that all parts are accessible for inspection, removal, and repair, and should be substantially supported.

Formal Interpretation

Question: Is it the intent of Section 3-17.4.2 to require guards against mechanical injury on all outdoor water motor gongs?

Answer: No. It is the intent of the Committee the word "guarded" relates to the protection against birds, vermin, and debris.

3-17.4.3 On each alarm check valve used under conditions of variable water pressure, a retarding device shall be installed. Valves shall be provided in the connections to retarding chambers, to permit repair or removal without shutting off sprinklers; these valves shall be so arranged that they may be locked or sealed in the open position.

Retarding chambers are required when alarm devices connected to alarm valves would otherwise be subject to false operation from pressure surges from variable pressure supplies, such as city water or fire pumps. When the supply is a constant pressure source such as a gravity tank or pressure tank, retarding chambers are not required.

3-17.4.4 Alarm valves, dry-pipe, pre-action and deluge valves shall be fitted with an alarm by-pass test connection for electric alarm switch or water motor gong or both. This pipe connection shall be made on the water supply side of the system and provided with a control valve and drain for the alarm piping. A check valve shall be installed in the pipe connection from the intermediate chamber of a dry-pipe valve.

Local water flow alarms should be tested at least quarterly (*see NFPA 13A, Recommended Practice for the Care and Maintenance of Sprinkler Systems, Section 4-5 and 4-6.1*). The required bypass from the water side of the system provides the facility to conduct these tests without tripping the system which, in addition to being costly and time consuming, would tend to introduce foreign material and scale into the piping.

3-17.4.5   A control valve shall be installed in connection with pressure-type contactor or water-motor-operated alarm devices and such valves shall be of the type which will clearly indicate whether they are open or closed and be so constructed that they may be locked or sealed in the open position. The control valve for the retarding chamber on alarm check valves of wet-pipe systems may be accepted as complying with this paragraph.

The valve being discussed is located in the piping between the connection from the alarm valve, a dry-pipe valve, a pre-action valve, or a deluge valve and the local water flow alarm device(s). This piping is normally subject to only atmospheric air pressure and, lacking an indicating type valve, there would be no means of visually ascertaining if the valve were open and that water could reach the alarm actuators in the event of fire.

3-17.5* Attachments — Mechanically Operated.   For all types of sprinkler systems employing water-motor-operated alarms, an approved ¾-in. strainer shall be installed at the alarm outlet of the waterflow detecting device except that when a retarding chamber is used in connection with an alarm valve, the strainer shall be located at the outlet of the retarding chamber unless the retarding chamber is provided with an approved integral strainer in its outlet. Water-motor-operated devices shall be protected from the weather, and shall be properly aligned and so installed as not to get out of adjustment. All piping to these devices shall be galvanized or brass or other approved corrosion resistant material of a size not less than ¾ in.

A-3-17.5   Water-motor-operated devices should be located as near as practicable to the alarm valve, dry-pipe valve, or other waterflow detecting device. The total length of the pipe to these devices should not exceed 75 ft (22.9 m) nor should the water-motor-operated device be located over 20 ft (6.1 m) above the alarm device or dry-pipe valve.

Formal Interpretation

Question: Is it the intent of Section 3-17.5 to require water-motor-operated alarm devices drain piping be of corrosion-resistant material?

Answer: No.

The strainer and the pipe of corrosion-resistant material are required to protect against obstruction of the small orifice through which water enters the water motor.

3-17.6 Attachments — Electrically Operated.

3-17.6.1   Electrically operated alarm attachments forming part of an auxiliary, central station, proprietary or remote station signaling system shall be installed in accordance with the following applicable NFPA standards.

(a) NFPA 71, *Central Station Signaling Systems,*

(b) NFPA 72B, *Auxiliary Protective Signaling Systems,*

(c) NFPA 72C, *Remote Station Protective Signaling Systems,*

(d) NFPA 72D, *Proprietary Protective Signaling Systems.*

3-17.6.2* The circuits of electrical alarm attachments forming part of a local sprinkler water flow alarm system need not be supervised.

Exception: If the local sprinkler water flow alarm system is part of a required local fire alarm system it shall be installed in accordance with NFPA 72A, Standard for Local Protective Signaling Systems.

A-3-17.6.2 Switches which will silence electric alarm sounding devices by interruption of electrical current are not desirable; however, if such means are provided, then the electrical alarm sounding device circuit should be arranged so that when the sounding device is electrically silenced, that fact shall be indicated by means of a conspicuous light located in the vicinity of the riser or alarm control panel. This light shall remain in operation during the entire period of the electrical circuit interruption.

Electrically operated water flow alarm attachments must be installed in accordance with NFPA 72A except as modified in this section.

3-17.6.3 Waterflow detecting devices, including the associated alarm circuits, shall be tested by an actual waterflow through use of a test connection. (*See 3-17.7.*)

3-17.6.4 Outdoor electric alarm devices shall be of a type specifically listed for outdoor use, and the outdoor wiring shall be in approved conduit, properly protected from the entrance of water in addition to the requirements of 3-17.6.1.

3-17.7 Drains. Drains from alarm devices shall be so arranged that there will be no danger of freezing, and so that there will be no overflowing at the alarm apparatus at domestic connections or elsewhere with the sprinkler drains wide open and under system pressure. (*See 3-11.4.*)

4

SPACING, LOCATION AND POSITION OF SPRINKLERS

4-1 General Information.

4-1.1* Basic Requirements.

A-4-1.1 The installation requirements are specific for the normal arrangement of structural members. There will be arrangements of structural members not specifically detailed by the requirements. By applying the basic principles, layouts for such construction can vary from specific illustrations, provided the maximum specified for the spacing of sprinklers (Section 4-2) and position of sprinklers (Section 4-3) are not exceeded.

All needless ceiling sheathing, hollow siding, tops of high shelving, partitions, or decks should be removed. Sheathing of paper and similar light flammable materials is particularly objectionable.

4-1.1.1* The basic requirements for spacing, location and position of sprinklers are specified in this chapter and are based on the following principles:

(a) Sprinklers installed throughout the premises,

(b) Sprinklers located so as not to exceed maximum protection area per sprinkler,

(c) Sprinklers positioned and located so as to optimize performance with respect to activation time and distribution,

(d) And as specified herein.

Exception No. 1:   See 4-4.3 and 4-4.4 for locations from which sprinklers may be omitted.

Exception No. 2:   Special sprinklers may be installed in accordance with 4-1.1.3.

Exception No. 3:   When sprinklers are specifically tested and test results prove deviations from clearance requirements to structural members offer no obstruction to spray discharge they may be positioned and located accordingly.

Exception No. 4:   Clearance between sprinklers and ceilings may exceed the maximum specified in Section 4-3 and 4-5.4 provided that, for the conditions of occupancy protected, tests or calculations show comparable sensitivity and performance of the sprinklers to those installed in conformance with Section 4-3.

Formal Interpretation

Question: Background: Construction - Structural steel frame with concrete floors. The floor to floor height is approximately 15½ ft. Approximately 9 ft above each floor, a gypsum deck has been poured to form an "interstitial space." Noncombustible ductwork, plumbing, electrical conduit, sprinkler piping and open cable trays run in the interstitial space. Access to the space is by five stairways, with access doors each being approximately 3 ft × 3 ft.

Question: Is it the intent of Section 4-1 of NFPA 13 that sprinkler protection of the interstitial space described above be provided to consider the building completely protected by automatic sprinklers?

Answer: No. The space described is essentially noncombustible with no occupancy and no combustible services [with the possible exception of the cable trays which could be protected in accordance with 4-4.4.4(b) if necessary] and so could be treated as a noncombustible concealed space. Use of the space for storage or the introduction of combustibles would require provision of sprinklers to maintain classification as completely protected by automatic sprinklers.

Formal Interpretation

Question: Is it the intent of Section 4-1.1.1 to require installation of a sprinkler head in every room in a building including (a) Shower Rooms, (b) Clothes Closets?

Answer: Yes. The intent is to require sprinklers in every area except where specifically excluded (for example, see Section 4-4.4.1), or except where the authority having jurisdiction permits omission of the sprinklers.

Formal Interpretation

Question: Is it the intent of 4-1.1.1 to require automatic sprinklers or equivalent automatic protection in rooms designated for the specific use of electrical equipment including buss ducts and circuit breaker panels?

Answer: Yes.

Formal Interpretation

Question: Does 4-1.1.1 require that a fire pump room containing a fire pump for a sprinkler system for a completely sprinklered building be protected by automatic sprinklers even if the fire pump is in a separate building?

Answer: Yes. Wherever the fire pump is located, it would be considered a part of the building and therefore required to be protected by automatic sprinklers.

To be effective, automatic sprinklers must be located in all areas of combustible construction or contents, in such a way that the sprinklers may respond promptly to a developing fire and be able to distribute their discharge without obstruction. This chapter addresses the spacing, location, and position of sprinklers so they may be effective. In noncombustible areas with noncombustible contents, sprinklers may be omitted. These spaces include chases, spaces above noncombustible ceilings in noncombustible buildings (*see* 4-4.4) and similar areas. They could also include areas in noncombustible buildings where introduction of combustible contents is precluded by building usage.

A-4-1.1.1 This standard contemplates full sprinkler protection for all areas. Other NFPA standards which mandate sprinkler installation may not require sprinklers in certain areas. The requirements of this standard should be used insofar as they are applicable. The authority having jurisdiction should be consulted in each case.

The position and clearance rules are specific but may be modified if it can be demonstrated by test that there is no impairment of fire control capability. In some cases, this may imply no impairment of water discharge. Some tests have indicated comparable fire control capability even with minor water discharge interference.

4-1.1.2 Residential sprinklers shall be installed in conformance with their listing and the positioning requirements of NFPA 13D, Installation of Sprinkler Systems for One- and Two-Family Dwellings and Mobile Homes.

The test procedure under which residential sprinklers have been listed differs from the procedure followed in listing sprinklers for this standard. Therefore, their use is confined to residential portions of occupancies and their spacing and location must be in accordance with NFPA 13D, *Standard for the Installation of Sprinkler Systems in One- and Two-Family Dwellings and Mobile Homes.* In other portions of such occupancies, sprinklers listed for general use and installed in accordance with the requirements of this standard are required.

4-1.1.3 Special sprinklers may be installed with protection areas or distances between sprinklers different than are specified in Sections 4-2 and 4-5 when found suitable for such use based on: fire tests related to the hazard category; tests to evaluate distribution, wetting of floors and walls, interference to distribution by structural elements; and tests to characterize response sensitivity, when installed in accordance with any special sprinkler listing limitation.

An example is the extended coverage horizontal sidewall sprinkler. The intent is to permit the testing laboratories to list new products with spacings greater than permitted in this chapter provided they can achieve the same degree of fire control.

Formal Interpretation

Question: May a horizontal sidewall sprinkler be installed at distances greater than 12 in. below the ceiling, when that sprinkler has been tested and listed by Underwriters Laboratories Inc. for distances greater than 12 in. below the ceiling, and the sprinkler is installed in accordance with that listing?

Answer: Yes. This is reflected in 4-1.1.1 Exception No. 4 of NFPA 13 which, while not specifically addressing sidewall sprinklers, was intended to apply to all sprinklers.

4-1.2* When partial sprinkler installations are installed, the requirements of this standard shall be used insofar as they are applicable. The authority having jurisdiction shall be consulted in each case.

The standard recognizes that, although it is intended to apply to fully sprinklered buildings, some codes and ordinances require sprinklers in only certain areas or occupancies to be installed according to this standard. In these circumstances, the authority having jurisdiction must be consulted to determine what additional requirements may be necessary to compensate for the lack of sprinklers in the remainder of the building. These may include fire separations between sprinklered and unsprinklered areas and increased water supply.

A-4-1.2   Installation of sprinklers throughout the premises is necessary for protection of life and property. In some cases partial sprinkler installations covering hazardous sections and other areas are specified in codes or standards or are required by authorities having jurisdiction, for minimum protection to property or to provide opportunity for safe exit from the building.

When buildings or portions of buildings are of combustible construction or contain combustible material, standard fire barriers should be provided to separate the areas which are sprinkler protected from adjoining unsprinklered areas. All openings should be protected in accordance with applicable standards and no sprinkler piping should be placed in an unsprinklered area unless the area is permitted to be unsprinklered by this standard.

Water supplies for partial systems should be adequate and designed with due consideration to the fact that in a partial system more sprinklers may be opened in a fire which originates in an unprotected area and spreads to the sprinklered area than would be the case in a completely protected building. Fire originating in nonsprinklered area may overpower the partial sprinkler system.

When sprinklers are installed in corridors only, sprinklers should be spaced up to the maximum of 15 ft (4.5 m) along the corridor, with one sprinkler opposite the center of any door or pair of adjacent doors opening onto the corridor, and with an additional sprinkler spaced inside each adjacent room above the door opening. When the sprinkler in the adjacent room provides full protection for that space, an additional sprinkler is not required in the corridor adjacent to the door.

Tests conducted at the National Bureau of Standards support this type of partial installation.

4-1.3 Definitions.

4-1.3.1 Smooth Ceiling Construction.   The term smooth ceiling construction as used in this standard includes:

(a) Flat slab, pan-type reinforced concrete, concrete joist less than 3 ft (0.9 m) on centers.

(b) Continuous smooth bays formed by wood, concrete or steel beams spaced more than 7½ ft (2.9 m) on centers — beams supported by columns, girders or trusses.

(c) Smooth roof or floor decks supported directly on girders or trusses spaced more than 7½ ft (2.9 m) on centers.

(d) Smooth monolithic ceilings of at least ¾ in. (19 mm) of plaster on metal lath or a combination of materials of equivalent fire-resistive rating attached to the underside of wood or bar joists.

(e) Open web-type steel beams regardless of spacing.

(f) Smooth shell-type roofs, such as folded plates, hyperbolic paraboloids, saddles, domes and long barrel shells.

(g) In (b) through (f) above, the roof and floor decks may be noncombustible or combustible. Item (b) would include standard mill construction.

(h) Suspended ceilings of noncombustible construction.

(i) Suspended ceilings of combustible construction where there is a full complement of sprinklers in the space immediately above such a ceiling and the space is unfloored and unoccupied.

(j) Smooth monolithic ceilings with fire resistance less than that specified under item (d) attached to the underside of wood or bar joists.

(k) Combustible suspended ceilings arranged other than as specified under item (i).

In general, smooth ceiling construction is such that it does not incorporate supporting members that are less than 7½ ft (2.3 m) on center or have members that would interfere with the distribution of water from sprinklers.

4-1.3.2 Beam and Girder Construction.   The term beam and girder construction as used in this standard includes noncombustible and combustible roof or floor decks supported by wood beams of 4 in. (102 mm) or greater nominal thickness or concrete or steel beams spaced 3 to 7½ ft (0.9 to 2.3 m) on centers and either supported on or framed into girders. [When supporting a wood plank deck, this includes semi-mill and panel construction and when supporting (with steel framing) gypsum plank, steel deck, concrete, tile, or similar material, would include much of the so-called noncombustible construction.]

Beam and girder construction is similar to smooth ceiling, but with supporting members at 3 to 7 ½ ft (0.9 to 2.3 m) centers which could interfere with the distribution of water from sprinklers.

4-1.3.3 Bar Joist Construction.   The term bar joist construction refers to construction employing joists consisting of steel truss-shaped members. This definition includes noncombustible and combustible roof and floor decks supported on bar joists.

Also known as open web steel joist. These members offer little resistance to the travel of heat under a ceiling, unlike solid beams. Obstruction to water distribution is minimal. (*See 4-2.4.4 and 4-2.4.5.*)

4-1.3.4 Panel Construction.   The term panel construction as used in this standard includes ceiling panels formed by members capable of trapping heat to aid the operation of sprinklers and limited to a maximum of 300 sq ft (27.9 m^2) in area. Beams spaced more than 7½ ft (2.3 m) apart and framed into girders qualify for panel construction provided the 300 sq ft (27.9 m^2) area limitation is met.

This is a version of beam and girder construction forming panels of up to 300 sq ft (27.9 m^2).

4-1.3.5 Standard Mill Construction.   The term standard mill construction as used in this standard refers to heavy timber construction as defined in NFPA 220, *Standard on Types of Building Construction.*

4-1.3.6 Semi-Mill Construction.   The term semi-mill construction as used in this standard refers to a modified standard mill construction where greater column spacing is used and beams rest on girders.

4-1.3.7 Wood Joist Construction.   The term wood joist construction refers to wood members of rectangular cross section, which may vary from 2 to 4 in. (51 to 102 mm) nominal width and up to 14 in. (356 mm) nominal depth, spaced up to 3 ft (0.9 m) on centers, and spanning up to 40 ft (12 m) between supports, supporting a floor or roof deck. Wood members less than 4 in. (102 mm) nominal thickness spaced more than 3 ft (0.9 m) on centers and open wood trusses with spacing up to 3 ft (0.9 m) on centers are also considered as wood joist construction.

The common type of light combustible construction utilizing nominal 2-in. (51-mm) wide ''beams'' on edge to support a wood deck, without sheathing. Sheathed joist is classified as smooth

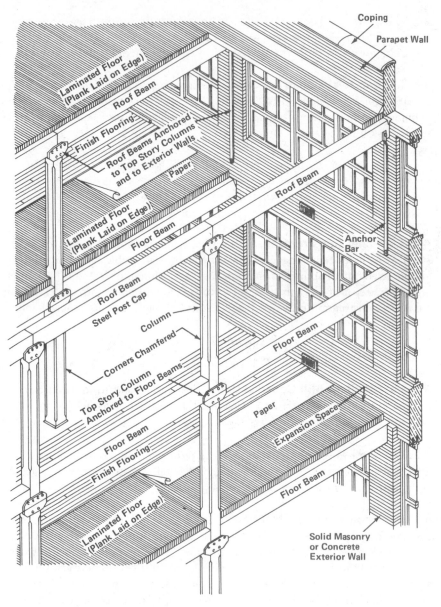

Figure 4.1.   Heavy timber construction of the laminated floor and beam type. (National Forest Products Association).

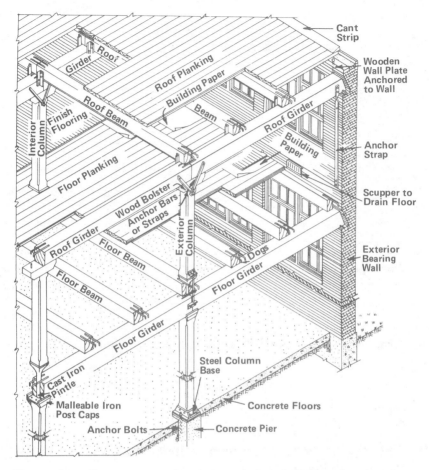

Figure 4.2. Components of a heavy timber building showing floor framing and identifying components of a type known as semimill.

ceiling [*see 4-1.3.1 (j)*]. Other constructions utilizing light wood beams less than 4 in. (102 mm) thick are considered wood joist construction regardless of the spacing of the beams.

4-1.3.8* Composite Wood Joist Construction.   The term composite wood joist construction refers to wood beams of I cross section constructed of wood flanges and solid wood web, supporting a floor or deck. Composite wood joists may vary in depth up to 48 in. (1.2 m), may be spaced up to 48 in. (1.2 m) on centers, and may span up to 60 ft (18 m) between supports. Joists channels shall be fire stopped to the full distance between joists with material equivalent to the joist construction and be limited to 300 sq ft (27.9 m²) channel area.

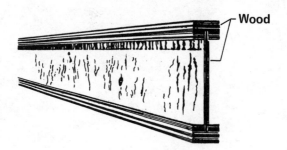

Figure A-4-1.3.8   Typical Composite Wood Joist Construction.

The fire stopping between the composite wood joists to form the 300 sq ft (27.9 m²) pocket is to trap the heat and accelerate sprinkler operation.

4-1.3.9 High-Piled Storage.   High-piled storage is defined as solid piled storage in excess of 12 ft (3.7 m) in height or palletized or rack storage in excess of 12 ft (3.7 m) in height. See Appendix D for availability of information for sprinkler protection of high-piled storage.

4-2 Spacing and Location of Upright and Pendent Sprinklers.   *(See also Sections 4-3 and 4-4.)*

The provisions of this section apply to upright and pendent sprinklers. The special provisions applicable to sidewall sprinklers are given in Section 4-5.

B-4 Spacing, Location, and Position of Upright and Pendent Sprinklers.

B-4-1.1   Cutting holes through partitions, either solid or slatted, to allow sprinklers on one side thereof to distribute water to the other side is not effective.

B-4-1.2   When wood cornices on masonry buildings face an exposure they should be replaced with a parapet, or the projecting woodwork should be cut away and metal flashing extended to cover the exposed edge of planking, or suitable sprinkler protection should be provided.

4-2.1 Distance Between the Branch Lines and Between Sprinklers on the Branch Lines.

4-2.1.1   For Light Hazard Occupancies, the distance between branch lines and between sprinklers on the branch lines shall not exceed 15 ft (4.6 m).

4-2.1.2*   For Ordinary Hazard Occupancies, the distance between the branch lines and between sprinklers on branch lines shall not exceed 15 ft (4.6 m).

The maximum distance of 15 ft (4.6 m) between sprinklers on branch lines or between branch lines is reduced to 12 ft (3.7 m) in buildings used for high-piled storage. If the protection area per sprinkler is permitted to exceed 100 sq ft (9.3 m^2) by the Exception to 4-2.2.4, then the reduction from 15 to 12 ft (4.6 to 3.7 m) need not be applied.

A-4-2.1.2   For examples of sprinkler layouts under smooth ceiling construction, refer to Figures A-4-2.1.2(A) and A-4-2.1.2(B).

The two figures which follow illustrate the spacing rules.

Formal Interpretation

Question No. 1: Is it the intent of 4-2.1.2 to permit 15-ft spacing for high-piled storage for systems designed in accordance with NFPA 231 for a density below 0.25 gpm/sq ft as noted in the Exception of 4-2.2.4?

Question No. 2: Does the Exception for hydraulically designed systems with densities below 0.25 gpm noted in paragraph 4-2.2.4 of NFPA 13 allow designers to exceed 12 ft-0 in. spacing for high piled stock in buildings with bays other than 25 ft-0 in. as noted in paragraph 4-2.1.2 of NFPA 13?

Answer: Yes. When sprinkler systems are hydraulically designed for densities below 0.25 gpm/sq ft, the distance between the branch lines or between sprinklers on branch lines shall not exceed 15 ft provided the sprinkler area coverage does not exceed 130 sq ft.

4-2.1.3   For Extra Hazard Occupancies, the distance between the branch lines and between sprinklers on the branch lines shall not exceed 12 ft (3.7 m).

4-2.1.4   In buildings used for high-piled storage (as defined in 4-1.3.9) the distance between the branch lines and between sprinklers on the branch lines shall not exceed 12 ft (3.7 m) except, in bays 25 ft (7.6 m) wide, a spacing of 12 ft 6 in. (3.8 m) between branch lines is permitted.

4-2.1.5 Distance from Walls.

4-2.1.5.1   The distance from the walls to the end sprinklers on the branch lines shall not exceed one-half of the allowable distance between sprinklers on

Flat Slab or Pan Type Reinforced Concrete

Maximum Spacing: 130 Sq Ft per Sprinkler
LXS = 130 or less

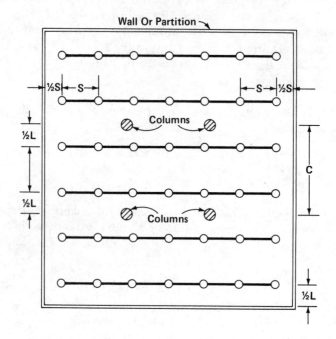

KEY

C = Column spacing.
L = Distance between branch lines, limit 15 ft.
S = Distance between sprinklers on branch lines, limit 15 ft.

Examples

C	L	S (Max)	C	L	S (Max)
21 ft 8 in.	10 ft 10 in.	12 ft 0 in.	21 ft 6 in.	10 ft 9 in.	12 ft 1 in.
24 ft 2 in.	12 ft 1 in.	10 ft 9 in.			

For SI Units: 1 in. = 25.4 mm; 1 ft = 0.3048 m; 1 ft^2 = 0.0929 m^2.

Figure A-4-2.1.2(A)   Layout of Sprinklers under Smooth Ceiling Construction
— Ordinary Hazard Occupancy.

Continuous Smooth Bays with Beams Supported on Columns

Maximum Spacing: 130 Sq Ft per Sprinkler
L × S = 130 or less

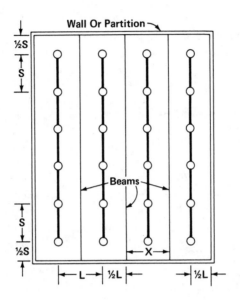

KEY

L = Distance between branch lines, limit 15 ft.
S = Distance between sprinklers on branch lines, limit 15 ft.
X = Width of bay.

Examples

X	L	S (Max)	X	L	S (Max)
10 ft 10 in.	10 ft 10 in.	12 ft 0 in.	10 ft 9 in.	10 ft 9 in.	12 ft 1 in.
12 ft 1 in.	12 ft 1 in.	10 ft 9 in.			

For SI Units: 1 in. = 25.4 mm; 1 ft = 0.3048 m; 1 ft^2 = 0.0929 m^2.

Figure A-4-2.1.2(B)   Layout of Sprinklers under Smooth Ceiling Construction — Ordinary Hazard Occupancy.

the branch lines. The distance from the walls to the end branch lines shall not exceed one-half the allowable distance between the branch lines. For exception relating to small rooms, refer to 4-4.20.

The allowable distance between sprinklers on the branch lines is determined by the actual distance between the branch lines and the permissible protection area per sprinkler (*see 4-2.2*). To minimize the amount of piping, branch lines are usually spaced as far apart as possible, while maintaining even spacing in the bay or room. If the spacing is not even, then the greatest distance between lines should be used to determine the allowable distance between sprinklers on the lines. If the distance of a line from a wall exceeds one-half of the distance between lines, then twice the distance of the line from the wall should be used in determining the allowable distance between sprinklers on the lines. For example, if the distance from the branch line to the wall was 7 ft (2.1 m) and the distance between branch lines was 13 ft (4.0 m), twice the distance from the branch line to the wall would be 14 ft (4.3 m), which should be used to determine the distance between sprinklers on the branch lines on the first line parallel to the wall. Other lines could use the 13- or the 14-ft dimension, dependent on whether symmetry was desired. If an ordinary hazard system with a protection area of 130 sq ft (12.1 m^2) were being installed, the distance between sprinklers on the branch lines would be 130 ÷ 14 = 9.3 ft (2.8 m). See 7-4.3.1.2(a).

4-2.1.5.2   Sprinklers shall be located a minimum of 4 in. (102 mm) from a wall.

Dead air spaces in corners can effect a sprinkler's operation time. The 4-in. limitation ensures that the sprinkler will operate properly.

4-2.2 Protection Area Limitations. (*See Section 3-3.*)

4-2.2.1 Light Hazard Occupancy.

4-2.2.1.1   Under smooth ceiling construction and under beam and girder construction [as defined in 4-1.3.1 items (a) through (i), and 4-1.3.2] the protection area per sprinkler shall not exceed 200 sq ft (18.6 m^2). For hydraulically designed sprinkler systems the protection area per sprinkler may be increased to 225 sq ft (20.9 m^2).

The permitted maximum area of 200 sq ft (18.6 m^2) with pipe schedule systems may be increased to 225 sq ft (20.9 m^2) if the system is hydraulically designed for these ceiling types.

4-2.2.1.2* Under open wood joist construction (as defined in 4-1.3.7) or under open composite wood joist construction (as defined in 4-1.3.8) of 16 in. (406 mm) nominal or less depth, the protection area per sprinkler shall not exceed 130 sq ft (12.1 m^2).

Because the nature of open wood joist construction, with numerous small pockets, tends to impede the flow of heat, and because this construction is readily subject to fire damage, the protection area is reduced to 130 sq ft (12.1 m^2).

Joists Above Girders or Framed into Girders;
Branch Lines Uniformly Spaced Between Girders

Maximum Spacing: 130 Sq Ft per Sprinkler
L × S = 130 or less

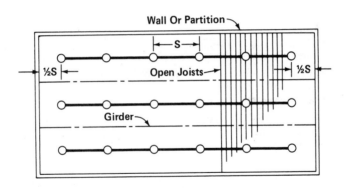

KEY
L = Distance between branch lines, limit 15 ft.
S = Distance between sprinklers on branch lines, limit 15 ft.
Y = Maximum distance between girders.

Examples

Y	L	S (Max)	Y	L	S (Max)
10 ft 9 in.	10 ft 9 in.	12 ft 1 in.	10 ft 10 in.	10 ft 10 in.	12 ft 0 in.
			12 ft 1 in.	12 ft 1 in.	10 ft 9 in.

For SI Units: 1 in. = 25.4 mm; 1 ft = 0.3048 m; 1 ft^2 = 0.0929 m^2.

Figure A-4-2.2.1.2 Layout of Sprinklers under Open Wood Joist Construction — Light and Ordinary Hazard Occupancies.

Figure A-4-2.2.1.2 is an illustration of the spacing rules for open wood joist construction, which are the same for both Light and Ordinary Hazard.

4-2.2.1.3   For other types of construction the protection area per sprinkler shall not exceed 168 sq ft (15.6 m²).

Formal Interpretation

Question: Series L, M, H, 60 and 60 open web joists as manufactured by the "Trus Joist Corporation" are trusses with structural wood cord members less than 4 in. nominal thickness and tubular steel web members.

If this type of truss is spaced more than 3 ft on centers in a combustible attic area, not accessible for storage (light hazard), would the protection area per sprinkler be limited to 168 sq ft as referenced in Section 4-2.2.1.3?

Answer: Yes.

4-2.2.2 Ordinary Hazard Occupancy.   For all types of construction the protection area per sprinkler shall not exceed 130 sq ft (12.1 m²).

4-2.2.3 Extra Hazard Occupancy.   The protection area per sprinkler shall not exceed 90 sq ft (8.4 m²) for any type of building construction, except protection area per sprinkler shall not exceed 100 sq ft (9.3 m²) where the system is hydraulically designed.

With the increase of fire load and/or heat release rate contemplated with an Extra Hazard Occupancy, the maximum spacing is reduced to 90 sq ft (8.4 m²) per sprinkler unless the system is hydraulically designed, in which case it may be 100 sq ft (9.3 m²). The increased spacing for the hydraulically designed system equates the spacing to that permitted in storage occupancies which have density requirements similar to those required for Extra Hazard Occupancies.

4-2.2.4 High-Piled Storage.   In buildings used for high-piled storage (as defined in 4-1.3.9) the protection area per sprinkler shall not exceed 100 sq ft (9.3 m²).

Exception: Sprinkler spacing may exceed 100 sq ft (9.3 m²) but shall not exceed 130 sq ft (12.1 m²) in systems hydraulically designed in accordance with NFPA 231, Indoor General Storage, and 231C, Rack Storage of Materials, for densities below 0.25 gpm per sq ft [(10.2 L/min)/m²)].

4-2.3* Location of Sprinklers and Branch Lines with Respect to Structural Members.

A-4-2.3   The arrangement of branch lines depends upon such construction features as the distance between girders or trusses, columns of mushroom type reinforced concrete, and beams of standard mill construction. Each space or bay should usually be treated as a unit, installing the same number of branch lines uniformly in each space. When single branch lines will suffice, they should be placed midway in each bay or space. The arrangement of branch lines also depends upon the structural members available and suitable for the attachment of hangers and upon the need for properly locating sprinkler deflectors in accordance with 4-2.4 and Section 4-3.

The direction in which branch lines are usually run in the common types of ceiling construction and framing is shown in Table A-4-2.3.

Table A-4-2.3

Type of Ceiling	Location of Branch Lines
Smooth Continuous:	
Concrete mushroom	Either direction
Concrete pan-type or flat	Either direction
Sheathed (ceiling attached to bottom of beams, wood joists, or bar joists):	
Girders beneath sheathing	Across the beam or joists
No girders beneath sheathing	Whichever direction facilitates easy and proper hanging
Bays more than 7½ ft (2.3 m) wide:	
Formed by beams supported on columns	Parallel to beams
Formed by beams supported on girders or trusses	Either across beams or parallel to beams in the bays above girders or trusses
Supported directly on girders	Parallel to girders
Supported directly on trusses	Either direction, parallel to or through trusses
Beam and Girder:	
Wood or steel beams spaced 3 to 7 ½ ft (0.9 to 2.3 m) apart	Across beams
Open Bar Joist	Across the joists or trusses (either through or under them)
Open Joist (wood, steel or concrete)	Across joists

4-2.3.1   Sprinklers may be located under beams, in bays, or combination of both, but the locations must meet the provisions outlined in 4-2.4 and Section 4-3.

The rules which follow give permissible locations for sprinklers either under beams or under a roof or ceiling. A combination of these two positions is permissible provided the clearance and position rules are followed.

4-2.3.2   Where there are two sets of joists under a roof or ceiling and there is no flooring over the lower set, sprinklers shall be installed above and below the lower set of joists where there is a clearance of from 6 in. to 12 in. (152 mm to 305 mm) between the top of the lower joist and bottom of the upper joist. (*See Figure 4-2.3.2.*)

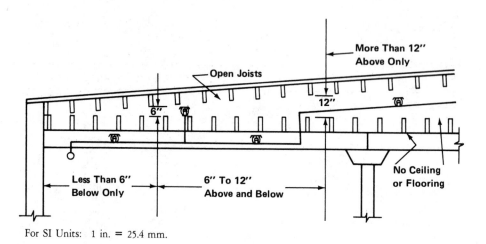

For SI Units: 1 in. = 25.4 mm.

Figure 4-2.3.2   Arrangement of Sprinklers under Two Sets of Open Joists — No Sheathing on Lower Joists.

Formal Interpretation

Question: Under paragraph 4-2.3.2 and its respective diagram, in spacing sprinklers above the lower set of joists, is it not the intent to measure the distance from the point where the space between joists (and/or wood trusses) is 6 in. and greater and not from the wall or where the top and bottom chords of the truss intersect?

Answer: No. This paragraph addresses only the special conditions stated in the beginning of the paragraph, and does not apply to wood trusses. Under the conditions stated in 4-2.3.2, the first sprinkler between the two sets of joists should be located at the point where the clear space between the two sets of joists is 6 in. In wood truss construction, sprinklers should be spaced from the end of the truss.

This is a special rule for a special condition. The first sprinkler above the lower set of joists should be located where the clear space is 6 in. (152 mm). The sprinkler below the lower set of joists should be spaced evenly in the space from the wall to the point where the clearance is 12 in. (305 mm). Sprinklers are not required below the lower set of joists beyond the point where clearance between the sets of joists exceed 12 in. (305 mm).

4-2.4 Clearance Between Sprinklers and Structural Members.

4-2.4.1 Trusses.   Sprinklers shall be at least 2 ft (0.6 m) laterally from truss members (web or chord) more than 4 in. (102 mm) wide, and at least 1 ft (0.3 m) laterally from truss members 4 in. (102 mm) or less in width. When sprinkler lines run above or through trusses, the sprinklers may be located on center line of truss, provided chord members are not more than 8 in. (203 mm) wide, and the deflector is at least 6 in. (152 mm) above the chord member. When sprinklers are located laterally beside chord members, clearances between the chord members and the sprinkler deflectors shall be in accordance with 4-2.4.6.

4-2.4.2 Girders.   When sprinkler lines are located perpendicular to and above girders, sprinklers shall be at least 3 ft 9 in. (229 mm) from girders except that they may be located directly above girders with the top flange not more than 8 in. (203 mm) wide, in which case the deflectors shall be at least 6 in. (152 mm) above the top of the girder.

4-2.4.3   When sprinkler deflectors are in accordance with Table 4-2.4.6, the girders may be disregarded in the spacing of the branch lines.

4-2.4.4 Open Web-Type Steel Beams.   (*See Figure 4-2.4.4.*) When branch lines are run across and through openings of open web-type steel beams, sprinklers may be spaced bay and beam provided:

(a) The distance between sprinklers and between branch lines conforms to 4-2.1,

(b) Sprinklers in the beam openings are located within 1 in. (25 mm) horizontally of the opening center line,

(c) The branch line is located within 1 in. (25 mm) horizontally of the opening center line, and

(d) Sprinklers on alternate lines are staggered.

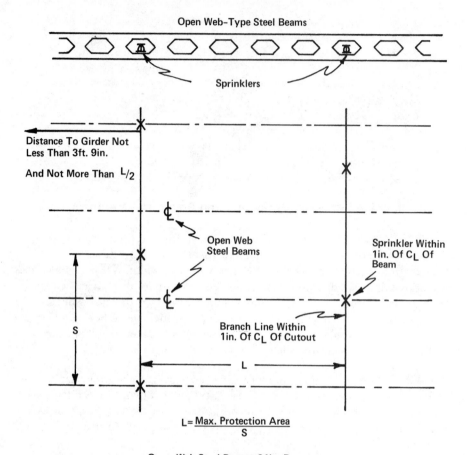

Open Web–Type Steel Beams

Sprinklers

Distance To Girder Not Less Than 3ft. 9in.

And Not More Than $L/2$

Open Web Steel Beams

Sprinkler Within 1in. Of C_L Of Beam

Branch Line Within 1in. Of C_L Of Cutout

S

L

$$L = \frac{\text{Max. Protection Area}}{S}$$

Open-Web Steel Beams, 21in. Deep, With Branch Line Through Beams

S= Spacing Of Sprinklers On Branch Lines
L= Distance Between Branch Lines
Lx S= Maximum Protected Area Per Sprinkler

For SI Units: 1 in. = 25.4 mm; 1 ft = 0.3048 m.

Figure 4-2.4.4   Location of Branch Lines and Sprinklers.

4-2.4.5 Bar Joists.   Sprinklers shall be at least 3 in. (76 mm) laterally from web members of open bar joists which do not exceed ½ in. (13 mm) or at least 6 in. (152 mm) laterally from web members which do not exceed 1 in. (25 mm). When the dimensions of the web member exceed 1 in. (25 mm), see 4-2.4.1.

4-2.4.6 Beams.   Deflectors of sprinklers in bays shall be at sufficient distances from the beams, as shown in Table 4-2.4.6 and Figure 4-2.4.6, to avoid obstruction to the sprinkler discharge pattern. Otherwise the spacing of sprinklers on opposite sides of the beams shall be measured from the beam and the distance shall not exceed ½ of the allowable distance between sprinklers.

Table 4-2.4.6   Position of Deflector when Located above Bottom of Beam

Distance from Sprinkler to Side of Beam	Maximum Allowable Distance Deflector above Bottom of Beam
Less than 1 ft	0 in.
1 ft to less than 2 ft	1 in.
2 ft to less than 2 ft 6 in.	2 in.
2 ft 6 in. to less than 3 ft	3 in.
3 ft to less than 3 ft 6 in.	4 in.
3 ft 6 in. to less than 4 ft	6 in.
4 ft to less than 4 ft 6 in.	7 in.
4 ft 6 in. to less than 5 ft	9 in.
5 ft to less than 5 ft 6 in.	11 in.
5 ft 6 in. to less than 6 ft	14 in.

For SI Units: 1 in. = 25.4 mm; 1 ft = 0.3048 m.

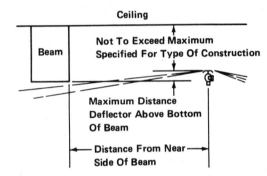

Figure 4-2.4.6   Position of Deflector, Upright, or Pendent, When Located Above Bottom of Beam.

It should be possible to exceed the distances above ceiling obstruction tabulated in Table 4-2.4.6 if the trajectory and resultant discharge pattern of a particular sprinkler has been verified by test.

Formal Interpretation

Question: Does Table 4-2.4.6 apply to the layout shown in Figure 1?

Answer: No. Section 4-2.4.6 offers the option as an alternate to Table 4-2.4.6 of spacing the sprinklers from the beam at a distance of not more than ½ the allowable distance between sprinklers. Under those conditions, the sprinklers may be positioned other than as specified in Table 4-2.4.6.

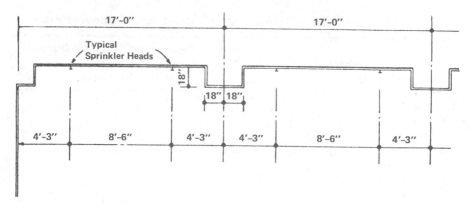

Figure 1.   Typical ceiling cross section

Formal Interpretation

Question No. 1: Does Table 4-2.4.6 apply to valances at stock fixtures approximately 3 ft from a wall and up to 5 ft deep from the ceiling?

Answer: No.

Question No. 2: If the answer to Question 1 is "no," what sections of NFPA 13 would apply?

Answer: Sections 4-4.11 and 4-4.13 more closely approximate the situation referred to.

4-2.4.7* Position of Deflectors.   Deflectors of sprinklers shall be parallel to ceilings, roofs, or the incline of stairs, but when installed in the peak of a pitched roof they shall be horizontal. Low-pitched roofs having slopes not greater than 1 in. per ft (83 mm/m) may be considered as level in the application of this rule and sprinklers may be installed with deflectors horizontal.

A-4-2.4.7   On sprinkler lines larger than 2 in., consideration should be given to the distribution interference caused by the pipe which can be minimized by installing sprinklers on riser nipples or installing sprinklers in the pendent position.

By having the sprinkler deflector parallel to the ceiling, a minimum of obstruction to the discharge pattern is achieved.

4-2.5 Clear Space Below Sprinklers.

4-2.5.1   A minimum of 18 in. (457 mm) clearance shall be maintained between top of storage and ceiling sprinkler deflectors. For in-rack sprinklers, the clear space shall be in accordance with NFPA 231C, *Rack Storage of Materials.*

The discharge pattern of upright and pendent sprinklers is a half paraboloid pattern filled with spray. In order to permit the distribution of water over the area which the sprinkler has been designed to protect, there must be no obstruction to the spray pattern. The rack storage fire tests and other tests with solid piled storage have shown that sprinklers are effective with 18-in. (457-mm) clearance.

4-2.5.2* The clearance from sprinklers to privacy curtains, free-standing partitions or room dividers shall be not less than the distances given in Table 4-2.5.2 as measured in Figure 4-2.5.2.

A-4-2.5.2 The distances given in Table 4-2.5.2 were determined through tests in which privacy curtains with either a solid fabric or close mesh ¼ in. (0.6 mm) top panel were installed. For broader mesh top panels [e.g., ½ in. (12.5 mm)] the obstruction of the sprinkler spray is not likely to be severe and the authority having jurisdiction may not need to apply the requirements in 4-2.5.2.

Table 4-2.5.2   Minimum Horizontal and Vertical Distances for Sprinklers

Horizontal Distance	Minimum Vertical Distance Below Deflector
6 in.	3 in.
9 in.	4 in.
12 in.	6 in.
15 in.	8 in.
18 in.	9½ in.
24 in.	12½ in.
30 in.	15½ in.
Greater than or equal to 36 in.	18 in.

For SI Units: 1 in. = 25.4 mm.

The distances specified in Table 4-2.5.2 were derived from NBS Report NBSIR 80-2097, *Full-Scale Fire Tests with Automatic Sprinklers in a Patient Room.*

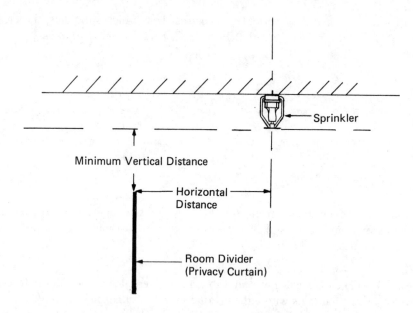

Figure 4-2.5.2   Standard Sprinkler Installed Near Privacy Curtain.

4-3 Position of Upright and Pendent Sprinklers.

4-3.1 Smooth Ceiling Construction (*as defined in 4-1.3.1*). Deflectors of sprinklers shall be located 1 in. to 10 in. (25 mm to 254 mm) below combustible ceilings or 1 in. to 12 in. (25 mm to 305 mm) below noncombustible ceilings. The operating elements of sprinklers shall be located below the ceiling.

Exception No. 1:   Deflectors of sprinklers under beams shall be located 1 in. to 4 in. (25 mm to 102 mm) below beams, and not more than 14 in. (356 mm) below combustible ceilings or not more than 16 in. (406 mm) below noncombustible ceilings.

Exception No. 2:   Special ceiling-type pendent sprinklers (concealed, recessed and flush types) may have the operating element above the ceiling and the deflector located nearer the ceiling when installed in accordance with their listing.

The rules establish minimum and maximum distances for sprinkler deflectors below the ceiling construction. The minimum of 1 in. (25 mm) is to permit installation and removal of upright sprinklers. The maximum distances are based on operating sensitivity and ceiling cooling considerations. Sprinklers under beams should not be more than 4 in. (102 mm) below the beam in order to be influenced by heat as it spills from one bay to another under the beam. As beam spacing forming pockets is decreased, resulting in smaller pocketed areas, the maximum distance below the ceiling increases. See summary in Appendix C.

4-3.2 Beam and Girder Construction (*as defined in 4-1.3.2*).

4-3.2.1   Deflectors of sprinklers in bays shall be located 1 in. to 16 in. (25 mm to 406 mm) below combustible or noncombustible roof or floor decks.

This section applies when the sprinkler is not located under a beam.

4-3.2.2   Deflectors of sprinklers under beams shall be located 1 in. to 4 in. (25 mm to 102 mm) below beams and not more than 20 in. (508 mm) below combustible or noncombustible roof or floor decks.

This section applies when the sprinkler is located under a beam.

4-3.2.3   Deflectors of sprinklers under concrete tee construction with stems spaced less than 7½ ft (2.3 m) but more than 3 ft (0.9 m) on centers shall, regardless of the depth of the tee, be located at or above a plane 1 in. (25 mm) below the level of the bottom of the stems of the tees and comply with Table 4-2.4.6.

The sprinkler is to be located as high as possible, taking into account the obstruction which could be offered by the leg of the concrete tee. This section is based on sensitivity tests conducted by the sprinkler industry.

4-3.3 Open Bar Joist Construction (*as defined in 4-1.3.3*).   Deflectors of sprinklers shall be located 1 in. to 10 in. (25 mm to 254 mm) below combustible or not more than 12 in. (305 mm) below noncombustible roof or floor decks.

4-3.4 Panel Construction (*as defined in 4-1.3.4*).

Formal Interpretation

Question: Under so-called Berkley Construction with a plywood deck on 2 in. by 6 in. wood stiffeners at 2 ft-0 in. centers framed into Glu-lam beams, with fiberglass insulation batts between the stiffeners with aluminum foil stapled to the lower edge of the stiffeners supporting the insulation, should sprinkler deflectors be positioned as for open wood joist construction or for panel construction with the aluminum sheathing considered as the "ceiling?"

Answer: The sheathing could be considered the ceiling in panel construction.

4-3.4.1   Deflectors of sprinklers in bays formed by members, such as beams framed into girders, resulting in panels up to 300 sq ft (27.9 m²) shall be located 1 in. to 18 in. (25 mm to 457 mm) below combustible or noncombustible roof or floor decks.

> This section applies when the sprinkler is not located under a beam. The increased maximum [by 2-in. (51-mm)] over beam and girder is a reflection of the 300 sq ft (27.9 m²) limit in the size of the panel.

4-3.4.2   Deflectors of sprinklers under the members, such as under beams framed into girders, forming panels up to 300 sq ft (27.9 m²) shall be located 1 in. to 4 in. (25 mm to 102 mm) below such members and not more than 22 in. (559 mm) below combustible or noncombustible roof or floor decks.

> This section applies when the sprinkler is located under a beam.

4-3.5 Open Wood Joist Construction (*as defined in* 4-1.3.7).   In open joist construction with joists spaced 3 ft (0.9 m) or less on centers, sprinklers shall be located with deflectors 1 in. to 6 in. (25 mm to 152 mm) below the bottom of the joists. If open joists are spaced more than 3 ft (0.9 m) on centers, sprinklers shall be located with deflectors placed in accordance with 4-3.1 or 4-3.2.

> Heat travel across joists is impaired by the numerous narrow pockets which tend to direct the heat in the direction of the joist channels, so sprinkler deflector distance is limited to 6 in. (152 mm) to help sensitivity.

4-3.6 Location Under Sheathed or Suspended Ceiling Under Any Type of Construction.   The position of sprinklers under sheathed or suspended ceilings with any type of construction shall be the same as for smooth ceiling construction; see 4-3.1.

The spacing rules are summarized in Appendix C as follows:

Summary of Spacing Rules

Maximum Distance of Deflectors Below Ceiling (Inches)

Type of Construction	in bays		under beams	
	comb.	noncomb.	comb.	noncomb.
Smooth Ceiling	10	12	14	16
Beam and Girder	16	16	20	20
Panel up to 300 sq ft	18	18	22	22
Bar Joists	10	12	—	—

Minimum below ceiling is 1 in.
Minimum below beams 1 in., maximum 4 in. Do not exceed maximum below ceiling.
Minimum below open wood joists — 1 in., maximum — 6 in.

Maximum Coverage per Sprinkler:
Light Hazard   — 200 sq ft smooth ceiling and beam girder construction (225 sq ft if hydraulically calculated)
130 sq ft open wood joist
168 sq ft all other types of construction
Ordinary Hazard — 130 sq ft all types of construction except
100 sq ft high piled storage *(see 4-2.2.2)*
Extra Hazard   — 90 sq ft all types of construction (100 sq ft if hydraulically calculated)

Direction of Lines: Either direction to facilitate hanging except:
Across beams for beams on girders 3 ft to 7½ ft on centers
Across joists for wood joists (open or sheathed) and bar joists (through or under)

Maximum Spacing Between Lines and Sprinklers:
Light and Ordinary Hazard — 15 ft except 12 ft for high piled storage *(see 4-2.1.2)*
Extra Hazard   — 12 ft

See 3-16.2.5 and Section 4-5 for rules on Sidewall Sprinklers.

For SI Units: 1 in. = 25.4 mm; 1 ft = 0.3048 m; 1 ft^2 = 0.0929 m^2.

4-4* Locations or Conditions Involving Special Consideration.

The rules in this section have been developed to deal with a number of special conditions.

A-4-4 Special Occupancy Considerations.

(a) Subject to the approval of the authority having jurisdiction, sprinklers may be omitted in rooms or areas where sprinklers are considered undesirable because of the nature of the contents, or in rooms or areas of noncombustible construction with wholly noncombustible contents and which are not exposed by other areas. Sprinklers should not be omitted from any room merely because it is damp or of fire-resistive construction.

(b) It is not advisable to install sprinklers when the application of water, or of flame and water, to the contents may constitute a serious life or fire hazard, as in the manufacture or storage of quantities of aluminum powder, calcium carbide, calcium phosphide, metallic sodium and potassium, quicklime, magnesium powder, and sodium peroxide. The manufacture and storage of such materials should be confined to specially cut-off, unsprinklered rooms or buildings of fire-resistive construction.

Formal Interpretation

Question: Is it the intent of Section 4-4 to require sprinkler protection in walk-in type coolers and freezers in fully sprinklered buildings?

Answer: Yes.

Formal Interpretation

Question: Are automatic sprinklers required in the combustible concealed space that is developed in the installation of a Mansard Roof on a wood frame structure that requires automatic sprinklers because of occupancy, i.e., nursing homes?

Answer: Yes. Automatic sprinklers are required in Mansard Roofs as described in the material submitted to the Subcommittee.

4-4.1 Combustible Form Board.   When roof or floor decks consist of poured gypsum or concrete on combustible form board supported on steel supports, the position of sprinkler deflectors shall be the same as for noncombustible construction as stated in Section 4-3. When combustible form board is located above suspended ceilings or in concealed spaces, see 4-4.4.1.

4-4.2 Metal Roof Decks.   When roof decks are metal with combustible adhesives or vapor seal, the position of sprinklers shall be the same as for combustible construction.

The subject of combustible metal roof decks is dealt with in detail in the *NFPA Fire Protection Handbook*, 15th Edition, Chapter 5.

4-4.3 Spaces Under Ground Floors.   Sprinklers shall be installed in all spaces below combustible ground floors except that, by permission of the authority having jurisdiction, sprinklers may be omitted when all of the following conditions prevail:

(a) The space is not accessible for storage purposes or entrance of unauthorized persons and is protected against accumulation of windborne debris;

(b) The space contains no equipment such as steam pipes, electric wiring, shafting, or conveyors;

(c) The floor over the space is tight;

(d) No combustible or flammable liquids or materials that under fire conditions may convert into combustible or flammable liquids are processed, handled or stored on the floor above.

This is an exception to the general rule that all combustible spaces shall be sprinklered, but is subject to jurisdictional approval and recognizes that there can be maintenance problems with piping under a floor, such as lack of access and danger of freezing. The conditions indicated are intended to eliminate sources of ignition from the concealed space and to limit combustibles to the floor only.

4-4.4 Concealed Spaces.

Formal Interpretation

Question: In a typical franchise restaurant having a roof system of a plywood deck on open web joist at 32-in. centers having steel web members and wood top and bottom chords with a ceiling sheathing attached to the bottom chord, is it the intent of 4-4.4 of NFPA 13 to permit 200 sq ft spacing of sprinklers in the blind space and under the ceiling in the restaurant seating area, the piping being sized by the pipe schedules for light hazard occupancy Tables 3-5.2 and 3-5.3?

Answer: No. The spacing of sprinklers in the blind space is governed by 4-2.2.1.3 and is limited to 168 sq ft per spinkler. If the ceiling below conforms to 4-2.2.1.1 then 200 sq ft spacing would apply. Pipe schedule Tables 3-5.2 and 3-5.3 would be applicable.

Question: In the restaurant above, is it intended to permit the sprinklers in the kitchen, mechanical and storage areas which would be spaced at 130 sq ft maximum per sprinkler to be fed by the piping sized by Tables 3-5.2 and 3-5.3?

Answer: No. The pipe schedule for these sprinklers should be sized by Table 3-6.2(a), 3-6.2(b) or 3-5.3 as these areas are considered to be Ordinary Hazard Occupancies.

Formal Interpretation

Question: Is it the intent of Section 4-4.4 of NFPA 13 to require sprinkler heads in the exterior canopy space of a shopping center, the interior of which is fully sprinklered and which is separated from the exterior canopy space by a rated partition in compliance with the National Building Code? The soffit of said canopy space is accessible by the removal of noncombustible 2 ft by 4 ft panels in a lay-in system with hold-down clips.

Answer: Yes. The only exception is where the space is less than 50 sq ft in area as covered in 4-4.4.1 Exception No. 7.

Formal Interpretation

Question No. 1: Section 4-4.4 of NFPA 13 deals with the installation of sprinklers in concealed spaces. Does Section 4-4.4 require the installation of sprinklers in a concealed space formed between a noncombustible roof or floor assembly and a suspended noncombustible lay-in tile ceiling system with enclosing end walls of masonry construction if the only combustibles present are fire retardant treated wood studs and top caps forming the top of one-hour rated interior partitions? The partitions consist of gypsumboard and noncombustible insulation fill terminating immediately above the ceiling and the studs with top caps extending another 4 to 6 in. above the gypsumboard but not to the structure above. The entire building is provided with an automatic sprinkler system below the ceiling.

Answer: No. The Committee intent is to permit limited combustibles if the exposed surfaces have been demonstrated not to propagate fire in the form in which they are installed in the space.

Question No. 2: Does the response to No. 1 depend on whether or not the floor or roof/ceiling design has an hourly fire rating?

Answer: No.

4-4.4.1   All concealed spaces enclosed wholly or partly by exposed combustible construction shall be protected by sprinklers.

Because this section deals with the requirement for sprinklers in combustible concealed spaces and does not specifically say that sprinklers are not required in noncombustible spaces, it has been commonly misinterpreted. The intent is that sprinklers are not required if a concealed space is constructed of noncombustible materials, has no combustible surfaces, and has no storage or occupancy. If a building qualifies as noncombustible under the appropriate building code, the presence of the minor amounts of combustibles permitted by the code, if exposed in the concealed space, would not require the installation of sprinklers in the space as indicated in 4-4.4.4.

Formal Interpretation

Question: Would partitions constructed of 2 by 4 wood studs with ⅝-in. sheet rock on each side extending from floor through a 1-hour rated drop ceiling and up to a concrete slab above be considered "combustible construction" as contemplated under 4-4.4.1?

Answer: No. It is the opinion of the Committee that where a combustible member is protected on both sides by noncombustible material such as sheet rock or metal-lath and plaster, the partition is not considered "exposed combustible construction" in the intent of Section 4-4.4.1.

Exception No. 1: *Spaces formed by studs or joists with less than 6 in. (152 mm) between the inside or near edges of the studs or joists. (See Figure 4-2.3.2.)*

Exception No. 2: *Spaces formed by bar joists with less than 6 in. (152 mm) between the roof or floor deck and ceiling.*

Exception No. 3: *Spaces formed by ceilings attached directly to or to within 6 in. (152 mm) of wood joist construction.*

Exception No. 3 applies to sheathed joist construction or similar narrow spaces formed by attachment of a ceiling tight to or to within 6 in. (152 mm) of beams.

Exception No. 4: *Spaces formed by ceilings attached directly to the underside of composite wood joist construction, provided the joist channels are fire stopped into volumes each not exceeding 160 cu ft (4.53 m³) using materials equivalent to the joists.*

Exception No. 5: *Spaces entirely filled with noncombustible insulation.*

In some cases it might be economically advantageous to fill an unsprinklered combustible concealed space with noncombustible insulation rather than to install sprinklers.

Exception No. 6: *In wood joist construction and composite wood joist construction with noncombustible insulation filling the space from the ceiling up to the bottom edge of the joist of the roof or floor deck, provided that in composite wood joist construction, the joist channels are fire stopped into volumes each not exceeding 160 cu ft (4.53 m³) using materials equivalent to the joists.*

Exception No. 7: *Small spaces over rooms not exceeding 50 sq ft (4.6 m²) in area.*

Exception No. 8: *When the exposed surfaces have a flame spread rating less than 25 and the materials have been demonstrated not to propagate fire in the form in which they are installed in the space or when the BTU content of the facing and substrate of insulation material does not exceed 1000 BTU per sq ft (11 356 kJ/m²).*

Exception No. 8 is intended to permit fire-retardant treated lumber or materials classed as limited combustible and to allow the use of paper-coated insulation material.

4-4.4.2   Sprinklers in concealed spaces having no access for storage or other use may be installed on the basis of Light Hazard Occupancy.

4-4.4.3   When heat producing devices such as furnaces or process equipment are located in the joist channels above a ceiling attached directly to the underside of composite wood joist construction which would not otherwise require sprinkler protection of the spaces, the joist channel containing the heat producing devices shall be sprinklered by installing two sprinklers in each joist channel, one on each side, adjacent to the heat producing device. The temperature rating of the sprinklers shall be as prescribed in Table 3-16.6.1 and Figure 3-16.3(a).

4-4.4.4*   In concealed spaces having exposed combustible construction, or containing exposed combustibles, in localized areas, the combustibles shall be protected as follows:

(a) If the exposed combustibles are in the vertical partitions or walls around all or a portion of the enclosure, a single row of sprinklers spaced not over 12 ft (3.7 m) apart nor more than 6 ft (1.8 m) from the inside of the partition may be installed to protect the surface. The first and last sprinklers in such a row shall not be over 5 ft (1.5 m) from the ends of the partitions.

(b) If the exposed combustibles are in the horizontal plane, permission may be given to protect the area of the combustibles on a light hazard spacing and add a row of sprinklers not over 6 ft (1.8 m) outside the outline of the area and not over 12 ft (3.7 m) on center along the outline. When the outline returns to a wall or other obstruction, the last sprinkler shall not be over 6 ft (1.8 m) from wall or obstruction.

These rules are designed to accommodate localized combustibles in concealed spaces which would, except for the localized combustibles, not require sprinklers.

Formal Interpretation

Background: A roof construction consisting of 2½ in. permadeck on 2 by 4 wood nailers on top chord of open web steel joist, with a noncombustible suspended ceiling below. The only combustible material is the 2 by 4 wood nailer and only exposed surface is the 1¾ face on the side of the nailer.

Question 1: Does this type of construction qualify as noncombustible construction?

Answer: No. Permadeck is classed as a limited-combustible construction material under NFPA 220, *Standard on Types of Building Material.*

Question 2: Is it the intent of the Committee on Automatic Sprinklers to strictly classify all blind spaces with wood construction as combustible construction, even if the amount of combustible material is very restricted and less than the "exposed combustibles in localized areas" mentioned in 4-4.4.4 of NFPA 13?

Answer: No. The intent of the Committee is to permit fire-retardant treated lumber in unrestricted amounts, or minor amounts of untreated wood.

Question 3: Are upright sprinklers required in the concealed spaces above noncombustible suspended ceiling in construction as described above?

Answer: No.

A-4-4.4.4   When there is a limited amount of combustibles available to burn and there is a limited prospect of fire propagation, sprinklers may not be required.

4-4.5 Spacing of Sprinklers Under Pitched Roofs.

Formal Interpretation

Question 1: Is it the intent of Section 4-4.5 to:

(a) Define the area of protection covered by a sprinkler head as the distance between sprinklers and branch lines as measured on the slope, or

(b) To establish the location of the sprinkler head with respect to the peak of a pitched roof building with the area of protection covered being determined by the spacing being projected to the horizontal plane of the floor?

Answer: The intent is expressed by (a).

Question 2: If the answer to the above is (a), why would it be allowed by Section 4-4.6 to project the spacing to the horizontal plane of the floor for curved roof buildings?

Answer: Section 4-4.6 does not permit the spacing of sprinklers under a curved roof to be determined by their projected distance on the floor. It does however, in 4-4.6.1, make an exception with respect to the sprinkler nearest a side wall when the roof curves down to the floor line.

The provisions of 4-4.6.3 are also an exception with respect to extra hazard spacing, permitting the area at the roof to be increased to that permitted for ordinary hazard provided the projected spacing on the floor does not exceed that for extra hazard [see 7-4.3.1.2(a)]. The general requirement is that spacing be determined on the plane of the roof.

4-4.5.1   Branch lines parallel to peaks of pitched roofs and sprinklers on lines perpendicular to peaks shall be spaced throughout the distance measured along the slope. This will place a row of sprinklers either in the peak or one-half the spacing down the slope from the peak.

The spacing of sprinklers is to be measured on the slope of the roof, and the maximum distances and areas given in Section 4-2 are applied on this slope. Note, however, that for hydraulically calculated systems the density is calculated on the basis of floor area.

4-4.5.2   Under saw-toothed roofs, the row of sprinklers at the highest elevation shall be not more than 3 ft (0.9 m) down the slope from the peak.

This is illustrated in section B-B of Figure A-1-9.2(B).

Formal Interpretation

Question No. 1: Please define "saw-toothed" roofs.

Answer: Saw-toothed roofs have regularly spaced monitors of saw-tooth shape, with the nearly vertical side glazed and usually arranged for venting [see Figure A-1-9.2(B)].

Question No. 2: Why is the row of sprinklers limited to maximum of 3 ft down the slope from the peak?

Answer: Sprinklers are limited to a maximum of 3 ft down the slope from the peak because if they are further removed from the venting they may not operate.

Question No. 3: (a) Where a peaked roof has a corridor parallel with the peak (center line of corridor is center line of peak) and the walls of the corridor are extended above the ceiling to the roof, thus subdividing the peaked roof into a peak with adjoining right triangles, could this "shed roof" (two right triangle areas) ever be considered a "saw-toothed" roof?

(b) Would the 3 ft maximum distance from the peak apply here?

(c) If so, why?

Answer: (a) No. The construction described does not include the glazing and venting features of saw-toothed roofs.

(b) No. Paragraph 4-4.5.1 would apply.

(c) Does not apply.

4-4.5.3   In 4-4.5.1 or 4-4.5.2 sprinklers in or near the peak shall have deflectors not more than 3 ft (0.9 m) vertically down from the peak. [*See Figures 4-4.5.3(a) and (b).*]

Sprinklers are required to be within 3 ft (0.9 m) vertically of the peak but must also meet the position requirements of Section 4-3. The 3-ft (0.9-m) limitation reflects concern for sensitivity.

Exception: In a steeply pitched roof the distance from the peak to the deflectors may be increased to maintain a horizontal clearance of not less than 2 ft (0.6 m) from other structural members. [See Figure 4-4.5.3(c).]

If the roof is steeply pitched sensitivity is compromised by distribution considerations.

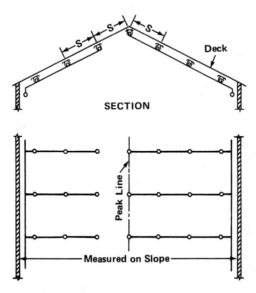

For SI Units: 1 in. = 25.4 mm.

Figure 4-4.5.3(a)   Sprinklers at Pitched Roofs; Branch Lines Run Up the Slope.

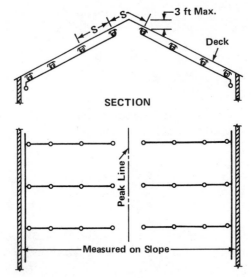

For SI Units: 1 in. = 25.4 mm.

Figure 4-4.5.3(b)   Sprinklers at Pitched Roofs; Branch Lines Run Up the Slope.

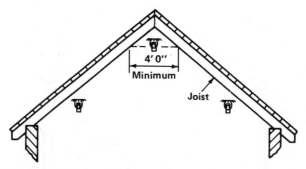

For SI Units: 1 in. = 25.4 mm; 1 ft = 0.3048 m.

Figure 4-4.5.3(c)   Desirable Horizontal Clearance for Sprinklers at Peak of Pitched Roof.

4-4.6 Spacing of Sprinklers Under Curved Roof Buildings.

4-4.6.1   When roofs are curved down to the floor line, the horizontal distance measured at the floor level from the side wall or roof construction to the nearest sprinklers shall not be greater than one-half the allowable distance between sprinklers in the same direction.

When the slope of the curved roof becomes steep near the side of the building, the consideration of spacing changes from on the plane of the roof to that of the floor for the location of the first sprinkler from the sidewall.

4-4.6.2 Deflectors of sprinklers shall be parallel with the curve of the roof or tilted slightly toward the peak of the roof. Deflectors of sprinklers shall be located as described for beam and girder construction or for the closest comparable type of ceiling construction.

4-4.6.3 When Extra Hazard Occupancy spacing of sprinklers is used under curved ceilings of other than fire-resistive construction, as in aircraft storage or servicing areas, the spacing as projected on the floor shall be not wider than required for Extra Hazard Occupancies, but in no case shall the spacing on the roof or ceiling be wider than required for Ordinary Hazard Occupancies.

The designer must consider the spacing under the curved ceiling on the slope and also the projected area on the floor.

4-4.7 Narrow Pocket.   Girders, beams or trusses forming narrow pockets of combustible construction along walls when of a depth which will obstruct the spray discharge pattern may require additional sprinklers positioned in accordance with Table 4-2.4.6.

4-4.8 Elevators, Stairs and Floor Openings.

4-4.8.1 Vertical Shafts.

4-4.8.1.1   One sprinkler shall be installed at the top of all shafts.

Sprinklers are to be provided at the top of all shafts used for elevators or stairs or other shafts open to more than one floor. Concealed combustible shafts must be sprinklered; concealed shafts of noncombustible construction and contents in a suitably rated enclosure do not require sprinklers.

4-4.8.1.2*   When vertical shafts have combustible sides, one sprinkler shall be installed at each alternate floor level. When a shaft having combustible surfaces is trapped, an additional sprinkler shall be installed at the top of each trapped section.

The additional sprinklers for shafts with combustible sides shall be placed to effectively wet the combustible surfaces. Some elevator and stair shafts are equipped with floor level trap doors; in this case, sprinklers are required at the ceiling of each level.

A-4-4.8.1.2   When practicable, sprinklers should be "staggered" at the alternate floor levels, particularly when only one sprinkler is installed at each floor level.

4-4.8.1.3   When accessible shafts have noncombustible surfaces, one sprinkler shall be installed near the bottom.

Refuse has a tendency to collect at the bottom of shafts. A properly located sprinkler should control a fire in such material.

4-4.8.1.4   When vertical openings are not protected by standard enclosures, sprinklers shall be so placed as to fully cover them. This necessitates placing sprinklers close to such openings at each floor level.

By placing sprinklers close to a ceiling opening, the floor area under the opening may be protected. Note that this is discussed under the heading of Vertical Shafts, implying small openings. Large openings are addressed in 4-4.8.2.3.

4-4.8.2* Stairways.

A-4-4.8.2   Floor or wall openings tending to create vertical or horizontal drafts, or other structural conditions that would delay the prompt operation of automatic sprinklers by preventing the banking up of the heated air from the fire, should be properly stopped in order to permit control of fire at any point by local sprinklers.

4-4.8.2.1   Stairways of combustible construction shall be sprinklered underneath whether risers are open or not.

4-4.8.2.2   Stairways of noncombustible construction with combustible storage beneath shall be sprinklered.

4-4.8.2.3*   When moving stairways, staircases, or similar floor openings are unenclosed, the floor openings involved shall be protected by draft stops in combination with closely spaced sprinklers.

The draft stops shall be located immediately adjacent to the opening, shall be at least 18 in. (457 mm) deep and shall be of substantially noncombustible material which will stay in place before and during sprinkler operation. Sprinklers, spaced not more than 6 ft (1.8 m) apart, shall be placed 6 to 12 in. (152 mm to 305 mm) from the draft stop on the side away from the opening to form a water curtain. Sprinklers in this water curtain shall be hydraulically designed to provide a discharge of 3 gpm per lineal foot [(37 L/min)/m] of water curtain, with no sprinklers discharging less than 15 gpm (56.8 L/min). The number of sprinklers calculated in this water curtain shall be the number in the length corresponding to the length parallel to the branch lines in the design area determined by 7-4.3.1. These sprinklers shall be added to the design area when considering the hydraulic design. Nominal ½ in. (12.7 mm) orifice closed head systems using sprinklers of Ordinary Temperature Classification are adequate for this purpose. When sprinklers are closer than 6 ft (1.8 m), cross baffles shall be provided in accordance with 4-4.19. When sprinklers in the normal pattern are closer than 6 ft (1.8 m) from the water curtain, it may be preferable to locate the water curtain sprinklers in recessed baffle pockets.

Exception: Openings having all horizontal dimensions between opposite edges at least 20 ft (6 m) and an area of at least 1000 sq ft (93 m²) such as those found in shopping malls, open atrium buildings or similar structures where all adjoining levels and spaces are protected with automatic sprinklers in accordance with this standard.

Fire experience has indicated that draft stops and closely spaced sprinklers are needed only with smaller sized openings.

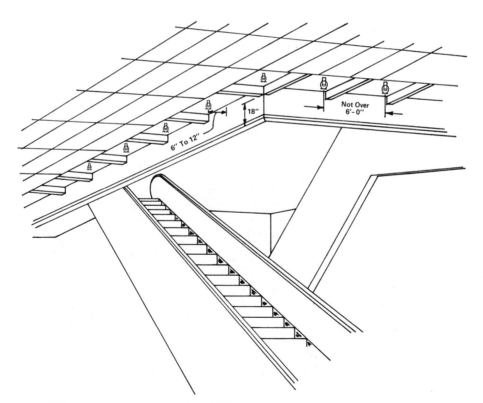

For SI Units: 1 in. = 25.4 mm; 1 ft = 0.3048 m.

Figure A-4-4.8.2.3   Sprinklers Around Escalators. (*See 4-4.19.*)

Formal Interpretation

Question: Is it the intent of 4-4.8.2.3 that the close spaced sprinklers used in combination with draft stops at moving stairways or large monumental staircases of similar unenclosed floor openings be closed sprinklers?

Answer: No. Close spaced sprinklers used in combination with draft stops and moving stairways, large monumental staircases or similar unenclosed floor openings may be either closed sprinklers or open sprinklers actuated by fixed temperature detection devices.

Use of deluge water curtains has become quite rare since the early 1960s. Privately conducted tests using closed spinklers indicated their effectiveness. Accidental discharge of deluge-type water curtains has resulted in considerable water damage as well as personal injury to persons on escalators when such false actuation has occurred.

4-4.8.2.4*   In noncombustible stair shafts, sprinklers shall be installed at the top and under the first landing above the lowest level. When the stair shaft serves two or more separate fire sections sprinklers shall also be installed at each floor landing.

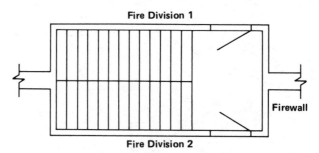

Figure A-4-4.8.2.4(a)   Noncombustible Stair Shaft Serving Two
Fire Sections.

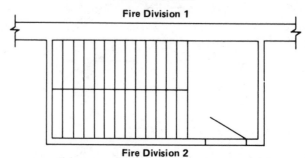

Figure A-4-4.8.2.4(b)   Noncombustible Stair Shaft Serving One Fire Section.

When a noncombustible stair shaft serves two fire-separated buildings or fire sections of one building as shown in Figure A-4-4.8.2.4(a) (that is, when the stair landing serves as a horizontal exit), sprinklers are required at each floor landing. If the stair serves only one fire section, then sprinklers are required only at the roof and under the lowest landing.

4-4.9* Building Service Chutes.   Building service chutes (linen, rubbish, etc.) shall be protected internally by automatic sprinklers. This will require a sprinkler above the top service opening of the chute, and above the lowest service opening and above service openings at alternate floor openings in buildings over two stories in height. The room or area into which the chute discharges shall also be protected by automatic sprinklers.

A-4-4.9   The installation of sprinklers at floor levels should be so arranged as to protect the sprinklers from mechanical injury, from falling materials, and not cause obstruction within the chute. This can usually be accomplished by recessing the sprinkler in the wall of the chute or by providing a protective deflector canopy over the sprinkler. Sprinklers should be placed so that there will be minimum interference of the discharge therefrom. (*See also 4-1.2.*) Sprinklers with special directional discharge characteristics may be advantageous.

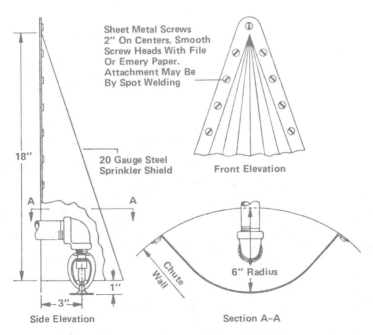

Figure 4.3.   Canopy for protecting sprinklers in building service chutes.

4-4.10* Exterior Canopies, Docks, and Platforms.

A-4-4.10   Small loading docks, covered platforms, ducts or similar small unheated areas may be protected by dry pendent sprinklers extending through the wall from wet sprinkler piping in an adjacent heated area, as shown in Figure A-4-4.10.

Where possible, the dry pendent sprinkler should extend down at a 45 degree angle. The width of the area to be protected should not exceed 7½ ft (2.3 m). Sprinklers should be spaced not over 12 ft (3.7 m) apart.

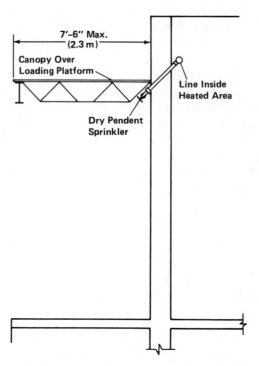

Figure A-4-4.10   Dry Pendent Sprinklers for Protection of Covered Platforms, Shipping Docks, and Similar Areas.

4-4.10.1   Sprinklers shall be installed under roofs or canopies over outside-loading platforms, docks, or other areas where combustibles are stored or handled.

Figure A-4-4.10 shows one method of protecting under roofs or canopies up to 7 ft, 6 in. (2.3 m) wide. Sprinklers are required under all such coverings where combustible goods are stored or handled.

4-4.10.2   Sprinklers shall be installed under exterior combustible roofs or canopies exceeding 4 ft (1.2 m) in width.

Exception: Sprinklers may be omitted where construction is noncombustible and areas under the canopies are not used for storage or handling.

Exterior canopies exceeding 4 ft (1.2 m) in width are to be sprinklered if of combustible construction whether goods are stored or handled under or not. If construction is noncombustible and the area under is essentially restricted to pedestrian use, sprinklers may be omitted.

4-4.10.3   Sprinklers shall be installed under exterior docks and platforms of combustible construction unless such space is closed off and protected against accumulation of debris.

This is an exception to the general requirement that sprinklers be installed in all areas of combustible construction but does not apply if windborne debris can accumulate and enhance the chance of a fire developing.

4-4.11* Decks.   Sprinklers shall be installed under decks and galleries which are over 4 ft (1.2 m) wide. Slatting of decks, walkways or the use of open gratings as a substitute for such sprinklers is not acceptable. Sprinklers installed under open gratings shall be of the listed intermediate level type or shielded from the discharge of overhead sprinklers.

There has to be some point at which ceiling sprinklers cannot be relied upon to control a fire under obstructions, such as shelves or decks, and this has been set at 4 ft (1.2 m). Because gratings or slatted constructions are frequently covered with goods in storage or by a light surface dust stop, sprinklers are required under such gratings and walkways.

A-4-4.11   Frequently, additional sprinkler equipment can be avoided by reducing the width of decks or galleries and providing proper clearances. Slatting of decks or walkways or the use of open grating as a substitute for automatic sprinklers thereunder is not acceptable. The use of cloth or paper dust tops for rooms forms obstruction to water distribution. If employed, the area below should be sprinklered.

Formal Interpretation

Question: Is subsection 4-4.11 intended to apply to steel racking with combustible materials stored on combustible shelves presenting a barrier (8 ft wide and 12 ft high, usually 2 or 3 shelves and of any length) to the proper operation of the sprinkler system?

Answer: No. It was not the intent of Section 4-4.11, when originally written, to apply to steel racking with combustible material stored on combustible shelves presenting a barrier to the proper operation of the sprinkler system.

Although it is not a mandatory requirement, the Committee wishes to call attention to B-4-2.4.8, Stock Fixtures, which states "Sprinklers should be installed in all stock fixtures which exceed 5 ft (1.5 m) in width, also in those which are less than 5 ft (1.5 m) but more than 2½ ft (0.8 m) in width unless bulkheaded with tight partitions. Sprinklers should be installed in any compartments which are larger than 5 ft (1.5 m) deep, 8 ft (2.4 m) long and 3 ft (0.9 m) high."

4-4.12 Library Stack Rooms.   For single tier stacks where 18-in. (457-mm) clearance can be provided between sprinkler deflectors and top of stacks, sprinklers shall be located without regard to stacks. For multitier stacks and for single-tier stacks where 18-in. (457-mm) clearance is not available between sprinkler deflectors and tops of stacks, branch lines shall be located in alternate aisles or in each aisle, depending on the arrangement of vertical shelf dividers. When vertical shelf dividers are incomplete, branch lines should be located in alternate aisles. If there are ventilation openings through floors or walkways, the location of branch lines shall be staggered in a vertical plane. When vertical shelf dividers are complete, so that lateral spread of sprinkler discharge will be prevented, branch lines shall be located in each aisle. (*See Figure 4-4.12.*)

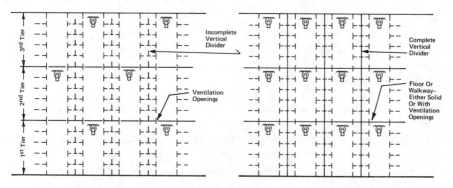

Figure 4-4.12   Sprinklers In Multitier Library Bookstacks.

While specific rules are given to cover a variety of conditions, judgment must be used to ensure that sprinklers are located so that a fire may be controlled.

4-4.13* Ducts.   Sprinklers shall be installed beneath ducts over 4 ft (1.2 m) wide unless ceiling sprinklers can be spaced in accordance with Table 4-2.4.6.

A-4-4.13   For ducts less than 4 ft (1.2 m) wide which obstruct distribution from ceiling sprinklers, see Appendix B-4-2.3.

Ducts are similar to decks (4-4.11) in their obstruction to sprinkler discharge, and the limit is again set at 4 ft (1.2 m), but if ceiling sprinklers can conform to Table 4-2.4.6, they need not be installed under the duct. In any case, the ceiling sprinklers must conform to Table 4-2.4.6.

Formal Interpretation

Question 1: Is it the intent of 4-4.13 (Ducts), of NFPA 13, to require sprinkler heads to be located below ducts greater than 4 ft wide in addition to the ceiling sprinkler heads located above the ducts when spacing requirements in accordance with Table 4-2.4.6 cannot be satisfied?

Answer: Yes.

Question 2: If the answer to the question above is "yes," could the ceiling sprinkler heads above the obstruction be omitted when the area of coverage above the obstruction protected by the omitted ceiling heads is noncombustible? Assume that sprinkler heads are located below the obstruction.

Answer: The Committee feels that this is a question that should properly be answered by the authority having jurisdiction who can properly assess all the circumstances.

4-4.14 Electrical Equipment.   When sprinkler protection is provided in generator and transformer rooms, hoods or shields installed to protect important electrical equipment from water shall be noncombustible.

Where it is feared that sprinkler discharge could damage electrical equipment, and shields are provided to deflect sprinkler water, they must be noncombustible. Omission of sprinklers solely due to presence of electrical equipment is not permitted. Frequently, generator and transformer rooms become storage rooms for combustible materials.

4-4.15* Open Grid Ceilings.   The following requirements are applicable to open grid ceilings in which the openings are ¼ in. (6.4 mm) or larger in least dimension, when the thickness or depth of the material does not exceed the least dimension of the openings and when such openings constitute at least 70 percent of the area of the ceiling material. Other types of open grid ceilings shall not be installed beneath sprinklers unless they are listed by a testing laboratory and are installed in accordance with the instructions contained in each package of the ceiling material. Ceilings made of highly flammable material may spread fire faster than sprinklers can control.

(a) In Light Hazard Occupancies when spacing of sprinklers of either standard or old style is not wider than 10 by 10 ft (3 X 3 m), a minimum clearance of at least 18 in. (457 mm) shall be provided between the sprinkler deflectors and the upper surface of the open grid ceiling. When spacing is wider than 10 by 10 ft (3 X 3 m) but not wider than 10 by 12 ft (3 X 3.7 m), a clearance of at least 24 in. (610 mm) shall be provided from standard sprinklers and at least 36 in. (914 mm) from old style sprinklers. When spacing is wider than 10 by 12 ft (3 X 3.7 m), a clearance of at least 48 in. (1219 mm) shall be provided.

(b) In Ordinary Hazard Occupancies, open grid ceilings may be installed beneath sprinklers only where such use is approved by the authority having jurisdiction, and shall be installed beneath standard sprinklers only. When sprinkler spacing is not wider than 10 by 10 ft (3 X 3 m), a minimum clearance of at least 24 in. (610 mm) shall be provided between the sprinkler deflectors and the upper surface of the open grid ceiling. When spacing is wider than 10 by 10 ft (3 X 3 m), a clearance of at least 36 in. (914 mm) shall be provided.

These requirements for clearance between the sprinkler deflectors and the top of the grid ceiling are to ensure that sprinkler discharge is not too severely impaired, for the grid ceiling will obstruct the discharge pattern. In addition to the height above the grid ceiling, other obstructions such as pipes or ducts must be considered.

A-4-4.15   The installation of open grid egg crate, louver, or honeycomb ceilings beneath sprinklers restricts the sidewise travel of the sprinkler discharge and may change the character of discharge.

4-4.16 Drop-out Ceilings.

4-4.16.1   Drop-out ceilings may be installed beneath sprinklers when ceilings are listed for that service and are installed in accordance with their listing. The authority having jurisdiction shall be consulted in all cases.

Drop-out ceilings are designed to shrivel and fall when heated by a fire, permitting the sprinkler which had been concealed above to come into play. Extreme care must be exercised in using only panels that are listed and tested for this use. When these ceilings are opaque so the fact that they are painted or have material stored above them is not easily detectable, their use should be discouraged. When any appreciable delay in sprinkler operation is undesirable, these ceilings become undesirable also.

4-4.16.2 Drop-out ceilings shall not be considered ceilings within the context of this standard.

Because the ceiling falls in the early stages of a fire, it need not be considered with respect to the position of sprinklers.

4-4.16.3 Piping installed above drop-out ceilings shall not be considered concealed piping. (*See 3-12.4, Exception No. 2.*)

Because piping is exposed to a fire, solder joints are not permitted in Ordinary Hazard Occupancies above drop-out ceilings.

4-4.16.4* Sprinklers shall not be installed beneath drop-out ceilings.

A-4-4.16.4 The ceiling tiles may drop before sprinkler operation. Delayed operation may occur because heat must then bank down from the deck above before sprinklers will operate.

Additionally, the danger exists for the drop-out ceiling tiles to hang up on a sprinkler and interfere with that sprinkler's proper operation.

4-4.17 Fur Vaults.

4-4.17.1 Sprinklers in fur storage vaults shall be located centrally over the aisles between racks and shall be spaced not over 5 ft (1.5 m) apart along the aisles.

Baffles are not required because old style sprinklers can be installed as close as 5 ft (1.5 m) apart without wetting the adjacent sprinklers.

4-4.17.2 When sprinklers are spaced 5 ft (1.5 m) apart along the sprinkler branch lines, pipe sizes may be in accordance with the following schedule:

1 in. pipe	4 sprinklers	2 in. pipe	20 sprinklers
1¼ in. pipe	6 sprinklers	2½ in. pipe	40 sprinklers
1½ in. pipe	10 sprinklers	3 in. pipe	80 sprinklers

This schedule applies to piping running in the same direction as the aisles.

4-4.17.3   Sprinklers shall be listed old style having orifice sizes selected to provide as closely as possible but not less than 20 gal per min (76 L/min) per sprinkler, based on the water pressure available.

Old style sprinklers are required because the test work which developed these requirements was done using old style sprinklers.

NOTE: See NFPA 81, *Standard on Fur Storage, Fumigation and Cleaning.* For tests of sprinkler performance in fur vaults see Fact Finding Report on Automatic Sprinkler Protection for Fur Storage Vaults of Underwriters Laboratories Inc., dated November 25, 1947.

4-4.18* Commercial-type Cooking Equipment and Ventilation Systems.

A-4-4.18   Automatic sprinklers protecting commercial-type cooking equipment and ventilation systems should be controlled by separate, readily accessible indicating-type control valves that are properly identified. *(See A-3-14.3.)*

The requirements with respect to protection of cooking equipment and ventilation systems were developed as a result of test work done by a task group of the National Fire Sprinkler Association and were introduced in 1968. Automatic sprinklers are effective for extinguishing fires in greases and cooking oils because the fine droplets of the water spray lower the temperatures below the point at which the fire will sustain itself.

4-4.18.1   In cooking areas protected by automatic sprinklers, sprinklers shall be provided to protect commercial-type cooking equipment and ventilation systems that are designed to carry away grease-laden vapors unless otherwise protected. *(See NFPA 96, Standard for Removal of Smoke and Grease-Laden Vapors from Commercial Cooking Equipment.)* Sprinklers shall be so located as to give complete coverage of cooking surfaces, within exhaust ducts, within exhaust hood plenum chamber, and under filters, if any.

Sprinklers are an effective way of controlling fires in cooking equipment, filters and ducts, if located to cover the cooking surfaces and all areas of an exhaust system to which fires could spread. Subject to acceptance by the authority having jurisdiction, if all cooking equipment is served by listed grease extractors, the sprinkler protection may be limited to the cooking surfaces. Some manufacturers of exhaust systems incorporating listed grease extractors provide listed built-in water spray fire protection for cooking surfaces in a pre-engineered package ready for connection to the sprinkler system.

4-4.18.2   Sprinklers with temperature classifications of intermediate, high or extra high will be required. Use of a temperature measuring device may be necessary to determine the appropriate temperature classification. Sprinkler systems shall be designed so that a cooking surface fire will operate sprinklers protecting the cooking surface prior to or simultaneously with sprinklers protecting the plenum chamber and ventilation ducts. This may be accomplished by installing sprinklers in the plenum chamber and ducts at least two temperature ratings higher than those protecting the cooking surfaces and not less than 325°F (163°C) or by use of thermal control valves.

It is important that the sprinklers over the cooking surface operate first, so that there is no chance of water from the sprinklers in the plenum cooling them and preventing operation. A thermal control valve is a thermally responsive valve which controls the supply of water to a number of open sprinklers so that all operate simultaneously.

4-4.18.3   Distance between sprinklers shall not exceed 10 ft (3 m) within and under exhaust hoods and in horizontal ducts. The first sprinkler in a horizontal duct shall be installed at the duct entrance.

4-4.18.4*   A standard ½-in. orifice pendent sprinkler with the frame parallel to the front edge of the deep fat fryer(s) shall be centered over each single or pair of fryers and shall be arranged to operate at not less than 30 psi (2.1 bars). A single sprinkler shall not protect more than 30 in. (762 mm) of deep fat fryer surface in any dimension. Sprinklers protecting deep fat fryers shall have their deflectors located at least 1 in. (25 mm) below the lower edge of the hood, and not less than 2 ft (0.6 m) nor more than 3 ft, 6 in. (1.1 m) above the deep fat fryer cooking surface. (*See Figure A-4-4.18.4.*)

Exception: Listed automatic spray nozzles may be used for the protection of deep fat fryers. The position, arrangement, location and water supply of individual spray nozzles shall be verified by approved test procedure.

Formal Interpretation

Question: Is it the intent of 4-4.18.4 that the sprinkler (or nozzle) be located over the front edge of the deep fat fryer?

Answer: No. The sprinkler is to be located as nearly as is practical over the center point of the fryer, with the plane of the frame arms supporting the deflector parallel to the front edge of the cooking equipment, unless test work indicates otherwise.

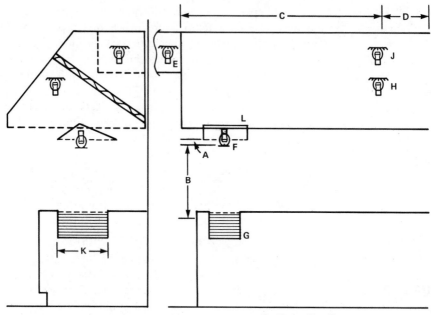

A Not Less Than 1 inch	G Deep Fat Fryer
B 2 feet to 3'6''	H Sprinkler Under Filter
C 5 feet Maximum	J Sprinkler Above Filter
D 10 feet Maximum to Next Sprinkler	K 30'' Maximum
E Sprinkler in Entrance to Horizontal Duct	L Shield Above Deep Fat Fryer
F Sprinkler Over Deep Fat Fryer	Sprinkler

For SI Units: 1 in. = 25.4 mm; 1 ft = 0.305 m.

Figure A-4-4.18.4 Typical Installation Showing Automatic Sprinklers Being Used for the Protection of Commercial Cooking Equipment and Ventilation Systems.

4-4.18.5   Other sprinklers shall be arranged so that their run-off does not fall into deep fat fryers. This may be accomplished by the use of a shield or unducted hood placed above the deep fat fryer. The shield or hood shall be placed above the sprinkler protecting the deep fat fryer and so located that it will not interfere with the sprinkler discharge.

This requirement is designed to prevent run-off of sprinkler water into a deep fat fryer. Application of water spray from a sprinkler does not cause boil-over problems associated with solid run-off.

4-4.18.6   One sprinkler shall be installed at the top of each vertical riser and an additional sprinkler shall be installed under any offset. Subject to the approval of the authority having jurisdiction, sprinklers may be omitted from a vertical riser located outside of a building provided the riser does not expose combustible material or the interior of a building and the horizontal distance between the hood outlet and the vertical riser is at least 25 ft (7.6 m).

Provided an offset is not over 5 ft (1.5 m), one sprinkler at the top of the lower vertical section will suffice.

4-4.18.7 Sprinklers and piping located at the top of a vertical riser, near the extremity of an exhaust duct, or in other areas subject to freezing, shall be properly protected against freezing by approved means.

4-4.18.8 Automatic sprinklers and spray nozzles used for protecting commercial-type cooking equipment and ventilation systems shall be replaced annually.

Exception: When automatic bulb type sprinklers or spray nozzles are used and annual examination shows no build-up of grease or other material on the sprinkler or spray nozzles.

The effect of grease and elevated temperature may seal the sprinkler orifice cap to the frame so that it will not release when the fusible element operates. Grease deposits may insulate and retard sprinkler operation. Because of the migration of solder components found in 360°F solder, the exception applies only to bulb-type sprinklers.

4-4.19 Baffles. Baffles (*except for in-rack sprinklers, see NFPA 231C, Standard on Rack Storage of Materials*) shall be installed whenever sprinklers are less than 6 ft (1.8 m) apart to prevent the sprinkler first opening from wetting adjoining sprinklers, thus delaying their operation. Baffles shall be located midway between sprinklers and arranged to baffle the actuating elements. Baffles may be of sheet metal about 8 in. (203 mm) wide and 6 in. (152 mm) high. When placed on branch line piping, the top of baffles shall extend 2 to 3 in. (51 to 76 mm) above the deflectors. (*See Figure A-3-15.5.4.*)

4-4.20 Small Rooms. Small room means a room with a smooth ceiling area not exceeding 800 sq ft (74.3 m²) of Light Hazard Occupancy classification.

Formal Interpretation

Question 1: Is it permissible to apply this rule to a hospital patient room?

Answer: Yes. This rule applies to any room of Light Hazard Occupancy which has a smooth ceiling and does not exceed 800 sq ft.

Question 2: Is it permissible to use multiple sprinklers in conjunction with the referenced sections in a pipe schedule system as long as the area per sprinkler does not exceed 200 sq ft?

Answer: Yes. This rule may be applied to either pipe schedule of hydraulically designed systems.

4-4.20.1* Within small rooms sprinklers may be located not over 9 ft (2.7 m) from any single wall; however, sprinkler spacing limitations of 4-2.1.1 and area limitations of 4-2.2.1.1 shall not be exceeded.

A-4-4.20.1 Example of sprinklers in small rooms for hydraulically designed and pipe schedule systems is shown in Figure A-4-4.20.1(a), and examples for hydraulically designed systems only are shown in Figures A-4-4.20.1(b), (c), and (d).

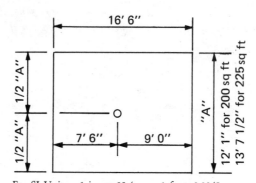

For SI Units: 1 in. = 25.4 mm; 1 ft = 0.3048 m.

Figure A-4-4.20.1(a).

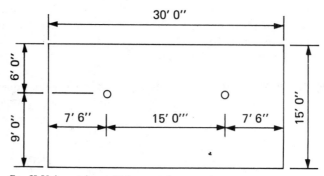

For SI Units: 1 in. = 25.4 mm; 1 ft = 0.3048 m.

Figure A-4-4.20.1(b).

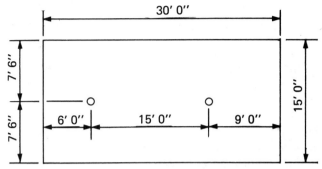

For SI Units: 1 in. = 25.4 mm; 1 ft = 0.3048 m.

Figure A-4-4.20.1 (c).

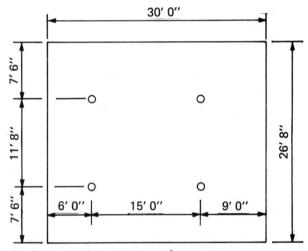

For SI Units: 1 in. = 25.4 mm; 1 ft = 0.3048 m.

Figure A-4-4.20.1 (d).

To accommodate ceiling patterns, lighting fixtures, diffusers, etc., in Light Hazard Occupancies with smooth ceilings, it is permissible to move the sprinklers a maximum of 1 ft, 6 in. (0.5 m) in one direction so that the maximum distance from one wall is 9 ft (2.7 m). In small rooms where other fixtures prevent the location of a sprinkler within the usual spacing rules this will save having to add additional sprinklers.

4-4.20.2 In hotels, sprinklers may be omitted from bathrooms not exceeding 55 sq ft (5.1 m^2) with noncombustible plumbing fixtures and with walls and ceilings surfaced with noncombustible materials.

Bathrooms which usually are surfaced with tile or other noncombustible material and contain very few combustible articles do not present a significant fire hazard.

B-4-2 Locations or Conditions Involving Special Consideration.

B-4-2.1 Overhead Doors.   When overhead doors form an obstruction to water distribution from sprinklers above, additional sprinkler protection may be required. When piping can be attached to the door structural framing, locate and space sprinklers under the doors in accordance with the rules for Ordinary Hazard Occupancy. When piping cannot be attached to the door structural framing, space sprinklers not over 12 ft (3.1 m) apart around the perimeter of the three accessible sides of the doors and at least 12 in. (305 mm) in from the edges of the doors. Deflectors should not be more than 10 in. (254 mm) below the doors in the open position. Sidewall sprinklers may be used when their distribution would be more effective than that from standard sprinklers. When doors are predominantly glass construction and when those doors, in an open position, will merely be over a traffic aisle, sprinkler protection is not necessarily required.

B-4-2.2 Tables.

B-4-2.2.1   Sprinklers should be installed under cutting, pressing, sewing machine and other work tables over 4 ft (1.2 m) wide. Sprinklers may be omitted under tables less than 5½ ft (1.7 m) but wider than 4 ft (1.2 m) if the tables are of temporary or semi-permanent nature, as determined by the authority having jurisdiction, and tight vertical partitions of galvanized iron or other noncombustible material are provided not over 10 ft (3 m) apart.

B-4-2.2.2   Partitions should be full width of table, extend from underside of table to floor and from front edge to back edge of table; should be substantially fastened to the underside of table and to floor; and should be reinforced with angle or channel iron uprights.

B-4-2.2.3   The outer edges of each partition should be smoothly finished (rounded if of metal) so as to prevent injury to employees.

B-4-2.2.4   Instructions should be obtained relative to the installation of "stops" under tables of unusual construction.

B-4-2.3 Obstructions.   Timbers, uprights, hangers, piping, lighting fixtures, ducts, etc., are likely to interfere with the proper distribution of water from sprinklers. Therefore, sprinklers should be so located or spaced that any interference is held to a minimum. The required clearance between such members and sprinklers is dependent upon the size of the obstruction to water distribution. The clearances should not be less than those specified between sprinklers and truss members in 4-2.4.1 and 4-2.4.5. (*See also 4-2.4.6.*)

B-4-2.4 Enclosures.

B-4-2.4.1 Sprinklers should be installed in enclosed equipment where combustible materials are processed or where combustible wastes or deposits may accumulate. Examples of such locations may include ovens, driers, dust collectors, conveyors, large ducts, spray booths, paper machine hood, paper machine economizers, parts of textile preparatory machines, and similar enclosures.

B-4-2.4.2 Sprinklers should be installed in small enclosed structures of combustible construction or containing combustible material. Examples of such locations may include penthouses, passageways, small offices, stock rooms, closets, vaults, or similar enclosures.

B-4-2.4.3 For small enclosures, pipes may be run outside the enclosures and sprinklers installed in approved dome-shaped covers about 10 in. (254 mm) in diameter. Where sprinklers can be nippled into the enclosure without forming an obstruction this should be done and dome-shaped covers omitted.

B-4-2.4.4 Sprinkler piping may be run above hoods over paper machines and similar equipment where dripping of condensation from sprinkler piping must be avoided and the sprinklers nippled through. The lower sprinklers under the hoods should be located outside of the line of the cylinders or rolls.

B-4-2.4.5 When enclosures are subject to freezing temperatures, special types of sprinkler protection should be provided. Manually operated systems should not be used.

B-4-2.4.6 The provision of other approved extinguishing systems does not permit the omission of sprinklers except in special situations, such as high value records or museum displays in vaults, commercial-type cooking equipment, computer or other electrical equipment enclosures. The authority having jurisdiction should be consulted.

B-4-2.4.7 Safe deposit or other vaults of fire-resistive construction will not ordinarily require sprinkler protection when used for the storage of records, files and other documents, when stored in metal cabinets.

B-4-2.4.8 Stock Fixtures. Sprinklers should be installed in all stock fixtures which exceed 5 ft (1.5 m) in width, also in those which are less than 5 ft (1.5 m) but more than 2½ ft (0.8 m) in width unless bulkheaded with tight partitions. Sprinklers should be installed in any compartments which are larger than 5 ft (1.5 m) deep, 8 ft (2.4 m) long and 3 ft (0.9 m) high.

B-4-2.4.9 Lighting fixtures of the pendent or surface mounted type may offer obstruction to discharge from sprinklers unless the clearances specified in Table 4-2.4.6 are provided.

4-4.21 Theater Stages.   Sprinklers shall be installed under the roof at the ceiling, in spaces under the stage either containing combustible materials or constructed of combustible materials; in all adjacent spaces and dressing rooms, storerooms, and workshops. When proscenium opening protection is required a deluge system shall be provided within 3 ft (0.9 m) of the stage side of the proscenium arch, with open sprinklers spaced up to a maximum of 6 ft (1.8 m) on center and designed to provide a discharge of 3 gpm/lineal foot [(37 L/min/)m] of water curtain, with no sprinkler discharging less than 15 gpm (56.8 L/min).

4-5* Spacing, Location and Position of Sidewall Sprinklers.   (*See 3-16.2.5.*)

A-4-5   The installation of sidewall sprinklers other than beneath smooth ceilings will require special consideration. Beams or other ceiling obstructions interfere with proper distribution and, when present, sidewall sprinklers should be spaced with regard to such obstructions.

The requirements of this section are for conventional sidewall sprinklers. Extended coverage sidewall sprinklers that have greater coverages than allowed in this section should be installed in accordance with their listing.

4-5.1 Distance Between Branch Lines and Sprinklers on Branch Lines.

4-5.1.1 Distance Between Branch Lines.   Rooms or bays having widths in excess of 15 ft up to 30 ft (4.6 m to 9.1 m) shall have sprinklers on two opposite walls or two opposite sides of bays with spacing as required in Section 4-5 and sprinklers regularly staggered. Additional branch lines shall be provided in rooms over 30 ft (9.1 m) in width except where special sprinklers are used (*see 4-1.1.3*).

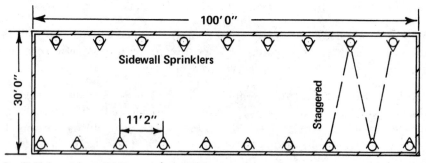

For SI Units: 1 in. = 25.4 mm; 1 ft = 0.3048 m.

Figure 4-5.1.1   **Spacing of Sidewall Sprinklers Under Combustible Smooth Ceilings, with Light Hazard Occupancy.**

4-5.1.2 Distance Between Sprinklers on Branch Lines.   Sidewall sprinklers shall be located not more than 10 ft (3 m) apart on walls for Ordinary Hazard Occupancies and not more than 14 ft (4.3 m) apart for Light Hazard Occupancies.

4-5.2 Protection Area Limitations for Light Hazard Occupancy.

4-5.2.1   With noncombustible smooth ceiling the protection area per sprinkler shall not exceed 196 sq ft (18.2 m^2) with the distance between sprinklers on lines not in excess of 14 ft (4.3 m).

4-5.2.2   With combustible smooth ceiling sheathed with plasterboard, metal, or wood lath and plaster, the protection area per sprinkler shall not exceed 168 sq ft (15.6 m^2) with the distance between sprinklers on lines not in excess of 14 ft (4.3 m). (*See Figure 4-5.1.1.*) When sheathing is combustible such as wood, fiberboard or other combustible material, the protection area per sprinkler shall not exceed 120 sq ft (11.1 m^2) with the distance between sprinklers on lines not in excess of 14 ft (4.3 m).

Exception: Noncombustible smooth ceiling spacing is permitted beneath a noncombustible smooth ceiling attached directly to the underside of a combustible sprinklered concealed space.

4-5.3 Protection Area Limitations for Ordinary Hazard Occupancy.

4-5.3.1   With noncombustible smooth ceiling the protection area per sprinkler shall not exceed 100 sq ft (9.3 m^2) with the distance between sprinklers on lines not in excess of 10 ft (3 m).

4-5.3.2   With combustible smooth ceiling sheathed with plasterboard, metal, wood lath and plaster, wood, fiberboard or other combustible material, the protection area per sprinkler shall not exceed 80 sq ft (7.4 m^2) with the distance between sprinklers on lines not in excess of 10 ft (3 m).

Exception: Noncombustible smooth ceiling spacing is permitted beneath a noncombustible smooth ceiling attached directly to the underside of a combustible sprinklered concealed space.

4-5.4* Position of Sidewall Sprinklers.   Sprinkler deflectors shall be at a distance from walls and ceilings not more than 6 in. (152 mm) or less than 4 in. (102 mm), unless special construction arrangements make a different position advisable for prompt operation and effective distribution.

Dead air spaces in corners can affect a sprinkler's operation time. The 4-in. (102 mm) limitation ensures that the sprinkler will operate properly.

Exception No. 1:   Horizontal-type sidewall sprinklers may be positioned 6 to 12 in. (152 to 305 mm) below noncombustible ceilings when listed for these positions.

Exception No. 2:   A horizontal-type sidewall sprinkler may be positioned less than 4 in. (102 mm) from the wall on which it is mounted.

The Exceptions for horizontal-type sidewall sprinklers permit: (1) the distance from the ceiling to be increased when the sprinkler has been listed for an increased distance not exceeding 12 in. (305 m), and (2) the minimum distance from the wall to be reduced.

A-4-5.4   Sidewall sprinklers should be placed to receive heat from a fire and at the same time most effectively distribute the water discharged by them. This is likely to be particularly important when heavy decorative molding is encountered near the junction of walls and ceilings.

5

TYPES OF SYSTEMS

5-1 Wet-Pipe Systems.

5-1.1* Definition.   A system employing automatic sprinklers attached to a piping system containing water and connected to a water supply so that water discharges immediately from sprinklers opened by a fire.

Wet-pipe systems are the most reliable and simple of all sprinkler systems since no equipment other than the sprinklers themselves need operate. Only those sprinklers which have been operated by heat over the fire will discharge water. All other types of systems included in this section utilize additional equipment for some specific purpose. This increase in sophistication also increases potential for failure. Therefore, wet-pipe systems are recommended where possible.

A-5-1.1   A dry-pipe, pre-action or deluge system may be supplied from a larger wet-pipe system, providing the water supply is adequate.

These auxiliary systems are connected to wet-pipe systems beyond the main control valve. They are considered a part of the wet-pipe system and must satisfy all rules regarding total sprinklers, system capacity, hydraulic calculations, etc., for that system. Each auxiliary system will require its own separate control valve, drains, etc. Alarms on the wet-pipe system may not be adequate for the auxiliary system and additional alarms for the auxiliary system may be needed.

5-1.2 Pressure Gages.   Approved pressure gages conforming to 2-9.2.2 shall be installed in sprinkler risers, above and below each alarm check valve.

5-1.3 Relief Valves.   A gridded wet-pipe system shall be provided with a relief valve not less than ¼ in. in size set to operate at pressure not greater than 175 psi (12.1 bars).

Exception No. 1:   When the maximum system pressure exceeds 165 psi (11.4 bars), the relief valve shall operate at 10 psi (0.7 bar) in excess of the maximum system pressure.

Exception No. 2:   When auxiliary air reservoirs are installed to absorb pressure increases.

The relief valves are installed to prevent excessive pressure buildup in gridded wet-pipe systems caused by increases in building temperature. These high pressures can cause a blowout or other system failure. Gridded systems have a much higher potential for this type failure because entrapped air can be completely evacuated if the piping arrangement has no dead end connections.

5-2 Dry-Pipe Systems.

5-2.1* Definition.   A system employing automatic sprinklers attached to a piping system containing air or nitrogen under pressure, the release of which as from the opening of a sprinkler permits the water pressure to open a valve known as a dry-pipe valve. The water then flows into the piping system and out the opened sprinklers.

Dry-pipe systems are installed in lieu of wet-pipe systems where piping is subject to freezing. Dry-pipe systems should not be used for the purpose of reducing water damage from pipe breakage or leakage since they operate too quickly to be of value for this purpose. Other systems, such as pre-action systems, may be suitable for this purpose where deemed necessary.

A-5-2.1   A dry-pipe system should be installed only where heat is not adequate to prevent freezing of water in all or sections of the system. Dry-pipe systems should be converted to wet-pipe systems when they become unnecessary because adequate heat is provided. Sprinklers should not be shut off in cold weather.

When two or more dry-pipe valves are used, systems should preferably be divided horizontally to prevent simultaneous operation of more than one system and resultant increased time delay in filling systems and discharging water plus receipt of more than one water flow alarm signal.

When adequate heat is present in sections of the dry-pipe system, consideration should be given to dividing the system into a separate wet-pipe system and dry-pipe system. Minimized use of dry-pipe systems is desirable where speed of operation is of particular concern.

Formal Interpretation

Question: Is it the intent of Section A-5-2.1 to recommend the use of a wet-pipe sprinkler system in the building in a cold climate where the heating system is turned off during non-working hours?

Answer: No. This section merely points out that wet-pipe systems are impractical in unheated spaces. It is not the intent of this section to recommend providing heat in any space to accommodate the sprinkler system nor to replace dry-pipe systems with wet-pipe systems.

B-5-3 Dry-Pipe Systems.

B-5-3.1 Eight-Inch Systems.   Where an 8-in. riser is employed in connection with a dry-pipe system, a 6-in. dry-pipe valve and a 6-in. gate valve between taper reducers may be used.

B-5-3.2 Dry-Pipe System Serving Several Remote Unheated Areas.   Where a single dry-pipe valve is used to supply piping and sprinklers located in several small unheated areas which are remote from each other, the dry-pipe valve and riser may be sized according to the number of sprinklers in the largest area. (*Also see 5-2.3.1.*)

5-2.2 Dry Pendent Sprinklers.   Automatic sprinklers installed in the pendent position shall be of the approved dry pendent type if installed in an area subject to freezing. The use of standard pendent sprinklers installed on return bends is permitted when both the sprinklers and the return bends are located in a heated area.

Dry pendent sprinklers are specially designed to prevent water from entering the drop pipe between sprinkler supply pipe and operating mechanism of the sprinkler. These sprinklers may also be used on wet-pipe systems where individual sprinklers are extended into spaces subject to freezing. Special application utilizing dry pendent sprinklers in upright and other positions should be in accordance with their listing.

5-2.3* Size of Systems.

Table A-5-2.3   Capacity of One Foot of Pipe
(Based on actual internal pipe diameters)

Nominal Diameter	Gal		Nominal Diameter	Gal	
	Sch 40	Sch 10		Sch 40	Sch 10
¾ in.	0.028	—	3 in.	0.383	0.433
1 in.	0.045	0.049	3½ in.	0.513	0.576
1¼ in.	0.078	0.085	4 in.	0.660	0.740
1½ in.	0.106	0.115	5 in.	1.040	1.144
2 in.	0.174	0.190	6 in.	1.501	1.649[1]
2½ in.	0.248	0.283	8 in.	2.66[3]	2.776[2]

For SI Units: 1 ft = 0.3048 m; 1 gal = 3.785 L.

[1]0.134 Wall Pipe
[2]0.188 Wall Pipe
[3]Schedule 30

5-2.3.1* Volume Limitations.   Not more than 500-gal (1893-L) system capacity for gridded systems or not more than 750-gal (2839-L) system capacity for nongridded systems shall be controlled by one dry-pipe valve.

Size of dry-pipe systems is limited by volume of piping in addition to area limitations of Chapter 3. The primary limitation is 500 gal (1893 L) capacity for gridded systems and 750 gal (2839 L) capacity for non-gridded systems. These volume limitations for a system may be exceeded if delivery to inspector's test pipe can be achieved in not more than 60 seconds. There is no 60-second limitation where these pipe volumes are not exceeded. Volume limitations, however, are imposed with the intent that water will be delivered to a sprinkler in a dry-pipe system as quickly as possible after it operates.

Exception No. 1*:   If check valves are installed in branches of the system to assist in more rapidly reducing the air pressure above the valve seat to the dry-pipe valve trip point, systems may exceed the above volume limitations, but no system branch shall have a capacity exceeding 400 gal (1514 L) for gridded systems nor 600 gal (2271 L) for nongridded systems, nor shall the total of a system branch plus common pipe exceed 500 gal (1893 L) for gridded systems nor 750 gal (2839 L) for nongridded systems. A hole ⅛ in. (3.2 mm) in diameter shall be drilled in the clapper of each check valve to permit equalization of air pressure among the various parts of the system. An approved indicating drain valve, connected by a bypass around each check valve, shall be provided as a means for draining the system. All check valves shall be located in heated enclosures to prevent the formation of ice.

Check valves shall not be installed in any piping where they may interfere with the hydraulic characteristics of the system, such as in the branch lines or cross mains of a gridded system.

A-5-2.3.1 Exception No. 1.   When check valves are installed in dry systems, the systems should deliver water to the inspector's test pipe in not more than 60 seconds starting at the normal air pressure.

The capacities of the various sizes of pipe given in Table A-5-2.3 are for convenience in calculating the air capacity of a system.

Exception No. 2:   Piping volume may exceed 500 gal (1893 L) for gridded systems or exceed 750 gal (2839 L) for nongridded systems if the system design is such that water is delivered to the system test pipe in not more than 60 seconds, starting at the normal air pressure on the system and at the time of fully opened inspection test connection.

5-2.4* Quick-Opening Devices.

Quick-opening devices, which are designed to hasten the tripping of a dry-pipe valve and thus speed delivery of water to operating sprinklers, consist primarily of accelerators and exhausters. Accelerators cause the dry-pipe valve to operate more quickly, thus allowing water pressure to expel air in the system at a faster rate and speed water movement to the open sprinklers. Exhausters cause air to be more quickly evacuated from sprinkler piping, thus shortening the time from sprinkler opening to dry-pipe valve operation and allowing more rapid movement of water to operating sprinklers after the dry-pipe valve has operated.

A-5-2.4   In the case of dry-pipe valves having relatively small priming chambers and in which the normal quantity of priming water fills, or nearly fills, the entire priming chamber, the objective contemplated by this rule will be met by requiring connection of the quick-opening device at a point on the riser above the dry-pipe valve, which will provide a capacity measure between the normal priming level of the air chamber and the connection of 1½, 2, and 3 gal (5.7, 7.6, and 11.4 L) for 4-, 5-, and 6-in. risers, respectively. Making the connection 24 in. (610 mm) above the normal priming water level will ordinarily provide this capacity.

5-2.4.1 When Required.   Dry-pipe valves shall be provided with an approved quick-opening device where system capacity exceeds 350 gal (1325 L) for gridded systems or capacity exceeds 500 gal (1893 L) for nongridded systems.

Even though quick-opening devices are not required where systems are of smaller capacity, they may be used whenever speed of operation is important, such as long extended piping systems and where life safety is an important factor.

5-2.4.2   The quick-opening device shall be located as close as practical to the dry-pipe valve. To protect the restriction orifice and other operating parts of the quick-opening device against submergence, the connection to the riser shall be above the point at which water (priming water and back drainage) is expected when the dry-pipe valve and quick-opening device are set, except where design features of the particular quick-opening device made these requirements unnecessary.

Special circumstances may require a quick-opening device, such as an exhauster, to be located at a point remote from the dry-pipe valve. Consideration should be given in these instances to protection against freezing of the device or its components. Remote devices should be accessible for servicing.

5-2.4.3   A soft disc globe or angle valve shall be installed in the connection between the dry-pipe sprinkler riser and the quick-opening device provided to accelerate operation of dry-pipe valve.

The shutoff valve between the dry-pipe sprinkler riser and the quick-opening device is necessary since these devices may require separate maintenance. It is highly undesirable to have the entire system out of service because of the condition of the quick-opening device. As a consequence of this ability to separately shut off these devices, they are frequently left out of service. Since system design has been based on proper operation of the quick-opening device, it should be maintained in operating condition at all times, and any necessary repairs should be expeditiously completed.

5-2.4.4   A check valve shall be installed between the quick-opening device and the intermediate chamber of the dry-pipe valve. If the quick-opening device requires pressure feedback from the intermediate chamber, a valve of the type which will clearly indicate whether it is opened or closed may be installed in place of that check valve. This valve shall be constructed so that it may be locked or sealed in the open position.

The shutoff valve for this type of quick-opening device is much the same as the shutoff valve at the connection to the sprinkler riser and should always be maintained open.

5-2.4.5   An approved antiflooding device shall be installed in the connection between the dry-pipe sprinkler riser and the quick-opening device, unless the particular quick-opening device has built-in antiflooding design features.

Quick-opening devices and their appurtenances, such as antiflooding devices, have in general had a poor service history. It is important that manufacturer's instructions be carefully followed when resetting each dry-pipe valve and quick-opening device. Many of the problems have been caused by not properly draining the equipment while resetting the dry-pipe valve after it has operated.

5-2.5* Location and Protection of Dry-Pipe Valve.

A-5-2.5   The dry-pipe valve should be located in an accessible place near the sprinkler system it controls.

This serves to reduce the amount of piping on a dry-pipe system, and to help limit system capacity. Long bulk mains from dry-pipe valves to system piping may use a large proportion of the system capacity and thus cause the actual sprinkler system to be quite small. Reduction of bulk pipe can be accomplished in large buildings by extending underground piping below floors to the dry-pipe valve. This type of arrangement introduces the problem of underground piping leaks under buildings and may dictate an alternate solution.

When exposed to cold, the dry-pipe valve should be located in an approved valve room or enclosure and, where this is not possible, in an underground pit acceptable to the authority having jurisdiction. Room should be of sufficient size to give at least 2½ ft (0.8 m) of free space at the sides and in front of, also above and below, the dry-pipe valve or valves, and this room, if feasible, should not be built until the valve is in position.

Size of enclosure should be governed by the number and arrangement of dry-pipe valves, so as to give ready access to these devices.

5-2.5.1   The dry-pipe valve and supply pipe shall be protected against freezing and mechanical injury.

5-2.5.2   Valve rooms shall be lighted and heated. The source of heat shall be of a permanently installed type.

5-2.5.3   The supply for the sprinkler in the dry-pipe valve enclosure shall be from the dry side of the system.

5-2.5.4   Protection against accumulation of water above the clapper shall be provided for a low differential dry-pipe valve. This may be an automatic high water level signaling device or an automatic drain device.

5-2.6* Cold Storage Rooms.

Dry-pipe systems installed in cold storage rooms require additional care and consideration since temperatures are below freezing and frequently in subzero ranges one hundred percent of the time. If a system trips accidentally, allowing the piping to fill with water, the nature of the occupancy may preclude heating this space to allow melting of ice and repair of system. It may be necessary in such systems to remove piping to a warm area where it would be thawed, emptied, and then reinstalled. Such operations are costly and result in long periods of systems out of service. Where conditions are particularly severe, it may be desirable to consider pre-action or other types of special systems. It is very important in cold storage plants for air supply to be dry enough to prevent accumulation of moisture in piping. This can be accomplished by taking air supply from the coldest freezer area, installation of air dryers, or using a moisture-free gas such as nitrogen.

A-5-2.6   Careful installation and maintenance, and some special arrangements of piping and devices as outlined in this section, are needed to avoid the formation of ice and frost inside piping in cold storage rooms which will be maintained at or below 32°F (0°C). Conditions are particularly favorable to condensation where pipes enter cold rooms from rooms having temperatures above freezing.

Whenever the opportunity offers, fittings such as specified in 5-2.6.1 and illustrated in Figures 5-2.6.1(A) and 5-2.6.1(B), as well as flushing connections specified in 3-8.2, should be provided in existing systems.

When possible, risers should be located in stair towers or other locations outside of refrigerated areas. This would reduce the probabilities of ice or frost formation within the riser (supply) pipe.

Cross mains should be connected to risers or feed mains with flanges. In general, flanged fittings should be installed at points which would allow easy dismantling of the system. Split ring or other easily removable types of hangers will facilitate the dismantling.

5-2.6.1 Fittings for Inspection Purposes.

5-2.6.1.1   Fittings for inspection purposes shall be provided whenever a cross main connects to a riser or feed main. This may be accomplished by a blind flange on a fitting (tee or cross) in the riser or cross main or a flanged removable section 24 in. (610 mm) long in the feed main as shown in Figure 5-2.6.1(A). Such fittings in conjunction with the flushing connections specified in 3-8.2 would permit examination of the entire lengths of the cross mains. Branch lines may be examined by backing the pipe out of fittings.

5-2.6.1.2   Whenever feed mains change direction, facilities shall be provided for direct observation of every length of feed main within the refrigerated area. This may be accomplished by means of 2-in. capped nipples or blind flanges on fittings.

5-2.6.1.3   Fittings for inspection purposes shall be provided whenever a riser or feed main passes through a wall or floor from a warm room to a cold room. This may be accomplished at floor penetrations by a tee with a blind flange in the cold room and at wall penetrations by a 24-in. (610-mm) flanged removable section in the warm room as shown in Figure 5-2.6.1(B).

5-2.6.2   A local low air-pressure alarm shall be installed on sprinkler systems supplying freezer sections.

5-2.6.3   Piping in cold storage rooms shall be installed with pitch, as outlined in 3-11.1.3.

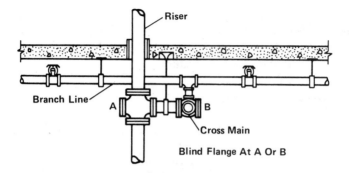

(a) Elevation At Riser And Cross Main

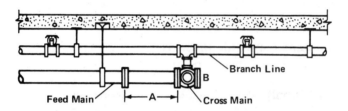

(b) Elevation At Feed Main And Cross Main

For SI Units: 1 in. = 25.4 mm; 1 ft = 0.3048 m.

Figure 5-2.6.1(A)   Fittings to Facilitate Examination of Feed Mains, Risers, and Cross Mains in Freezing Areas.

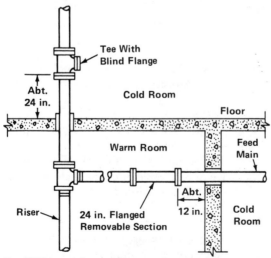

For SI Units: 1 in. = 25.4 mm; 1 ft = 0.3048 m.

Figure 5-2.6.1(B)   Fittings in Feed Main or Riser Passing through Wall or Floor from Warm Room to Cold Room.

5-2.6.4 The air supply for dry-pipe systems in cold storage plants shall be taken from the freezers of lowest temperature or through a chemical dehydrator. Compressed nitrogen gas from cylinders may be used in place of air in dry-pipe systems to eliminate introducing moisture.

5-2.7 Air Pressure and Supply.

Proper air pressure should be maintained on systems at all times. It is important to follow manufacturer's instructions regarding range of air pressures to be maintained. Low air pressures could result in accidental operation of the dry-pipe valve. High air pressures result in slower operation because this air must be exhausted from the system before water can be delivered to open sprinklers.

5-2.7.1 Maintenance of Air Pressure. Air or nitrogen pressure shall be maintained on dry-pipe systems throughout the year.

5-2.7.2* Air Supply. The compressed air supply shall be from a source available at all times and having a capacity capable of restoring normal air pressure in the system within 30 minutes, except for low differential dry-pipe systems where this time may be 60 minutes. Where low differential dry-pipe valves are used, the air supply shall be maintained automatically.

Standard dry-pipe valves normally have a differential in water pressure to air pressure at trip point of approximately 5.5:1. A low differential dry-pipe valve usually is an alarm check valve which has been converted to a dry-pipe valve. The differential in water pressure and air pressure at the trip point of these valves is approximately 1.1:1.

A-5-2.7.2 The compressor should draw its air supply from a place where the air is dry and not too warm. Moisture may cause trouble from condensation in the system.

5-2.7.3 Air Filling Connection. The connection pipe from the air compressor shall not be less than ½ in. and shall enter the system above the priming water level of the dry-pipe valve. A check valve shall be installed in this air line and a shutoff valve of renewable disc type shall be installed on the supply side of this check valve and shall remain closed unless filling the system.

5-2.7.4 Relief Valve. An approved relief valve shall be provided between compressor and controlling valve set to relieve at a pressure 5 psi (0.3 bar) in excess of maximum air pressure which should be carried in the system.

5-2.7.5 Shop Air Supply.   When the air supply is taken from a shop system having a normal pressure greater than that required for dry-pipe systems and an automatic air maintenance device is not used, the relief valve shall be installed between two control valves in the air line and a small air cock, which is normally left open, shall be installed in fitting below relief valve.

The relief valve on systems which use shop air supply should be set to prevent overpressurizing of the sprinkler system. As described above, high pressures are undesirable since they result in slower operation of the system.

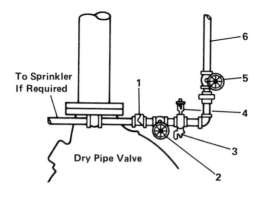

1. Check Valve
2. Control Valve (Renewable Disc Type)
3. Small Air Cock (Normally Open)

4. Relief Valve
5. Same as No. 2
6. Air Supply

Figure 5-2.7.5   Air Supply from Shop System.

5-2.7.6 Automatic Air Compressor.   When a dry-pipe system is supplied by an automatic air compressor or plant air system, any device or apparatus used for automatic maintenance of air pressure shall be of a type specifically approved for such service and capable of maintaining the required air pressure on the dry-pipe system. Automatic air supply to more than one dry-pipe system shall be connected to enable individual maintenance of air pressure in each system. A check valve or other positive back flow prevention device shall be installed in the air supply to each system to prevent air or water flow from one system to another.

It is not permitted to pump directly from an automatic air compressor through a fully opened supply pipe into the sprinkler system. An approved device must be installed between the automatic air supply and the dry-pipe valve. The restricted opening in air maintenance devices prevents the air supply system from adding air at too fast a rate, thus preventing or slowing operation of the dry-pipe valve. Automatic air compressors may require an air holding tank to prevent damage to the air compressor.

5-2.7.7 Air Pressure To Be Carried.   The air pressure to be carried shall be in accordance with the instruction sheet furnished with the dry-pipe valve, when available, or 20 psi (1.4 bars) in excess of the calculated trip pressure of the dry-pipe valve, based on the highest normal water pressure of the system supply. The permitted rate of air leakage shall be as specified in 1-11.3.2.

5-2.7.8   When used, nitrogen shall be introduced through a pressure regulator set to maintain system pressure in accordance with 5-2.7.7.

5-2.8 Pressure Gages.   Approved pressure gages conforming to 2-9.2.2 shall be connected:

(a) On the water side and air side of dry-pipe valve,

(b) At the air pump supplying the air receiver,

(c) At the air receiver,

(d) In each independent pipe from air supply to dry-pipe system, and

(e) At exhausters and accelerators.

5-3 Pre-Action and Deluge Systems.

   These systems are more technical in nature and require specialized knowledge and experience in design and installation. Manufacturer's specifications and listings must be strictly adhered to. Since the systems are somewhat more complex, failure to maintain them properly will seriously reduce their reliability.

5-3.1 Definitions.

Pre-Action System means a system employing automatic sprinklers attached to a piping system containing air that may or may not be under pressure, with a supplemental fire detection system installed in the same areas as the sprinklers; actuation of the fire detection system as from a fire opens a valve which permits water to flow into the sprinkler piping system and to be discharged from any sprinklers which may be open.

Deluge System means a system employing open sprinklers attached to a piping system connected to a water supply through a valve which is opened by the operation of a fire detection system installed in the same areas as the sprinklers; when this valve opens water flows into the piping system and discharges from all sprinklers attached thereto.

The primary difference between pre-action and deluge systems is that sprinklers in pre-action systems are closed and sprinklers in deluge systems are open. Operation of the detection system in a pre-action system only fills the pipe with water until a sprinkler opens. However, operation of the detection system in a deluge system results in flow from all sprinklers on the system. Deluge systems are normally used for high hazards requiring an immediate application of water over an entire hazard. Pre-action systems are normally used to protect properties where there is danger of serious water damage as a result of damaged automatic sprinklers or broken piping.

5-3.2* Description.   Pre-action and deluge systems are normally without water in the system piping and the water supply is controlled by an automatic valve operated by means of fire detection devices and provided with manual means for operation which are independent of the sprinklers. Systems may have equipment of the types described in (a) through (f) below. *(See 5-3.5.2.)*

(a) Automatic sprinklers with both sprinkler piping and fire detection devices automatically supervised,

(b) Automatic sprinklers with sprinkler piping and fire detection devices not automatically supervised,

(c) Open sprinklers with only fire detection devices automatically supervised,

(d) Open sprinklers with fire detection devices not automatically supervised,

(e) Combination of open and automatic sprinklers with fire detection devices automatically supervised,

(f) Combination of open and automatic sprinklers with fire detection devices not automatically supervised.

Automatic supervision consists of a means to give an alarm which indicates impairment of a device or system.

A-5-3.2   Pre-action and deluge systems may also have outside sprinklers for protection against exposure fire.

B-5 Types of Systems.

B-5-1 Pre-Action and Deluge Systems — Valves.

B-5-1.1   In hazardous locations, an approved indicating type valve or manual means for operation of pre-action or deluge valve should be installed in a location where access to the control valves is not likely to be prevented under fire emergency conditions.

B-5-1.2   With deluge systems, the deluge valve should be located as close as possible to the hazard protected, outside any fire or explosion hazard area.

5-3.3* General.

A-5-3.3   Conditions of occupancy or special hazards may require quick application of large quantities of water and in such cases deluge systems may be needed.

Fire detection devices should be selected to assure operation, yet guard against premature operation of sprinklers, based on normal room temperatures and draft conditions.

In locations where ambient temperature at ceiling is high, from heat sources other than fire conditions, heat-responsive devices should be selected which operate at higher than ordinary temperature and which are capable of withstanding the normal high temperature for long periods of time.

When corrosive conditions exist, materials or protective coatings which resist corrosion should be used.

To help avoid ice formation in piping because of accidental tripping of dry-pipe valves in cold storage rooms, a deluge automatic water control valve may be used on the supply side of the dry-pipe valve. When this combination is employed:

Dry systems may be manifolded to a deluge valve, the protected area not exceeding 40,000 sq ft (3716 m²). The distance between valves should be as short as possible to minimize water hammer.

The dry-pipe valves should be pressurized to 50 psi (3.4 bars) to reduce the possibility of dry-pipe valve operation from water hammer.

5-3.3.1   A supply of spare fusible elements for heat-responsive devices, not less than two of each temperature rating, shall be maintained on the premises for replacement purposes.

5-3.3.2   When hydraulic release systems are used, it is possible to water column the deluge valve or deluge-valve actuator if the heat-actuated devices (fixed temperature or rate-of-rise) are located at extreme heights above the valve. Refer to the manufacturer for height limitations of a specific deluge valve or deluge valve actuator.

5-3.3.3   All new pre-action or deluge systems shall be tested hydrostatically as specified in 1-11.2.1. In testing deluge systems, plugs shall be installed in fittings and replaced with open sprinklers after the test is completed, or automatic sprinklers may be installed and the operating parts removed after test is completed.

5-3.4 Location and Spacing of Fire Detection Devices.   Spacing of fire detection devices other than automatic sprinklers shall be in accordance with their listing by testing laboratories or in accordance with manufacturer's specifications. When automatic sprinklers are used as detectors, the distance between detectors and the area per detector shall not exceed the maximum permitted for suppression sprinklers as specified in 4-2.1 and 4-2.2; they shall be positioned in accordance with Section 4-3, but need not conform with the clearance requirements of 4-2.4. (*See NFPA 72E, Standard for Automatic Fire Detectors.*)

5-3.5 Pre-Action Systems.

5-3.5.1   All components of pneumatic, hydraulic or electrical pre-action systems shall be compatible.

5-3.5.2 Size of Systems.   Not more than 1,000 closed sprinklers shall be controlled by any one pre-action valve.

5-3.5.3 Supervision.   Sprinkler piping and fire detection devices shall be automatically supervised when there are more than 20 sprinklers on the system.

Formal Interpretation

Question 1: Is it the intent of 5-3.5.3, NFPA 13 (Supervision), to limit methods of supervision to pressurized air monitoring devices when there are more than 20 sprinklers on the system?

Answer: No.

Question 2: If the answer to Question 1 above is "no," would a downstream pressure sensing device for water flow be adequate to satisfy the supervision requirements of 5-3.5.3, NFPA 13 (Supervision)?

Answer: No. Supervision (electrical or mechanical) in this section refers to constant monitoring of piping and detection equipment to assure integrity of the system.

5-3.5.4 Pipe Schedule.   (*See Sections 3-5, 3-6, 3-7 and Chapter 7.*)

5-3.5.5 Pendent Sprinklers.   Automatic sprinklers on pre-action systems installed in the pendent position shall be of the approved dry pendent type if installed in an area subject to freezing.

5-3.6* Deluge Systems.   The fire detection devices or systems shall be automatically supervised when there are more than 20 sprinklers on the system.

A-5-3.6   Deluge systems are usually applied to severe conditions of occupancy. In designing piping systems the pipe sizes should be calculated in accordance with the standard for hydraulically designed sprinkler systems as given in Chapter 7.

When 8-in. piping is employed to reduce friction losses in a system operated by fire detection devices, a 6-in. pre-action or deluge valve and 6-in. gate valve between taper reducers may be used.

5-3.7 Devices for Test Purposes and Testing Apparatus.

5-3.7.1   When fire detection devices installed in circuits are located where not readily accessible, an additional fire detection device shall be provided on each circuit for test purposes at an accessible location and shall be connected to the circuit at a point which will assure a proper test of the circuit.

5-3.7.2   Testing apparatus capable of producing the heat or impulse necessary to operate any normal fire detection device shall be furnished to the owner of the property with each installation. Where explosive vapors or materials are present, hot water, steam or other methods of testing not involving an ignition source shall be used.

5-3.7.3 Pressure Gages.   Approved pressure gages conforming to 2-9.2.2 shall be installed as follows:

(a) Above and below pre-action valve and below deluge valve.

(b) On air supply to pre-action and deluge valves.

5-4 Combined Dry-Pipe and Pre-Action Systems.

5-4.1 General.

5-4.1.1* Definition.

A Combined Dry-Pipe and Pre-Action Sprinkler System.   A system employing automatic sprinklers attached to a piping system containing air under pressure with a supplemental fire detection system installed in the same areas as the sprinklers; operation of the fire detection system, as from a fire, actuates tripping devices which open dry-pipe valves simultaneously and without loss of air pressure in the system. Operation of the fire detection system also opens approved air exhaust valves at the end of the feed main which facilitates the filling of the system with water which usually precedes the opening of sprinklers. The fire detection system also serves as an automatic fire alarm system.

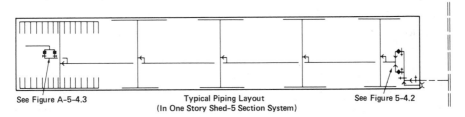

See Figure A-5-4.3   Typical Piping Layout   See Figure 5-4.2
(In One Story Shed–5 Section System)

Figure A-5-4.1.1   Typical Piping Layout for Combined Dry-Pipe and Pre-Action Sprinkler System.

A-5-4.1.1 When Installed.   Combined dry-pipe and pre-action systems may be installed when wet-pipe systems are impractical. They are intended for use in but not limited to structures where a number of dry-pipe valves would be required if a dry-pipe system were installed.

> Combined dry-pipe and pre-action systems are employed primarily where more than one dry-pipe system would be required by reason of area or capacity and it is not possible to install supply piping to each dry-pipe valve through a heated or protected space. An example of this use is piers and wharves. These are special application systems and should not be used merely because areas are large in an ordinary building.

5-4.1.2   Combined automatic dry-pipe and pre-action systems shall be so constructed that failure of the fire detection system shall not prevent the system from functioning as a conventional automatic dry-pipe system.

> Conventional automatic operation of the dry-pipe valve provides a fail-safe backup to the fire detection system. Since the primary purpose of the fire detection system is to cause piping to start filling

sooner, loss of the detection system will only result in a much slower operating mode. Although somewhat impaired, the system will be reasonably fail-safe.

5-4.1.3 Combined automatic dry-pipe and pre-action systems shall be so constructed that failure of the dry-pipe system of automatic sprinklers shall not prevent the fire detection system from properly functioning as an automatic fire alarm system.

Quite the opposite effect from Section 5-4.1.2 results when the pre-action system operates but the dry-pipe valve does not. There would be no water to the piping system and sprinklers with this type failure. There would be an alarm only.

5-4.1.4 Provisions shall be made for the manual operation of the fire detection system at locations requiring not more than 200 ft (61.0 m) of travel.

Manual operation of the fire detection system provides rapid means of filling the system with water from remote locations throughout the protected area. This can be used in special circumstances where there is danger of an incident and the system can be converted to a wet-pipe system prior to a fire.

5-4.1.5 Except as indicated in 5-2.2, automatic sprinklers installed in the pendent position shall be of the approved dry pendent type.

5-4.2 Dry-Pipe Valves in Combined Systems.

5-4.2.1 Where the system consists of more than 600 sprinklers or has more than 275 sprinklers in any fire area, the entire system shall be controlled through two 6-in. dry-pipe valves connected in parallel and shall feed into a common feed main. These valves shall be checked against each other. (*See Figure 5-4.2.*)

The primary purpose of providing two valves is to maintain service on the system while servicing the other dry-pipe valve. The second dry-pipe valve in parallel also provides an additional degree of safety, since one valve would serve as a backup to the other.

5-4.2.2 Each dry-pipe valve shall be provided with an approved tripping device actuated by the fire detection system. Dry-pipe valves shall be cross connected through a 1-in. pipe connection to permit simultaneous tripping of both dry-pipe valves. This 1-in. pipe connection shall be equipped with a gate valve so that either dry-pipe valve can be shut off and worked on while the other remains in service.

Although each dry-pipe valve serves as a backup valve, simultaneous operation of both valves is desirable to increase flow characteristics of the system.

5-4.2.3   The check valves between the dry-pipe valves and the common feed main shall be equipped with ½-in. bypasses so that a loss of air from leakage in the trimmings of a dry-pipe valve will not cause same to trip until the pressure in the feed main is reduced to the tripping point. A gate valve shall be installed in each of these bypasses so that either dry-pipe valve can be completely isolated from the main riser or feed main and from the other.

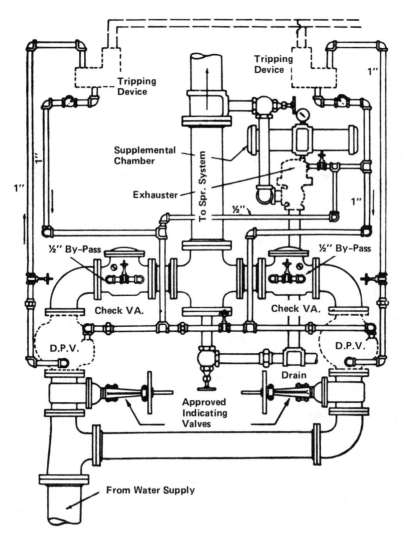

Figure 5-4.2   Header for Combined Dry-Pipe and Pre-Action Sprinkler System, Standard Trimmings Not Shown.

5-4.2.4   Each combined dry-pipe and pre-action system shall be provided with approved quick opening devices at the dry-pipe valves.

5-4.3* Air Exhaust Valves.   One or more approved air exhaust valves of 2-in. or larger size controlled by operation of a fire detection system shall be installed at the end of the common feed main. *(See Figure A-5-4.3.)* These air exhaust valves shall have soft seated globe or angle valves in their intakes; also, approved strainers shall be installed between these globe valves and the air exhaust valves.

These systems are of very large air capacity. The exhaust valves serve to evacuate air at remote ends of the piping and allow water to fill the system more rapidly. The air exhaust valves should be located where they may be tested and readily maintained.

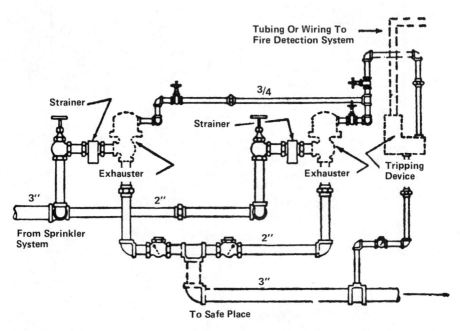

Figure A-5-4.3   Arrangement of Air Exhaust Valves for Combined Dry-Pipe and Pre-Action Sprinkler System.

5-4.4 Subdivision of System Using Check Valves.

5-4.4.1   Where more than 275 sprinklers are required in a single fire area, the system shall be divided into sections of 275 sprinklers or less by means of check valves. If the system is installed in more than one fire area or story, not more than 600 sprinklers shall be supplied through any one check valve. Each section shall have a 1¼-in. drain on the system side of each check valve supplemented by a drum drip.

This section provides a means to increase size of systems to effect a cost saving. Its use is usually determined by economic considerations.

5-4.4.2 Section drain lines and drum drips shall be located in heated areas or inside of thermostatically controlled electrically heated cabinets of sufficient size to enclose drain valves and drum drips for each section. Drum drips shall also be provided for all low points except that heated cabinets need not be required for 20 sprinklers or less.

5-4.4.3 Air exhaust valves at end of feed main and associated check valves shall be protected against freezing.

5-4.5 Time Limitation. The sprinkler system shall be so constructed and the number of sprinkler heads controlled shall be so limited that water shall reach the furthest sprinkler within a period of time not exceeding 1 minute for each 400 ft (122 m) of common feed main from the time the heat-responsive system operates. Maximum time permitted not to exceed 3 minutes.

Since these systems are a compromise for economic reasons, time restrictions should not be exceeded. Excessive time for delivery of water should be reviewed with the authority having jurisdiction.

5-4.6 System Test Pipe. The end section shall have a system test pipe as required for dry-pipe systems.

5-5 Antifreeze Systems.

5-5.1 Definition. Antifreeze system means a system employing automatic sprinklers attached to a piping system containing an antifreeze solution and connected to a water supply. The antifreeze solution, followed by water, discharges immediately from sprinklers opened by a fire.

5-5.2* Where Used. The use of antifreeze solutions SHALL be in conformity with any state or local health regulations.

State or local plumbing and health regulations may not allow the introduction of foreign materials to piping systems connected to public water. Where these regulations are in effect, the use of antifreeze and the type of solution permitted should be checked with local authorities. It may be permitted where piping arrangements make contamination of public water virtually impossible. In many instances antifreeze systems are the only practical solution to a fire protection problem in a small area of a building and deviations from codes will be allowed.

A-5-5.2   Antifreeze solutions may be used for maintaining automatic sprinkler protection in small unheated areas. Antifreeze solutions are recommended only for systems not exceeding 40 gallons.

Because of the cost of refilling the system or replenishing small leaks, it is advisable to use small dry valves where more than 40 gallons are to be supplied.

Propylene glycol or other suitable material may be used as a substitute for priming water, to prevent evaporation of the priming fluid, and thus reduce ice formation within the system.

5-5.3 Antifreeze Solutions.

5-5.3.1   When sprinkler systems are supplied by public water connections the use of antifreeze solutions other than water solutions of pure glycerine (C.P. or U.S.P. 96.5 percent Grade) or propylene glycol shall not be permitted. Suitable glycerine-water and propylene glycol-water mixtures are shown in Table 5-5.3.1.

Table 5-5.3.1   Antifreeze Solutions
To Be Used if Public Water Is Connected to Sprinklers

Material	Solution (by Volume)	Spec. Grav. at 60°F (15.6°C)	Freezing Point °F	°C
Glycerine	50% Water	1.133	−15	−26.1
C.P. or U.S.P. Grade*	40% Water	1.151	−22	−30.0
	30% Water	1.165	−40	−40.0
	Hydrometer Scale 1.000 to 1.200			
Propylene Glycol	70% Water	1.027	+ 9	−12.8
	60% Water	1.034	− 6	−21.1
	50% Water	1.041	−26	−32.2
	40% Water	1.045	−60	−51.1
	Hydrometer Scale 1.000 to 1.200 (Subdivisions 0.002)			

*C.P.   — Chemically Pure.
 U.S.P. — United States Pharmacopoeia 96.5%.

5-5.3.2   If public water is not connected to sprinklers, the commercially available materials indicated in Table 5-5.3.2 are suitable for use in antifreeze solutions.

Table 5-5.3.2   Antifreeze Solutions

To Be Used if Public Water Is Not Connected to Sprinklers

Material	Solution (by Volume)	Spec. Grav. at 60°F (15.6°C)	Freezing Point °F	°C
Glycerine	If glycerine is used, see Table 5-5.3.1			
Diethylene Glycol	50% Water	1.078	−13	−25.0
	45% Water	1.081	−27	−32.8
	40% Water	1.086	−42	−41.1
	Hydrometer Scale 1.000 to 1.120 (Subdivisions 0.002)			
Ethylene Glycol	61% Water	1.056	−10	−23.3
	56% Water	1.063	−20	−28.9
	51% Water	1.069	−30	−34.4
	47% Water	1.073	−40	−40.0
	Hydrometer Scale 1.000 to 1.120 (Subdivisions 0.002)			
Propylene Glycol	If propylene glycol is used, see Table 5-5.3.1			
Calcium Chloride 80% "Flake"	Lb CaCl$_2$ per gal of Water			
Fire Protection Grade*	2.83	1.183	0	−17.8
Add Corroison inhibi-	3.38	1.212	−10	−23.3
tor	3.89	1.237	−20	−28.9
of sodium bichromate	4.37	1.258	−30	−34.4
¼ oz per gal water	4.73	1.274	−40	−40.0
	4.93	1.283	−50	−45.6

*Free from magnesium chloride and other impurities.

5-5.3.3*   An antifreeze solution shall be prepared with a freezing point below the expected minimum temperature for the locality. The specific gravity of the prepared solution shall be checked by a hydrometer with suitable scale.

Verification of the hydrometer scale is very important. Commonly available hydrometers, such as those used for automobile service stations, are often not adequate.

A-5-5.3.3   Beyond certain limits, increased proportion of antifreeze does not lower the freezing point of solution. (*See Figure A-5-5.3.3.*) Glycerine, diethylene glycol, ethylene glycol and propylene glycol should never be used without mixing with water in proper proportions because these materials tend to thicken near 32°F (0°C).

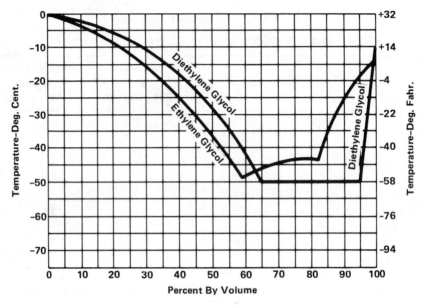

Figure A-5-5.3.3   Freezing Points of Water Solutions of Ethylene Glycol and Diethylene Glycol.

5-5.4* Arrangement of Supply Piping and Valves.

All permitted antifreeze solutions are heavier than water. At the point of contact (interface) the heavier liquid will be below the lighter liquid in order to prevent diffusion of water into the unheated areas. In most cases, this necessitates the use of a 5-ft (1.5-m) drop pipe or U-loop as illustrated in Figure 5-5.4. The preferred arrangement is to have the sprinklers below the interface between the water and the antifreeze solution.

If sprinklers are above the interface, a check valve with ⅟₃₂-in. (0.8-mm) hole in the clapper shall be provided in the U-loop. A water control valve and two small solution test valves shall be provided as illustrated in Figure 5-5.4. An acceptable arrangement of filling cup is also shown.

Figure 5-5.4 shows the most common arrangement of supply piping and valves. Actual field conditions may require modification of this arrangement. For example, if piping on the antifreeze system is at a higher elevation than the water supply, it is not practical to install a gravity-type fill cup, as shown on the drawing, at the supply

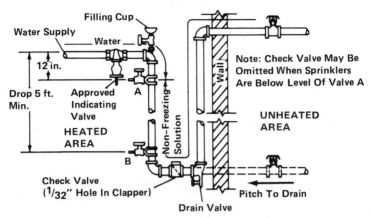

NOTE: The ⅟₃₂-in. (0.8-mm) hole in the check valve clapper is needed to allow for expansion of the solution during a temperature rise and thus prevent damage to sprinkler heads.

For SI Units: 1 in. = 25.4 mm; 1 ft = 0.3048 m.

Figure 5-5.4   Arrangement of Supply Piping and Valves.

location. It is generally relocated out to the high point of the piping system. In a few instances, it may be necessary to use the valves at A or B to pump antifreeze into the system if the fill cup is impractical.

B-5-2 Filling with Antifreeze Solutions.   With water supply valve closed and the system drained, fill the piping through the filling cup, using a suitable antifreeze solution of the proper concentration. Vent the air at the end sprinklers. Back out all sprinklers slightly until the liquid appears so that the piping will be completely filled and all air expelled. If the filling cup is not above the highest sprinklers, the piping may be filled through valve B by means of a small pump or through a filling cup installed at the highest branch sprinkler line. If the last-named method is used, the drop pipe should be filled through the filling cup shown in diagram. Then tighten the sprinkler heads and open valve A until the 12-in. (305-mm) section of pipe above this valve is empty and the level of the antifreeze solution in the drop pipe is at valve A. Close valve A. Close the filling connection valve and slowly open the supply valve wide.

A-5-5.4   To avoid leakage, the materials and workmanship should be excellent, the threads clean and sharp, and the joints tight. Use only metal-faced valves.

5-5.5* Testing.   Before freezing weather each year, the solution in the entire system shall be emptied into convenient containers and brought to the proper specific gravity by adding concentrated liquid as needed. The resulting solution may be used to refill the system.

If emptying of the antifreeze piping is impractical, tests may be conducted as indicated in A-5-5.5.

A-5-5.5   Tests should be made by drawing a sample of the solution from valve B two or three times during the freezing season, especially if it has been necessary to drain the building sprinkler system for repairs, changes, etc. A small hydrometer should be used so that a small sample will be sufficient. When water appears at valve B or when the test sample indicates that the solution has become weakened, empty the entire system and recharge as previously described.

5-6 Automatic Sprinkler Systems with Nonfire Protection Connections.

5-6.1 Circulating Closed Loop Systems.

It has been considered feasible for many years to make use of the sprinkler piping, which normally stands idle, for a more active purpose.

This section allows sprinkler piping to be used for additional functions other than fire protection, such as heating and air conditioning of a building. A basic criterion used in the development of this section of the standard is that the auxiliary functions be added in such a manner and with sufficient controls to ensure there would be no reduction in overall fire protection as a result of adding these auxiliary uses. It was determined this could be controlled more adequately with totally closed systems. This section and the limitation of a closed system is viewed as a first step. Future changes may result in extended use of sprinkler piping for additional purposes.

This section attempts to answer the need to improve economies of building materials by providing multiple use of sprinkler piping systems.

5-6.1.1 Definition.   A circulating closed loop is one with nonfire protection connections to automatic sprinkler systems in a closed loop piping arrangement for the purpose of utilizing sprinkler piping to conduct water for heating or cooling. Water is not removed or used from the system, but only circulated through the piping system.

These systems are frequently heat pump systems where water is circulated through heating and air conditioning equipment utilizing sprinkler pipe as the primary conductors. Heat is added to, or rejected from, the circulating water as dictated by requirements of the system. Water is neither added nor removed from the piping system.

5-6.1.2 System Components.

5-6.1.2.1 Basic Principle.   A circulating closed loop system is primarily a sprinkler system, and all provisions of this standard such as control valves, area limitation of a system, alarms, fire department connections, sprinkler spacing, etc., are to be satisfied.

Exception: Items as specifically detailed within 5-6.1.

5-6.1.2.2   Piping, fittings, valves and pipe hangers shall meet requirements specified in Chapter 3.

5-6.1.2.3   A dielectric fitting shall be installed in junction where dissimilar piping materials are joined, e.g., copper to steel.

Exception: Dielectric fittings are not required in junction where sprinklers are connected to piping.

5-6.1.2.4   It is not required that other auxiliary devices be listed for sprinkler service; however, these devices such as pumps, circulating pumps, heat exchangers, radiators, and luminaires shall be pressure rated at 175 or 300 psi (12.1 or 20.7 bars) (rupture pressure of 5 × rated water working pressure), to match required rating of sprinkler system components.

Although this section does not require testing and listing of the auxiliary equipment or components, they must be of materials acceptable in Chapter 3 and of sufficient strength to be equal to all other sprinkler system components.

5-6.1.2.5   Auxiliary devices shall incorporate materials of construction and be so constructed that they will maintain their physical integrity under fire conditions to avoid impairment to the fire protection system.

5-6.1.2.6   Auxiliary devices where hung from the building structure shall be supported independently from the sprinkler portion of the system, following recognized engineering practices.

5-6.1.3 Hydraulic Characteristics.   Piping systems for attached heating and cooling equipment shall have auxiliary pumps or an arrangement made to return water to the piping system in order to assure the following:

(a)* Water for sprinklers shall not be required to pass through heating or cooling equipment. At least one direct path shall exist for water flow from the sprinkler water supply to every sprinkler. Pipe sizing in the direct path shall be in accordance with design requirements of this standard.

This section is intended to ensure there is no reduction in flow to sprinklers regardless of operation or presence of the auxiliary equipment. There may be an enhancement of the sprinkler system, but no reduction.

A-5-6.1.3(a)   Outlets should be provided at critical points on sprinkler system piping to accommodate attachment of pressure gages for test purposes.

(b) No portions of the sprinkler piping shall have less than the sprinkler system design pressure regardless of the mode of operation of the attached heating or cooling equipment.

(c) There shall be no loss or outflow of water from the system due to or resulting from the operation of heating or cooling equipment.

(d) Shut-off valves and a means of drainage shall be provided on piping to heating or cooling equipment at all points of connection to sprinkler piping and shall be installed in such a manner as to make possible repair or removal of any auxiliary component without impairing the serviceability and response to the sprinkler system. All auxiliary components including strainer shall be installed on the auxiliary equipment side of the shut-off valves.

5-6.1.4 Water Temperature.

5-6.1.4.1 Maximum.   In no case shall maximum water temperature flowing through the sprinkler portion of the system exceed 120°F (49°C). Protective control devices listed for this purpose shall be installed to shut down heating or cooling systems when temperature of water flowing through the sprinkler portion of the system exceeds 120°F (49°C). When water temperature exceeds 100°F (37.8°C), intermediate or higher temperature rated sprinklers shall be used.

5-6.1.4.2 Minimum.   Precaution shall be taken to ensure that temperatures below 40°F (4.4°C) will not be permitted.

5-6.1.5 Obstruction to Discharge.   Automatic sprinklers shall not be obstructed by auxiliary devices, piping, insulation, etc., from detecting fire or from proper distribution of water.

5-6.1.6 Valve Supervision.   Position of all valves controlling sprinkler system (post indicator, main gate, sectional control) shall be supervised open by one of the following methods:

(a) Central station, proprietary, or remote station alarm service.

(b) Local alarm service, which will cause the sounding of an audible signal at a constantly attended point.

Supervision of all control valves is required in this type of system in consideration of the combined use characteristics of the system. Some repairs may be made by those not familiar with sprinkler systems and supervision is to lessen the chance of improper valve closure.

5-6.1.7 Signs.   Caution signs shall be attached to all controlling sprinkler valves. The caution sign shall be worded as follows:

"This valve controls fire protection equipment. Do not close until after fire has been extinguished. Use auxiliary valves when necessary to shut supply to auxiliary equipment. CAUTION: Automatic alarm will be sounded if this valve is closed."

5-6.1.8 Water Additives.   Materials added to water shall not adversely affect the fire fighting properties of the water and shall be in conformity with any state or local health regulations. Due care and caution shall be given to the use of additives which may remove or suspend scale from older piping systems. When additives are necessary for proper system operation, due care shall be taken to ensure additives are replenished after alarm testing or whenever water is removed from the system.

5-6.1.9 Water Flow Detection.   The supply of water from sprinkler piping through auxiliary devices, circulatory piping, and pumps shall not under any condition or operation, transient or static, cause false sprinkler water flow signals.

5-6.1.9.1   Sprinkler water flow signal shall not be impaired when water is discharged through opened sprinkler or through system test pipe while auxiliary equipment is in any mode of operation (on, off, transient, stable).

5-6.1.10* Working Plans.   Working plans shall be prepared and submitted in accordance with Section 1-9. Special symbols shall be used and explained for auxiliary piping, pumps, heat exchangers, valves, strainers and the like, clearly distinguishing those devices and piping runs from those of the sprinkler system. Model number, type and manufacturer's name shall be identified for each piece of auxiliary equipment.

5-6.1.11 Testing.

5-6.1.11.1   All sprinkler system and auxiliary system components shall be hydrostatically tested in accordance with 1-11.3.

5-6.1.11.2   Sprinkler system discharge tests shall be conducted using system test pipes described in 3-9.1. Pressure gages shall be installed at critical points and readings taken under various modes of auxiliary equipment operation. Water

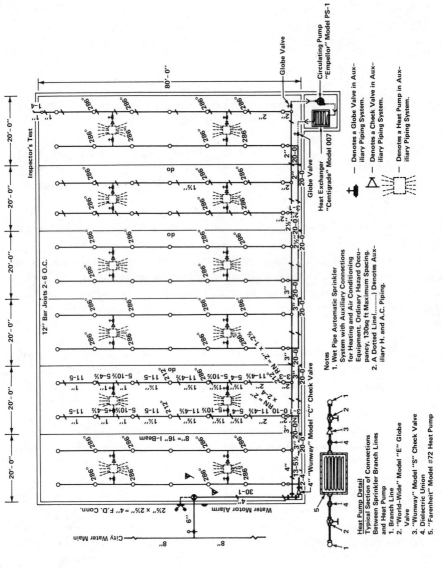

Figure A-5-6.1.10(a).

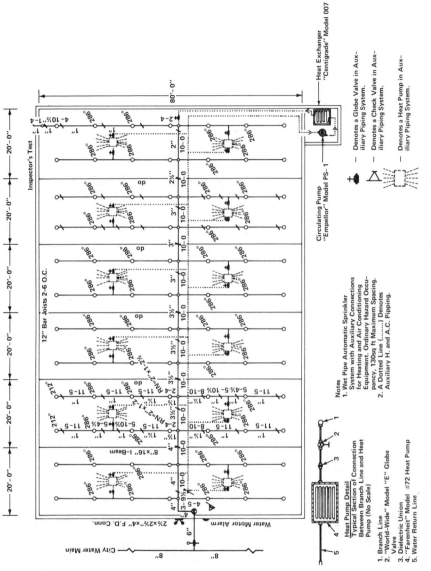

Figure A-5-6.1.10(b).

flow alarm signals shall be responsive to discharge of water through system test pipes while auxiliary equipment is in each of the possible modes of operation.

5-6.1.12 Contractor's Material and Test Certificate. Additional information shall be appended to the Contractor's Material and Test Certificate described in Section 1-12 as follows:

(a) Certification that all auxiliary devices, such as heat pumps, circulating pumps, heat exchangers, radiators, and luminaires have a pressure rating of 175 or 300 psi (12.1 or 20.7 bars).

(b) All components of sprinkler system and auxiliary system have been pressure tested as a composite system in accordance with 1-11.3, Hydrostatic Tests.

(c) Water flow tests have been conducted and water flow alarms have operated while auxiliary equipment is in each of the possible modes of operation.

(d) With auxiliary equipment tested in each possible mode of operation and with no flow from sprinklers or test connection, water flow alarm signals did not operate.

(e) Excess temperature controls for shutting down the auxiliary system have been properly field tested.

6

OUTSIDE SPRINKLERS FOR PROTECTION AGAINST EXPOSURE FIRES

6-1 Water Supply and Control.

6-1.1 Water Supply.

6-1.1.1* Sprinklers installed for protection against exposure fires shall be supplied from a standard water supply as defined in Chapter 2, or other supply such as manual valves, pumps or fire department connections when approved by the authority having jurisdiction.

A-6-1.1.1 The water supply should be capable of furnishing total demand for all exposure sprinklers operating simultaneously for protection against the exposure fire under consideration for a duration of not less than 60 minutes.

6-1.1.2 When automatic systems of sprinklers are installed, water supplies shall be from an automatic source.

6-1.1.3 When water supply feeds other fire protection appliances, it shall be capable of furnishing total demand for such appliances as well as the outside sprinkler demand.

6-1.1.4 When fire department connections are used for water supply, they shall be so located that they will not be affected by the exposing fire.

There are two purposes for the installation of sprinklers around the outer perimeter of a building which may be exposed to fire from an adjacent building or structure. The first is to prevent radiated or convected heat from a fire from entering the building through windows, doors, or other openings in the exposed walls. The second is to deter rupture or destruction of glazing or other weather protecting membrane and to deter ignition of or heat damage to combustible sheathing, eaves, cornices, or other exposed surfaces.

Rapidity of water application should be ensured by automatic actuation of the outside sprinkler system. Delayed application of water to heated glass surfaces may cause breakage from thermal shock. Delayed application of water to combustible or heat sensitive surfaces, such as plastic, may contribute to unnecessary fire or heat damage to these surfaces.

Water should be continuously delivered to the outside sprinkler system at the required design rate for the time or duration of exposure expected from fire in the adjacent building or structure or other exposure source such as outside flammable liquid tanks, etc. Time may be in excess of 60 minutes if fire department operations cannot be expected to quickly bring streams to play on the exposed surface and cannot promptly arrest the exposing fire.

The simultaneous water supply requirements expected for the fire department and for automatic sprinkler systems in the exposing and exposed building should be added to the expected demand of the outside sprinklers in order to determine that the water supply will be of adequate rate and duration.

6-1.2 Control.

6-1.2.1   Each system of outside sprinklers shall have an independent control valve. Where more than one valve is required, the division between sprinklers on each valve shall be vertical and not horizontal, except as noted in Appendix B-6-2.3.

6-1.2.2   Manually controlled open sprinklers shall be used only where constant supervision is present.

Division of sprinkler piping and sprinklers into systems should anticipate the extent of exposure from each adjacent building or structure if it should become involved in fire. Further division of sprinkler piping may be necessary if the surface to be protected is quite large. The water supply may limit the area that can be protected. The exposing fire may be assumed to be limited to a smaller area of heat radiation and convection than the whole of an adjacent building side if the exposing building is compartmented. Also, if the exposing building is sprinklered, it is not considered to cast exposure.

Flames from an adjacent building or structure will be convected upward in a rectangular or plume area. Each outside sprinkler sytem should be arranged to simultaneously discharge water over the corresponding rectangular or plume area, assuming that radiation will occur from ground level upward.

6-1.2.3   Automatic systems may be of the open or closed sprinkler head type. Closed sprinklers in areas subject to freezing shall be on dry-pipe or nonfreezing systems when not prohibited by local public health authorities.

6-1.2.4* Automatic systems of open sprinklers shall be controlled by the operation of fire detection devices designed for the specific application.

Where automatic actuation of outside sprinkler systems cannot be provided, the rapidity of operation should be ensured by manual operation of strategically located and identified valves. Responsible personnel who have been instructed in the operation and importance of these sytems should be constantly in attendance in the building or the premises.

A-6-1.2.4 Spacing of approved fire detection devices should not exceed 30 ft (9.1 m) apart on buildings of less than three stories in height, and not exceed 40 ft (12.1 m) apart on buildings three or more stories in height. On buildings in excess of eight stories in height, there shall be at least one line of fire detectors for each eight stories with fire detectors staggered. One line of fire detectors should be located close up under the cornice, eave, or outside parapet.

Approved fire detection devices may be of the fixed temperature, rate-anticipation or rate-of-rise type. They should be located on the exterior of the exposed building in such a manner as to rapidly receive and detect radiation or convected heat from the burning building or structure. Spacing and location of detectors should be closer than that specified if there is any obstruction to line-of-sight view of surfaces of exposing building or structure protected by the outside sprinkler system.

6-2 System Components.

6-2.1* Valves.

A-6-2.1 Valves should be so located as to be easily accessible.

6-2.1.1 Control valves shall be of the approved indicating type and shall be distinctively marked by letters not less than ½ in. (13 mm) high to clearly explain their use.

The control valves of automatically actuated outside sprinkler systems should be kept open to ensure that water will be immediately discharged from all operating sprinklers when radiation from fire impinges on the protected building's exterior surface. These valves should be supervised in accordance with 3-14.2.3 and identified in accordance with 3-14.3, for example:

"THIS VALVE SHOULD BE KEPT OPEN AT ALL TIMES. THIS VALVE CONTROLS THE WATER SUPPLY TO THE OUTSIDE SPRINKLER SYSTEM LOCATED ON THE EXTERIOR OF THE _____ WALL OF THIS BUILDING."

When a manually controlled outside sprinkler system is provided, the location of the system controlled by the valve and the importance of rapid opening of the valve should be included in the wording of the sign, for example:

"THIS VALVE MUST BE OPENED FULLY TO SUPPLY WATER TO SPRINKLERS LOCATED ON THE OUTSIDE OF THE _____WALL OF THIS BUILDING. IMMEDIATE OPERATION WILL RETARD SPREAD OF FIRE TO THIS BUILDING FROM THE ADJACENT BUILDING/STRUCTURE."

6-2.1.2 Drain Valve.   Each system of outside sprinklers shall have a separate drain valve installed on system side of each control valve. Drain valves shall be in accordance with 3-11.2, except that in no case shall valves be smaller than 1 in.

6-2.1.3 Check Valves.   When sprinklers run on two adjacent sides of a building, protecting against two separate and distinct exposures, with separate control valves for each side, the end lines shall be connected together with check valves located so that one sprinkler around the corner will operate. The intermediate pipe between the two check valves shall be arranged to drain. As an alternate solution, an additional sprinkler shall be installed on each system located around the corner from the system involved.

The "trapped" section of pipe and sprinklers located between two check valves must be drained automatically to prevent freezing of water in piping after system operation. This may be accomplished by providing a down-turned tee outlet and ⅜-in. orifice sprinkler located in the piping at a low point. If the ⅜-in. orifice sprinkler is not opened after sprinkler system operation, it must be removed to drain the piping and then replaced.

6-2.1.4   When one exposure affects two sides of the protected structure, the system shall not be subdivided between the two sides, but rather shall be arranged to operate as a single system.

Heat transfer from one building or structure to another is accomplished by convection or by radiation.

Radiation is direct along a line-of-sight. Convection is via fire gases, which may be flaming or deflected by drafts or wind. Additional sprinklers may be required to discharge on surfaces other than perpendicular to the line-of-sight to radiant heat sources. In these cases, outside sprinkler systems should be extended beyond the wall area directly opposite the exposing building or structure.

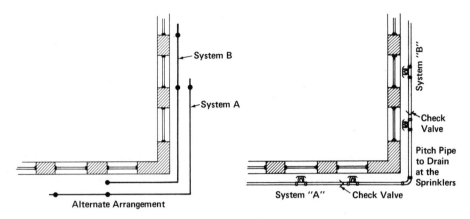

Figure 6-2.1.4   Arrangement of Check Valves.

6-2.2 Pipe and Fittings.   Approved corrosion-resistant pipe and fittings shall be used for the equipment as far back as the control valve on the water supply.

Outside sprinkler systems will be exposed to the elements of weather. The exterior of the piping will be subject to corrosion. The interior of the piping will be subject to corrosion if the sprinklers are open as in a manually or automatically controlled deluge system. Pipe scale formed in the interior of the system could plug the sprinklers when the system is activated. Galvanized pipe and fittings are normally used for this application.

6-2.3 Strainers.   An approved strainer shall be provided in the riser or feed main which supplies sprinklers having orifices smaller than ⅜ in. (9.5 mm).

Scale, mud, or other debris may be transmitted from the water supply by the rapidly moving water of a deluge or open sprinkler system. Sprinklers with orifices smaller than ⅜ in. diameter may become plugged. Strainers should be of an approved type, located adjacent to the control valve and capable of being flushed without the necessity of stopping water flow while the outside sprinkler system is operating.

6-2.4 Gage Connections.   Pressure gage shall be installed just below control valve of each system.

6-3 Sprinklers.   Only sprinklers of such type as are approved for window, cornice, sidewall or ridge pole service shall be installed for such use except where adequate coverage by use of other types of approved sprinklers and/or nozzles has been demonstrated. Sprinklers may be of small orifice [¼ in., ⁵⁄₁₆ in. and ⅜ in. (6.4 mm, 7.9 mm and 9.5 mm)] or large orifice [½ in., ⅝ in. and ¾ in. (12.7 mm, 15.9 mm and 19.1 mm)].

Many sprinkler designs approved for use in outside sprinkler systems may be equipped with heat-sensitive fusible elements. This will permit installation of wet-pipe systems in non-freezing climates or dry-pipe systems in areas where freezing of water may occur during winter months.

Other sprinkler designs approved for use in outside sprinkler systems may not be equipped with heat-sensitive fusible elements. Outside sprinkler systems that utilize these types of open sprinkler must be of the deluge type.

The principle of fire protection afforded by outside sprinkler systems is based on the ability of water to carry away heat which is transmitted from fire in an adjacent building or structure. Water is generally transparent to radiated heat but will tend to mix with convected fire gases when transmitted through the water. It is necessary that the water issuing from an outside sprinkler system be forcefully applied to the surface; that the water remain in contact while wetting and running down the surface to be protected; and that the water not be blown away or stream away without wetting all surfaces to be protected. Heat radiated through the water to the surface being protected must be absorbed from the surface by the water. Water spray adjacent to the surface will tend to diffuse the radiant energy and reduce its effect on the surface being protected.

B-6 Outside Sprinklers.

B-6-1 Type.

B-6-1.1 Small orifice sprinklers will normally be used where exposure is light or moderate, area of coverage is small, or where one horizontal line of window sprinklers is installed at each floor level.

B-6-1.2 Large orifice sprinklers should be used where exposure is severe, or where one horizontal line of sprinklers is used to protect windows at more than one floor level.

B-6-2 Window Sprinklers.

B-6-2.1 When exposure hazard is light or moderate, and only one horizontal line of sprinklers is installed, the sprinklers should have ⅜-in. (9.5-mm) orifices. Where conditions require more than one line of sprinklers, the sprinklers should have orifices as shown in the following table:

Table B-6-2.1

	2 Lines	3 Lines	4 Lines	5 Lines	6 Lines
Top line	⅜ in.	⅜ in.	⅜ in.	⅜ in.	⅜ in.
Next below	⁵⁄₁₆ in.	⁵⁄₁₆ in.	⅜ in.	⅜ in.	⅜ in.
Next below		¼ in.	⁵⁄₁₆ in.	⁵⁄₁₆ in.	⁵⁄₁₆ in.
Next below			¼ in.	⁵⁄₁₆ in.	⁵⁄₁₆ in.
Next below				¼ in.	¼ in.
Next below					¼ in.

For SI Units: 1 in. = 25.4 mm.

The definition of exposure severity is discussed in Appendix A of NFPA 80A-1980, *Recommended Practice for Protection of Buildings from Exterior Fire Exposures*. Table 2-2.4(a) of NFPA 80A provides a guide for assessing severity of the radiation from the burning, exposing building. Other factors, such as separation distance between buildings and structures, also affect the rates of heat radiation and water application required for the exposed surface.

B-6-2.2 When there are over six horizontal rows of windows, sprinklers over the first story may be omitted. Sprinklers may also be omitted over second story windows if field test indicates wetting of all surfaces.

B-6-2.3 If over six lines are used, system should be divided horizontally with independent risers.

B-6-2.4 Large orifice sprinklers may be used for protecting windows in two or three stories from one line of sprinklers. This will be determined by window and wall construction, such that all parts of the windows and frames will be thoroughly wetted by a single line of sprinklers.

B-6-2.5 For buildings not over three stories in height, one line of sprinklers will often be sufficient, located at the top story windows. For buildings more than three stories in height, a line of sprinklers may be used in every other story beginning at the top. With an odd number of stories, the lowest line can protect the first three stories. When several lines are used, the orifice should be decreased one size for each successive line below the top. In no case should an orifice less than ½ in. (12.7 mm) be used.

Run-down of water applied to windows at upper levels is expected to cumulatively protect the lower level(s) of windows. The designer of the outside sprinkler system should be assured that water will flow over all obstacles in the path to lower level(s) of windows and that water will "sheet" across frames and glazing of lower windows.

B-6-2.6   For windows not exceeding 5 ft (1.5 m) wide protected by small orifice sprinklers, one sprinkler should be placed at center near top, so located that water discharged therefrom will wet the upper part of the window, and by running down over the sash and glass wet the entire window. This may ordinarily be accomplished by placing one sprinkler in the center with deflector about on a line with the top of the upper sash and 7 in., 8 in. and 9 in. (178 mm, 203 mm and 229 mm) in front of the glass, with windows 3 ft, 4 ft and 5 ft (0.9 m, 1.2 m and 1.5 m) wide respectively. When windows are over 5 ft (1.5 m) wide, or where mullions interfere, two or more sprinklers should be used.

B-6-2.7   When windows are 3 ft (0.9 m) or less in width, a size smaller orifice than required by B-6-2.1 may be used, but in no case smaller than ¼ in. (6.4 mm).

B-6-2.8   For windows up to 5 ft (1.5 m) wide protected by large orifice sprinklers, use one ½-in. (12.7-mm) sprinkler at center of each window. For windows from 5 ft to 7 ft (1.5 m to 2.1 m) wide, use ⅝-in (15.9-mm) sprinkler at center of each window. For windows from 7 ft to 9 ½ ft (2.1 m to 2.9 m) wide, use one ¾-in. (19.1-mm) sprinkler at center of each window. For windows from 9 ½ ft to 12 ft (2.9 m to 3.7 m) wide, use two ½-in (12.7-mm) sprinklers at each window.

B-6-2.9   Large orifice wide deflector sprinklers should be placed with deflectors 2 in. (50.8 mm) below top of sash and 12 in. to 15 in. (305 mm to 581 mm) out from the glass. When face of glass is close to exterior wall, cantilever brackets or similar type hangers may be used to maintain the window sprinklers 12 in. to 15 in. (505 mm to 381 mm) out from the glass.

The designer may utilize water spray patterns furnished by the sprinkler manufacturer to be assured that the location of the sprinkler and the orifice sizes selected will provide adequate density and complete coverage of the surface to be protected. The addition of a wetting agent to the water will assist in obtaining complete coverage intended in design.

B-6-3 Cornice Sprinklers.

B-6-3.1   The discharge orifice should be at least ⅜ in. (9.5 mm) in diameter except, when exposure is severe, ½-in. or ⅝-in. (12.7-mm or 15.9-mm) cornice sprinklers should be installed.

B-6-3.2   Sprinklers should not be over 8 ft (2.4 m) apart, except as noted in B-6-3.6. Projecting beams or other obstructions may make additional sprinklers necessary.

B-6-3.3   For cornices with bays up to 8 ft (2.4 m) wide, sprinklers should be placed in center of each bay. With cornices with bays from 8 to 10 ft (2.4 to 3.0 m) wide, sprinkler orifices should be increased one size.

B-6-3.4   Cornice sprinklers should be located with deflectors approximately 8 in. (203 mm) below the roof plank.

B-6-3.5   When wood cornices are 30 in. (762 mm) or less above the windows, cornice sprinklers may be supplied by the same pipe used for window sprinklers.

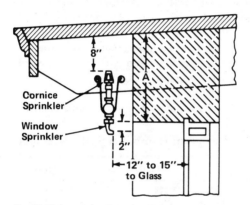

For SI Units: 1 in. = 25.4 mm.

Figure B-6-3.5   Location of Window and Cornice Sprinklers.

Cornice sprinklers may be of the open type, thus necessitating a piping system which relies on deluge operation — either automatically or manually controlled. The water spray patterns developed by the sprinkler manufacturers should be used to establish adequate placement and orifice size.

B-6-3.6   Where the overhang of the cornice is not over 1 ft (0.3 m), window sprinklers should be used and be spaced as follows:

⅜-in. and ½-in. sprinklers . . . . Not over 5 ft apart

⅝-in. sprinklers . . . . Not over 7 ft apart

¾-in. sprinklers . . . . Not over 9 ft apart

For SI Units: 1 in. = 25.4 mm; 1 ft = 0.3048 m.

B-6-3.7   The window sprinklers should be placed above the pipe near the outer edge of the cornice with deflectors not over 3 in. (76 mm) down from the cornice and at such an angle as to throw the water upward and inward.

B-6-3.8   With an overhang of more than 1 ft (0.3 m), cornice sprinklers should be used.

6-4 Piping System.

6-4.1*   Pipe sizes of lines, risers, feed mains, and water supply shall be hydraulically calculated in accordance with Chapter 7 to furnish a minimum of 7 psi (0.5 bar) at any sprinkler with all sprinklers facing the exposure operating, or pipe sizes shall be in accordance with 6-4.2 and 6-4.3.

A-6-4.1   Hydraulic calculations should include all other fire protection systems or devices, such as inside sprinklers and hydrants, to determine that there is no danger of impairing their operation.

6-4.2   Branch line sizes on pipe schedule systems shall be as follows:

Table 6-4.2   Maximum Number of Sprinklers Supplied on Line

Size of Pipe Inches	Orifice Size — In. (mm)						
	¼ (6.4)	³⁄₁₆ (7.9)	⅜ (9.5)	⁷⁄₁₆ (11.1)	½ (12.7)	⅝ (15.9)	¾ (19.1)
1	4	3	2	2	1	1	1
1¼	8	6	4	3	2	2	1
1½		9	6	4	3	3	2
2				5	4	4	3

6-4.3   Risers and feed main sizes on pipe schedule systems shall be as follows for central feed risers:

Pipe Size	Number of Sprinklers		
	⅜ in. (9.5 mm) or smaller orifice	½ in. (12.7 mm) orifice	¾ in. (19.1 mm) orifice
1½	6	3	2
2	10	5	4
2½	18	9	7
3	32	16	12
3½	48	24	17
4	65	33	24
5	120	60	43
6		100	70

The piping should be sized and the water supply characteristic should be established for wet-pipe, dry-pipe, and deluge-type outside sprinkler systems. Such calculations may indicate the need for larger water supplies, the division of systems, or the up-sizing of pipe in order to simultaneously deliver water from all operating sprinklers to adequately absorb both radiated and convected heat from an adjacent burning structure.

6-5 Testing and Flushing.

6-5.1 Tests.

6-5.1.1   All piping shall be tested hydrostatically as specified in 1-11.3.

6-5.1.2   Operating tests shall be made of the system when completed, except where such tests may risk water damage.

6-5.2 Flushing.   Flushing shall be conducted in accordance with 1-11.2.

Operating tests should reveal that all surfaces to be protected from radiant and convective heat are uniformly wetted with adequate discharge pressure from the most remote sprinkler. Water shall be so applied that the effects of wind and drafts will not seriously detract from the intended water discharge and patterns of the sprinklers and from the wetting of all surfaces to be protected.

7

HYDRAULICALLY DESIGNED
SPRINKLER SYSTEMS

7-1 General.

7-1.1 Definition.

7-1.1.1 A hydraulically designed sprinkler system is one in which pipe sizes are selected on a pressure loss basis to provide a prescribed density [gal per min per sq ft $(L/min)/m^2$] distributed with a reasonable degree of uniformity over a specified area. This permits the selection of pipe sizes in accordance with the characteristics of the water supply available. The stipulated design density and area of application will vary with occupancy hazard.

To hydraulically design a sprinkler system, the following must be known:

1) The design density and the assumed area of sprinkler operation. This information is contained in Table 2-2.1(B) for light, ordinary (Groups 1, 2 & 3) and extra (Groups 1 & 2) hazard occupancy. Other NFPA standards that prescribe density and area of operation are NFPA 231, *Standard for General Storage*; NFPA 231C, *Standard for Rack Storage of Materials*; NFPA 231D, *Standard for Storage of Rubber Tires*; and NFPA 409, *Standard on Aircraft Hangars*. Some of the major insurance companies also have publications that specify density and area of application for special hazards for properties insured by them.

2) The characteristics of the water supply available. The types of acceptable water supplies as stated in Chapter 2 are:

a) Connections to water works systems — Information on the characteristics of the water supply may be obtained from the water department, fire department, insurance companies, or by conducting a water flow test as described in Section B-2-1.

b) Gravity tanks — The available pressure is determined by the height of the tank and the friction loss, based on the required flow, is easily calculated. So the determination of the water supply characteristics is a simple matter.

For more information, refer to NFPA 22, *Standard for Water Tanks for Private Fire Protection*.

c) Pumps — The fire pump curve determines the characteristics of the water supply. When pumps take suction from tanks, the elevation of the bottom of the tank with relationship to the pump must be considered. Where suction is from water mains, the fire pump curve is added to the water supply curve for the water main.

For more information, refer to NFPA 20, *Standard for the Installation of Centrifugal Fire Pumps.*

d) Pressure tanks — In this case the designer establishes the characteristics required and chooses a tank of proper volume and establishes the air pressure from the procedure given in A-2-6.3.

For more information, refer to NFPA 22, *Standard for Water Tanks for Private Fire Protection.*

This standard does not require that a factor of safety be applied to the water supply characteristics curve. A prudent designer may choose to leave some margin below the curve, particularly when the water works system comprises the sole supply. This margin would depend on the accuracy of the flow test information and the possibility of a reduction in the water pressure or volume, in the foreseeable future, due to changes in the water works system.

7-1.1.2*   The design basis for such a system or addition to an existing system supersedes the rules in the sprinkler standard governing pipe schedules, except that all systems continue to be limited by area, and pipe sizes shall be no less than 1-in. nominal for ferrous piping and ¾-in. nominal for copper tubing. The size of pipe, number of sprinklers per branch line, and number of branch lines per cross main are otherwise limited only by the available water supply. However, sprinkler spacing and all other rules covered in this and other applicable standards shall be observed.

A-7-1.1.2   When additional sprinkler piping is added to an existing system, the existing piping does not have to be increased in size to compensate for the additional sprinklers, provided the new work is calculated and the calculations include that portion of the existing system as may be required to carry water to the new work.

In 1940 the use of ¾-in. steel pipe was eliminated from this standard in order to improve water discharge at end sprinklers and to reduce the danger of clogging. This restriction still prevails in all systems including hydraulically designed systems.

The number of sprinklers per branch line is limited (*see 3-5.1, 3-6.1 and 3-7.1*) on pipe schedule systems, and prior to 1966 the number of branch lines on a cross main was limited to fourteen. These rules do not apply to hydraulically designed systems.

7-1.2*   The installer shall properly identify a hydraulically designed automatic sprinkler system by a permanently attached placard indicating the location, and the basis of design (discharge density over designed area of discharge, including gallons per minute and residual pressure demand at base of riser). Such signs shall be placed at the controlling alarm valve, or dry-pipe valve, for the system containing the hydraulically designed layout.

A-7-1.2   Embossed plastic tape, pencil, ink, crayon, etc. should not be considered permanent markings. The pressure values should be rounded to the nearest psi (0.1 bar) and discharge flow to the nearest 5 gpm (20 L/min) increment. The placard should be secured to the riser with durable wire, chain or equivalent.

This system as shown on .... company

print no. ....dated ....

for ....

at .... contract no.............

is designed to discharge at a rate of ....gpm

(L/min) per sq ft of floor area over a maximum area of

.... sq ft (m²) when supplied with

water at a rate of .... gpm (L/min)

at .... psi (bars) at the base of the riser.

Figure A-7-1.2   Sample Nameplate.

The information contained on the placard is of vital importance in accessing the capability of the system in controlling fires as the occupancy of the building changes or the water supply weakens over the years. This information is necessary in hydraulically designing revisions or additions to the system.

7-2 Information Required.

7-2.1 Basic Design Information.   Basic design criteria for hydraulically designed sprinkler systems shall be obtained from this or other applicable standards. Where no standards exist, the authority having jurisdiction shall be consulted.

7-2.2 Sprinkler System Requirements.   The following information shall be included when applicable:

(a) Area of water application . . . . sq ft

(b) Minimum rate of water application (density) . . . . . gpm/sq ft

(c) Area per sprinkler. . . . . sq ft

(d) Allowance for inside hose and outside hydrants . . . . . gpm

(e) Allowance for in-rack sprinklers . . . . gpm

7-2.3* Water Supply Information.   The following information shall be included: water flow data with existing or proposed water supply, dead end or circulating.

(a) Location and elevation of static and residual test gage with relation to the riser reference point

(b) Flow location

(c) Static pressure, psi

(d) Residual pressure, psi

(e) Flow, gpm

(f) Date

(g) Time

(h) Test conducted by or information supplied by. . . .

A-7-2.3   Designers should consult with the authority having jurisdiction on the water supply to be used in system calculations prior to system design and calculation.

7-2-4 Information Required on the Drawings.

7-2.4.1   In addition to the requirements of Section 1-9, the drawings shall also contain the information mentioned in the remainder of 7-2.4.

7-2.4.2 Hydraulic Reference Points.   Reference points may be shown by a number and/or letter designation and shall correspond with comparable reference points shown on the hydraulic calculation sheets.

Hydraulic reference points are important to the designer, as they help ensure that parts of the system are not overlooked. They are essential to the plan checker, particularly in the case of gridded systems. Good coordination between calculations and plans is essential should it be necessary to re-evaluate the system several years after design or should additions to the system become necessary.

7-2.4.3 Sprinklers.   Description of sprinklers used.

7-2.4.4 System Design Criteria.   The minimum rate of water application (density), the design area of water application, in-rack sprinkler demand, and the water required for hose streams both inside and outside shall be included.

The system design criteria must be on the drawings for the benefit of the plan reviewer and for future reference. This would include —
   a) Density
   b) Area of water application
   c) In-rack sprinkler demand — if applicable
   d) Inside hose demand — if applicable
   e) Outside hose demand — this would be zero when applying 2-2.1.2.6.

7-2.4.5 Actual Calculated Requirements.   The total quantity of water and the pressure required shall be noted at a common reference point for each system.

The actual calculated demand is normally referenced to the base of the riser in buildings containing several systems. In single system buildings, the reference point may be at the base of the riser or the point of connection to the city main.

7-2.4.6 Elevation Data.   Relative elevations of sprinklers, junction points and supply or reference points shall be noted.

7-3 Data Sheets and Abbreviations.

7-3.1 General.   Hydraulic calculations shall be prepared on form sheets that include a summary sheet, detailed work sheets, and a graph sheet. (*See copy of typical forms, Figures A-7-3.3 and A-7-3.4.*)

7-3.2 Summary Sheet. The summary sheet shall contain the following information, when applicable:

(a) Date

(b) Location

(c) Name of owner and occupant

(d) Building number or other identification

(e) Description of hazard

(f) Name and address of contractor or designer

(g) Name of approving agency

(h) System design requirements

1. Design area of water application . . . . sq ft

2. Minimum rate of water application (density) . . . . gpm per sq ft

3. Area per sprinkler . . . . sq ft

(i) Total water requirements as calculated including allowance for inside hose and outside hydrants

(j) Water supply information.

7-3.3* Detailed Work Sheets. Detailed work sheets (*for sample work sheet, refer to Figure A-7-3.3*) or computer printout sheets shall contain the following information:

(a) Sheet number

(b) Sprinkler description and discharge constant (K)

(c) Hydraulic reference points

(d) Flow in gpm

(e) Pipe size

(f) Pipe lengths, center to center of fittings

(g) Equivalent pipe lengths for fitting and devices

Contract No. _____   Sheet No. _____ of _____

Name & Location _____

Nozzle Type & Location	Flow in GPM (L/min)	Pipe Size in.	Fitting & Devices	Pipe Eqiv. Length	Friction Loss psi/ft. (bars/m)	Req. Psi. (bars)	Normal Pressure	Notes
	q			lgth.		Pt	Pt	
				ftg.		Pf	Pv	
	Q			tot.		Pe	Pn	
	q			lgth.		Pt	Pt	
				ftg.		Pf	Pv	
	Q			tot.		Pe	Pn	
	q			lgth.		Pt	Pt	
				ftg.		Pf	Pv	
	Q			tot.		Pe	Pn	
	q			lgth.		Pt	Pt	
				ftg.		Pf	Pv	
	Q			tot.		Pe	Pn	
	q			lgth.		Pt	Pt	
				ftg.		Pf	Pv	
	Q			tot.		Pe	Pn	
	q			lgth.		Pt	Pt	
				ftg.		Pf	Pv	
	Q			tot.		Pe	Pn	
	q			lgth.		Pt	Pt	
				ftg.		Pf	Pv	
	Q			tot.		Pe	Pn	
	q			lgth.		Pt	Pt	
				ftg.		Pf	Pv	
	Q			tot.		Pe	Pn	
	q			lgth.		Pt	Pt	
				ftg.		Pf	Pv	
	Q			tot.		Pe	Pn	
	q			lgth.		Pt	Pt	
				ftg.		Pf	Pv	
	Q			tot.		Pe	Pn	
	q			lgth.		Pt	Pt	
				ftg.		Pf	Pv	
	Q			tot.		Pe	Pn	
	q			lgth.		Pt	Pt	
				ftg.		Pf	Pv	
	Q			tot.		Pe	Pn	
	q			lgth.		Pt	Pt	
				ftg.		Pf	Pv	
	Q			tot.		Pe	Pn	
	q			lgth.		Pt	Pt	
				ftg.		Pf	Pv	
	Q			tot.		Pe	Pn	
	q			lgth.		Pt	Pt	
				ftg.		Pf	Pv	
	Q			tot.		Pe	Pn	
	q			lgth.		Pt	Pt	
				ftg.		Pf	Pv	
	Q			tot.		Pe	Pn	
	q			lgth.		Pt	Pt	
				ftg.		Pf	Pv	
	Q			tot.		Pe	Pn	
	q			lgth.		Pt	Pt	
				ftg.		Pf	Pv	
	Q			tot.		Pe	Pn	

Figure A-7-3.3   Sample Work Sheet.

(h) Friction loss in psi per ft of pipe

(i) Total friction loss between reference points

(j) In-rack sprinkler demand

(k) Elevation head in psi between reference points

(l) Required pressure in psi at each reference point

(m) Velocity pressure and normal pressure if included in calculations

(n) Notes to indicate starting points, reference to other sheets or to clarify data shown

(o)* Sketch to accompany gridded system calculations to indicate flow quantities and directions for lines with sprinklers operating in the remote area. [*See Figure A-7-3.3(o).*]

The worksheet on page 291 is just one of several used in the sprinkler industry today. Many of the computer print-outs are in this same form.

Figure A-7-3.3(o) depicts a gridded system that has been calculated. The dotted lines outline the calculated area and the numbers in circles are reference points. The arrows on the pipe indicate the direction of flow and the numbers alongside indicate quantity in gallons per minute.

A sketch of this type must accompany hydraulic calculations for gridded systems to facilitate checking.

7-3.4* Graph Sheet.   Water supply curves and system requirements, plus hose and in rack sprinkler demand when applicable, shall be plotted on semi-logarithmic graph paper ($Q^{1.85}$) so as to present a graphic summary of the complete hydraulic calculation.

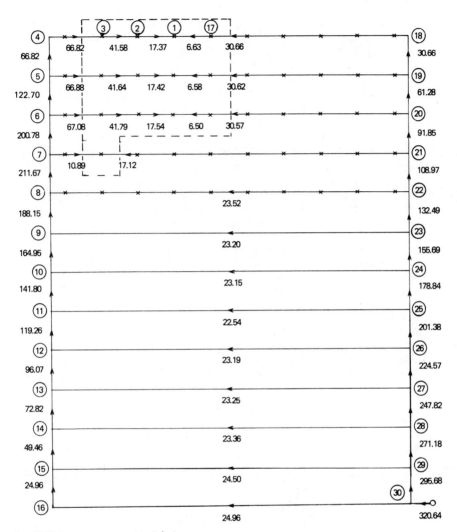

For SI Units: 1 gpm = 3.785 L/min.

Figure A-7-3.3(o)   Flow Quantities and Direction.

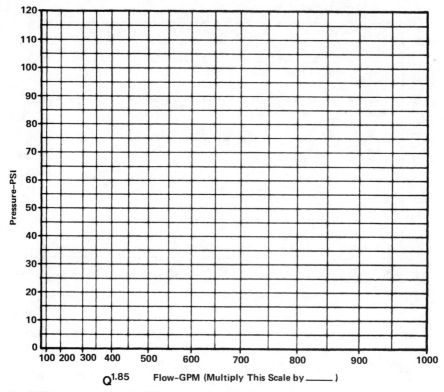

For SI Units: 1 gpm = 3.785 L/min; 1 psi = 0.0689 bar.

Figure A-7-3.4   Sample Graph Sheet.

This graph sheet is used to plot the characteristics of the water supply and the sprinkler system demand. [*See Figure A-7-3.4(d)*]. It is referred to as hydraulic graph paper. The scale along the abscissa must be to the 1.85 power since in the Hazen and Williams formula pressure (P) in pounds per square inch is proportional to the flow (Q) in gallons per minute to the 1.85 power.

7-3.5 Abbreviations and Symbols.   The following standard abbreviations and symbols shall be used on the calculation form:

Symbol or Abbreviation	Item
p	Pressure in psi
gpm	U.S. Gallons per minute
q	Flow increment in gpm to be added at a specific location

Q	Summation of flow in gpm at a specific location
P_t	Total pressure in psi at a point in a pipe
P_f	Pressure loss due to friction between points indicated in location column
P_e	Pressure due to elevation difference between indicated points. This can be a plus value or a minus value. Where minus, the (—) shall be used; where plus, no sign need be indicated
P_v	Velocity pressure in psi at a point in a pipe
P_n	Normal pressure in psi at a point in a pipe
E	90° Ell
EE	45° Ell
Lt.E	Long Turn Elbow
Cr	Cross
T	Tee — flow turned 90 degrees
GV	Gate Valve
BV	Butterfly Valve
Del V	Deluge Valve
ALV	Alarm Valve
CV	Swing Check Valve
WCV	Butterfly (Wafer) Check Valve
St	Strainer
psi	Pounds per square inch
v	Velocity of water in pipe in feet per second

7-4 Calculation.

7-4.1 Formulas.

7-4.1.1 Friction Loss Formula.   Pipe friction losses shall be determined on the basis of Hazen and Williams formula.

$$p = \frac{4.52 \ Q^{1.85}}{C^{1.85} \ d^{4.87}}$$

where p is the frictional resistance in pounds pressure per square inch per foot of pipe, Q is the gallons per minute flowing and d is the actual internal diameter of pipe in inches with C as the friction loss coefficient.

$$\text{For SI Units: } P_m = 6.05 \times \frac{Q_m^{1.85}}{C^{1.85} \ d_m^{4.87}} \times 10^5$$

P_m is the frictional resistance in bars per meter of pipe, Q_m is the flow in L/min and d_m is the actual internal diameter in mm with C as the friction loss coefficient.

This is the most popular of the exponential formulas and was developed by experiment and experience. The friction coefficient (C) in this formula is constant for a specific type or roughness of pipe, and is independent of velocity. The "C" factors are shown in Table 7-4.3.1.4.

Formal Interpretation

Question: Is it acceptable to calculate sprinkler systems without any limit on flow velocities?

Answer: Yes, NFPA 13 does not limit the velocity of water in pipe.

7-4.1.2 Velocity Pressure Formula.   Velocity pressure shall be determined on the basis of the formula

$P_v = 0.001\ 123\ Q^2/D^4$
P_v = velocity pressure psi.

where:

Q = flow in gpm.
D = the inside diameter in inches.

For SI units: 1 in. = 25.4 mm; 1 gal = 3.785 L; 1 psi = 0.0689 bar.

NFPA 15, *Standard for Water Spray Fixed Systems*, defines velocity pressure as "a measure of energy required to keep the water in a pipe in motion." The basic formula for velocity pressure is:

$$P_V = 0.433\ \frac{V^2}{2_g}$$

Where: Pv = Velocity pressure in PSI
V = Velocity in FPS
g = 32.2 FPS/S

By substituting $Q \div A$ for V in this formula, and using proper units, the more usable formula that appears in the text is found.

7-4.1.3   Normal pressure P_n shall be determined on the basis of the formula

$$P_n = P_t - P_v$$

where:

P_t = total pressure in psi (bars)
P_v = velocity pressure in psi (bars)

See A-7-4.3.1.7

7-4.1.4 Hydraulic Junction Points. For gridded systems only, pressures at hydraulic junction points shall balance within 0.5 psi (0.03 bar). The highest pressure at the junction point shall be carried into the calculations.

The value of 0.5 PSI was selected as a reasonable degree of accuracy for balancing at hydraulic junction points. This is easily obtained by most computer programs without an excessive number of iterations.

7-4.2 Equivalent Pipe Lengths of Valves and Fittings.

7-4.2.1 Table 7-4.2 shall be used to determine the equivalent length of pipe for fittings and devices unless manufacturer's test data indicate other factors are appropriate. For saddle type fittings having friction loss greater than that shown in Table 7-4.2, the increased friction loss shall be included in hydraulic calculations.

The fitting and valve losses shown in Table 7-4.2 are calculated values that have been rounded to a whole number for convenience. The exception is the swing check valve where several makes of valves were averaged to produce the numbers in the tables. Where possible, the designer should use the friction loss value for the specific check valve that is to be installed.
Some 1-in. saddle type fittings have a friction loss equivalent of 21 ft of pipe as opposed to the 5 ft shown in the table. An appreciable error would be introduced into the calculations if this increased loss is ignored.

7-4.2.2 Use Table 7-4.2 with Hazen and Williams C = 120 only. For other values of C, the values in Table 7-4.2 shall be multiplied by the factors indicated below:

Value of C	100	120	130	140	150
Multiplying Factor	0.713	1.00	1.16	1.33	1.51

(This is based upon the friction loss through the fitting being independent of the C factor available to the piping.)

The loss through a fitting is constant and as the parenthetical note indicates is independent of the C-Factor. This may be illustrated for a 2 in. tee (equivalent length 10) using C = 100 and 120 and a flow of 100 gpm.
$P_f (100) = 10 \times 0.713 (2.63 \times 10^{-5} \times 100^{1.85}) = 0.94$ psi
$P_f (120) = 10 \times (1.87 \times 10^{-5} \times 100^{1.85}) = 0.94$ psi
The numbers in the brackets are the Hazen and Williams formula for 2-in. schedule 40 pipe with all constants combined in one multiple.

Table 7-4.2   Equivalent Pipe Length Chart

Fittings and Valves	¾ in.	1 in.	1¼ in.	1½ in.	2 in.	2½ in.	3 in.	3½ in.	4 in.	5 in.	6 in.	8 in.	10 in.	12 in.
45° Elbow	1	1	1	2	2	3	3	3	4	5	7	9	11	13
90° Standard Elbow	2	2	3	4	5	6	7	8	10	12	14	18	22	27
90° Long Turn Elbow	1	2	2	2	3	4	5	5	6	8	9	13	16	18
Tee or Cross (Flow Turned 90°)	3	5	6	8	10	12	15	17	20	25	30	35	50	60
Butterfly Valve	—	—	—	—	6	7	10	—	12	9	10	12	19	21
Gate Valve	—	—	—	—	1	1	1	1	2	2	3	4	5	6
Swing Check*	—	5	7	9	11	14	16	19	22	27	32	45	55	65

Fittings and Valves Expressed in Equivalent Feet of Pipe.

For SI Untis: 1 ft = 0.3048 m.

*Due to the variations in design of swing check valves, the pipe equivalents indicated in the above chart to be considered average.
NOTE: This table applies to all types of pipe listed in Table 7-4.3.1.4.

7-4.2.3 Specific friction loss values or equivalent pipe lengths for alarm valves, dry-pipe valves, deluge valves, strainers and other devices shall be made available to the authority having jurisdiction.

7-4.3* Calculation Procedure.

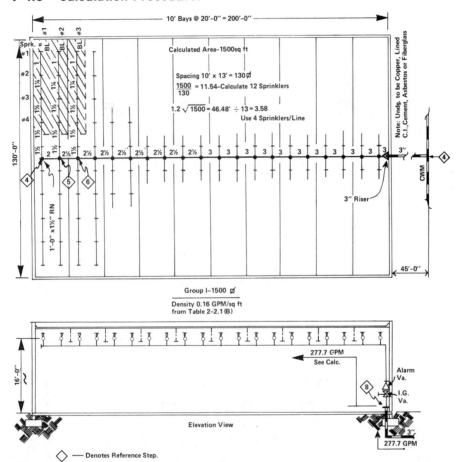

For SI Units: 1 ft = 0.3048 m; 1 ft² = 0.0929 m²; 1 gpm = 3.785 L/min.

Figure A-7-4.3(a).

This figure shows the floor plan and elevation of the sprinkler system in a small building (130 ft × 200 ft) which has been hydraulically designed. The density of 0.16 gpm/sq ft and the calculated area of 1500 sq ft are both obtained from Table 2-2.1(B) for an ordinary hazard Group 1 occupancy.

To determine the number of sprinklers to be calculated, it is necessary to divide the calculated area by the coverage per sprinkler which, in this case, is 1500 ÷ 130 = 11.54 sprinklers, which is rounded off to 12. The number of sprinklers to calculate per branch line is determined by dividing $1.2 \sqrt{A}$ (*see 7-4.3.1*) by the distance between sprinklers on the line which, in this case, is $1.2 \sqrt{1500} \div 13 = 3.58$. Therefore, use 4 sprinklers.

The calculated area is the hydraulically most remote rectangle encompassing four sprinklers on the branch line and three branch lines to equal the total required of twelve. This area is indicated by cross-hatching, and since this system is symmetrical about the cross main, this area may be located as shown or could be in the lower left corner of the building.

HYDRAULIC CALCULATIONS

FOR

ABC COMPANY

_____

_____

CONTRACT NO. *4001*

DATE *1-7-75*

DESIGN DATA—

OCCUPANCY CLASSIFICATION *ORD. GR. 1*

DENSITY *0.16* GPM/SQ. FT.

AREA OF APPLICATION *1500* SQ. FT.

COVERAGE PER SPRINKLER *130* SQ. FT.

NO. OF SPRINKLERS CALCULATED *12*

TOTAL WATER REQUIRED *650* GPM.
INCLUDING HOSE STREAMS.

NAME OF CONTRACTOR_____

NAME OF DESIGNER_____

AUTHORITY HAVING JURISDICTION_____

Figure A-7-4.3(b).

This sheet is normally used as the cover sheet for the calculations and contains the information required by 7-3.2.

CONTRACT NAME _GROUP I 1500 Φ_   SHEET _2_ OF _3_

STEP NO.	NOZZLE IDENT. AND LOCATION	FLOW IN G.P.M.	PIPE SIZE	PIPE FITTINGS AND DEVICES	EQUIV. PIPE LENGTH	FRICTION LOSS P.S.I./FOOT	PRESSURE SUMMARY	NORMAL PRESSURE	D=0.16 GPM/Φ NOTES K=5.65	REF. STEP
1	1 BL-1	q	1		L 13.0	C=120	Pt 13.6	Pt	q=130X.16=20.8	
		Q 20.8			F		Pe	Pv		
					T 13.0	.140	Pf 1.8	Pn		
2	2	q 22.2	1¼		L 13.0		Pt 15.4	Pt	q=5.65√15.4	
		Q 43.0			F		Pe	Pv̇		
					T 13.0	.141	Pf 1.8	Pn		
3	3	q 23.4	1½		L 13.0		Pt 17.2	Pt	q=5.65√17.2	
		Q 66.4			F		Pe	Pv		
					T 13.0	.149	Pf 1.9	Pn		
4	4 DN RN	q 24.7	1½	2T-16	L 20.5		Pt 19.1	Pt	q=5.65√19.1	4
		Q 91.1			F 16.0		Pe	Pv		
					T 36.5	.267	Pf 9.7	Pn		
5	CM TO BL-2	q	2		L 10.0		Pt 28.8	Pt	K=91.1/√28.8	5
		Q 91.1			F		Pe	Pv	K=16.98	
					T 10.0	.079	Pf .8	Pn		
6	BL-2 CM TO BL-3	q 92.4	2½		L 10.0		Pt 29.6	Pt	q=16.98√29.6	6
		Q 183.5			F		Pe	Pv		
					T 10.0	.122	Pf 1.2	Pn		
7	BL-3 CM	q 94.2	2½		L 70.0		Pt 30.8	Pt	q=16.98√30.8	
		Q 277.7			F		Pe	Pv		
					T 70.0	.262	Pf 18.3	Pn		
8	CM TO F/S	q	3	E 5	L 119.0		Pt 49.1	Pt	Pe=15 x .433	8
				AV15	F		Pe 6.5	Pv		
		Q 277.7		GV 1	T 140.0	.091	Pf 12.7	Pn		
9	THRU UNDERGROUND TO CITY MAIN	q	3	E 5	L 50.0	C=150	Pt 68.3	Pt	COPPER	9
				GV 1	F 32.0	TYPE "M"	Pe	Pv	21 X 1.51 = 32	
		Q 277.7		T 15	T 82.2	.069	Pf 5.7	Pn		
		q			L		Pt 74.0	Pt		
					F		Pe	Pv		
		Q			T		Pf	Pn		
		q			L		Pt	Pt		
					F		Pe	Pv		
		Q			T		Pf	Pn		
							Pt			

Figure A-7-4.3(c).

In step number one, sprinkler one and branch line (BL) one are identified. The flow for the first sprinkler is determined by multiplying the density by the coverage per sprinkler, which in this case is 0.16 gpm/ft^2 × 130 ft^2 = 20.8 gpm. The pipe size is determined by trial and error. The fitting directly connected to a sprinkler is not included in the calculations [7-4.3.1.4(d)]; therefore, nothing is shown under the "Pipe Fittings and Device" column. The equivalent pipe length is the center to center distance between sprinklers, which in this case is 13 ft. The friction loss (psi/ft) is determined by using the Hazen and Williams formula. When using Schedule 40 pipe and C = 120 the formula becomes:

$P = 5.10 × 10^4 × Q^{1.85} = 0.140$ psi/ft

By multiplying 0.14 psi/ft × 13 ft = 1.8 psi the friction loss from sprinkler one to sprinkler two is determined.

The total pressure (P_t) is determined by using the formula $P_t = (Q ÷ K^2)$ which in this case is $(20.8 ÷ 5.65)^2 = 13.6$ psi.

The pressure at sprinkler two is 13.6 psi plus 1.8 psi friction loss or 15.4 psi. The flow (Q) from sprinkler two is equal to K × $\sqrt{p}$, 5.65 × $\sqrt{15.4}$, or 22.2 gpm. This flow is added to sprinkler one and same procedures followed as above. In step four, the four sprinklers have been calculated, and it is necessary to carry this back to the cross main. The pipe length (L) included is 13.0 ft between sprinklers + 6.5 ft for starter piece plus 1.0 ft for riser nipple for a total of 20.5 ft. The tees at the top and bottom of the riser nipple are included and are equal to 8 ft of pipe each. (*See Figure 7-4.2.*) A "K-Factor" is established for branch line one by dividing the flow (Q) by the square root of the pressure (P) which in this case is 91.1 divided by the square root of 28.8, or 16.98. This "K-Factor" is used in predicting the flow in subsequent branch lines that are identical to branch line one. In Step 8, the pressure due to elevation (Pe) of 6.5 psi is added at one point. This can be done only if the branch lines are basically level as they are assumed to be in this case. In buildings with pitched roofs or branch lines at different elevations, the pressure due to elevation (Pe) would have to be applied where it occurs. In Step 9, please note that copper pipe is used so the "C-Factor" is 150 (*see Table 7-4.3.1.4*) and the fitting and device equivalent lengths are totalled and multiplied by 1.51 (*see 7-4.2.2*).

The city water supply curve is based upon 90 psi static pressure and a 60-psi residual pressure with 1000 gpm flowing.

The system demand curve is started at the 6.5 psi point on the ordinate to allow the pressure due to elevation, Pe, and continued to a point of 277.7 gpm at 74.0 psi. The system demand curve then represents the friction loss in the system. The horizontal dashed line drawn from the above demand point to the city water supply curve represents the amount of water available for hose streams. In this case about 450 gpm are available and only 250 gpm are required. This represents a factor of safety of approximately 7 psi. If this factor of safety is warranted (*see discussion on 7-1.1.1*) the design is complete. If not, some pipe sizes may be reduced to absorb the 7 psi. If the system demand point falls above the city water supply curve, some of the piping must be enlarged so the friction loss will decrease. Other factors that could be used to reduce the pressure to a point at or below the supply curve are sprinkler orifice size and reduced sprinkler spacing. These would normally be used only in higher hazard occupancies.

7-4.3.1*   For all systems the design area shall be the hydraulically most demanding "rectangular area" having a dimension parallel to the branch lines equal to 1.2 times the square root of the area of sprinkler operation corresponding to the density used. This may include sprinklers on both sides of the cross main. Any fractional sprinkler shall be carried to the next higher whole sprinkler.

   The 1966 edition of this standard was the first to include a Chapter 7. The following is from that edition: "The design area shall be the hydraulically most remote area and shall include all sprinklers on both sides of the cross main."
   This meant that, if the system shown in Figure A-7-4.3(a) was to be calculated, the ten sprinklers on the end of the system (five on each side of the cross main) plus two on the next line would be calculated.
   The first mention of gridded systems appeared in the 1975 edition as follows: "7-4.3.1 - Exception No. 2: For gridded systems, the design area shall be the hydraulically most remote area which approaches a square." This definition was adopted since it did not seem logical to calculate all of the sprinklers on one line of a grid as had been done on "Tree" systems since 1966. In the 1978 edition, the wording of this paragraph was drastically revised as follows:
   7-4.3.1* For all systems, the design area shall be the hydraulically most demanding "rectangular area" having a dimension parallel to the branch lines equal to, or greater than, 1.2 times the square root of the area of sprinkler operation corresponding to the density used. Any fractional sprinkler shall be carried to the next higher whole sprinkler.

   This revision recognized that the sprinkler operating area should be the same regardless of the configurations of the sprinkler system piping.

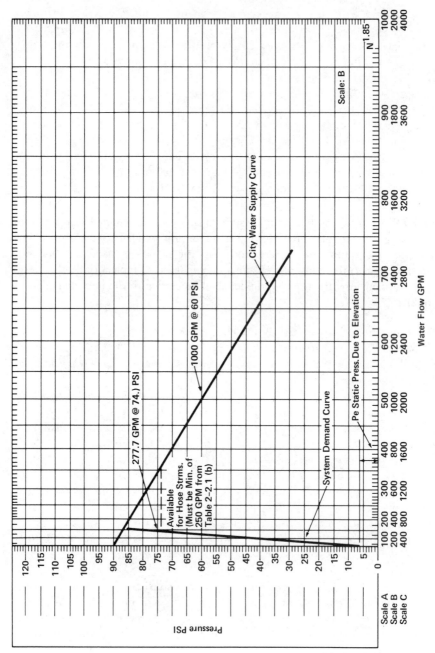

Figure A-7-4.3 (d).

The square area was changed to a rectangular area and the 1.2 $\sqrt{A}$ factor added to ensure that the long side of the rectangle would be along the branch line. In cases of sprinkler spacing of 13 ft on the line and 10 ft between branch lines and where the design area included twelve sprinklers, some designers were calculating three sprinklers on the line (3 × 13 = 39) over four lines (4 × 10). This was almost a perfect square, but the design would probably be inadequate if four sprinklers operated on the branch line; hence, the change.

Exception No. 1: Where the design area under consideration consists of a corridor protected by one row of sprinklers, the maximum number of sprinklers that need be calculated is 5.

Exception No. 1 has been in the standard for several years; however, in 1978 the number of sprinklers was reduced from seven to five. It seemed highly unlikely that seven sprinklers in a corridor in a fully sprinklered building would be opened by one fire — hence, the change. This exception would only be used when the design is based on the area of the largest room with automatic or self-closing doors as described in 2-2.1.2.8. Then the designer would have to calculate the largest room as well as the five sprinklers in the corridor to determine which requires the higher water supply.

Exception No. 2: In systems having branch lines with an insufficient number of sprinklers to fulfill the 1.2 $\sqrt{A}$ requirement, the design area shall be extended to include sprinklers on adjacent branch lines supplied by the same cross main.

Exception No. 2 was added in 1980 for clarification. Figure A-7-4.3.1(b) System D shows an example.

Exception No. 3: Where the design area is based on the largest room as per 2-2.1.2.8 [including the exception to subsection (a)] the above dimensional requirements do not apply.

Exception No. 4: Where the design area under consideration consists of a building service chute supplied by a separate riser the maximum number of sprinklers that need be calculated is 3.

Figure A-7-4.3.1(a)   Example:

Assume a remote area of 1,500 sq ft with sprinkler coverage of 120 sq ft

$$\frac{1500}{120} = 12.5 \quad \text{Therefore calculate 13 sprinklers}$$

$$\sqrt{1500} = 38.73 \times 1.2 = \frac{46.48}{12} = 3.87 \quad \text{Use 4 sprinklers on a line}$$

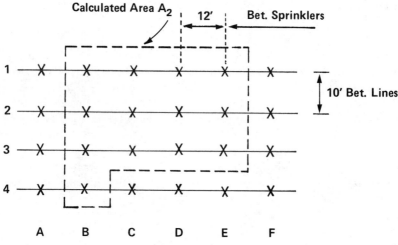

For SI Units: 1 ft = 0.3048 m; 1 ft^2 = 0.0929 m^2.

Figure A-7-4.3.1(a)   Example of Determining the Number of Sprinklers to Be Calculated.

NOTE: Extra sprinkler (or sprinklers) on branch line 4 may be placed in any adjacent location from B to E at the designer's option.

This figure is an example of the number of sprinklers to calculate on a gridded system. In order for this figure to apply to a tree or loop system, the branch line would terminate at the last sprinkler on the right in the calculated area and the cross main would be to the left. Also, the note would only apply to grids. The extra sprinkler on line four would have to be "B" for tree and loop systems as it is closest to the supply and would be hydraulically most demanding.

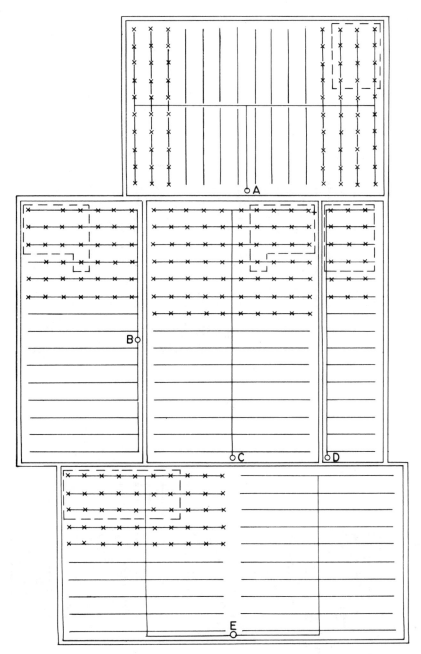

Figure A-7-4.3.1(b)   Example of Hydraulically Most Demanding Area.

System A shows design area for twelve sprinklers with three branch lines and four sprinklers per branch line. Systems B and C indicate the location of the extra sprinkler on the fourth branch line.

System D illustrates 7-4.3.1, Exception 2.

System E illustrates that sprinklers on both sides of the cross main may be required to fulfill the required design area and the $1.2 \sqrt{A}$ restriction on the sprinklers on the branch line.

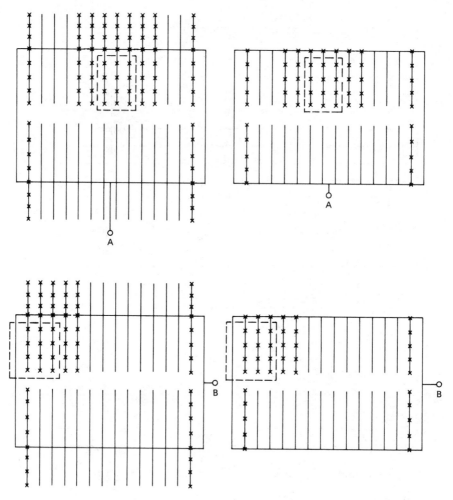

Figure A-7-4.3.1(c)   Example of Hydraulically Most Demanding Area.

This figure illustrates the design area for looped systems with various riser locations and branch line configurations.

7-4.3.1.1* For gridded systems, the designer shall verify that the hydraulically most demanding area is being used. A minimum of two additional sets of calculations shall be submitted to demonstrate peaking of demand area friction loss when compared to areas immediately adjacent on either side along the same branch lines.

In the 1975 standard, the design area was shown as a square area located along the far cross main (not the supply cross main). It was quickly learned that this was not necessarily the hydraulically most remote area so this section was inserted to force the designer to move the design area along the branch line to prove that, in fact, the hydraulically most remote area had been chosen.

Exception: Computer programs which show the peaking of the demand area friction loss shall be acceptable based on a single set of calculations.

Figure A-7-4.3.1.1   Example:

If Area A_2 is selected as the most remote area, submit calculations to show that Areas A_1 and A_3 are subject to less friction loss.

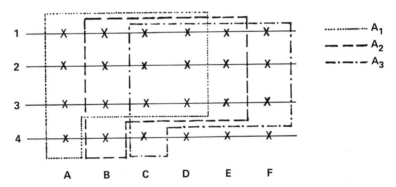

Figure A-7-4.3.1.1   Example of Determining the Most Remote Area for a Gridded System.

7-4.3.1.2* System piping shall be hydraulically designed using design densities and areas of operation in accordance with Table 2-2-1 (B) as required for the occupancies involved.

(a)* The density shall be calculated on the basis of floor area. The area covered by any sprinkler for use in hydraulic design and calculations shall be determined as follows:

The first sentence in (a) has been in the standard for several years and applies to cases where the sprinklers are installed under a sloping roof. The area of coverage per sprinkler is based on the projected area on the floor. Even though the one line in Figure A-7-4.3.1.2 is only 2 ft from the wall, the area of coverage is assumed to be 10 ft × 12 ft or 120 sq ft.

A-7-4.3.1.2(a)   Example. Design Area for Sprinkler 1 (see Figure A-7-4.3.1.2).

1. Along Branch Lines —	2 ft, 10 ft	(0.6 m, 3.0 m)
	S=10	(S=3.0)
2. Between Branch Lines —	2 ft, 12 ft	(0.6 m, 3.7 m)
	L=12	(L=3.7)
3. Design Area —	10×12=120 sq ft	(3.0×3.7)=11.1 m²

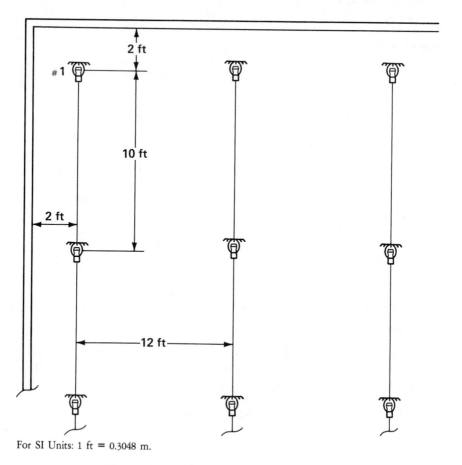

For SI Units: 1 ft = 0.3048 m.

Figure A-7-4.3.1.2   Sprinkler Design Area.

1. Along Branch Lines. Determine distance to next sprinkler (or to wall in case of end sprinkler on branch line) upstream and downstream. Choose larger of either twice the distance to the wall or distance to the next sprinkler. Call this "S."

2. Between Branch Lines. Determine perpendicular distance to branch lines (or to wall in case of the last branch line) on each side of branch on which the subject sprinkler is positioned. Choose the larger of (1) the larger distance to the next branch line, or (2) in the case of the last branch line, twice the distance to the wall. Call this "L."

3. Design Area for Sprinkler = S × L.

Exception: This does not apply to small rooms. (See 4-4.20.)

As stated in the exception to (a) 3, this does not apply to small rooms. In small rooms the area of coverage per sprinkler is based on the area of the room divided by the number of sprinklers in the room.

(b)* When sprinklers are installed above and below a ceiling or in a case where more than 2 areas are supplied from a common set of branch lines, the branch lines and supplies shall be calculated to supply the largest water demand.

A-7-4.3.1.2(b)   This subsection contemplates a ceiling constructed to reasonably assure that a fire on one side of the ceiling will operate sprinklers on one side only. When a ceiling is sufficiently open or of such construction that operation of sprinklers above and below ceiling may be anticipated, the operation of such additional sprinklers should be considered in the calculations.

Formal Interpretation

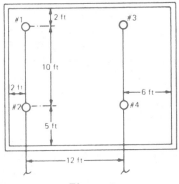

Figure 8.

Question: When hydraulically calculating sprinkler head number one in Figure 8 above, should the area covered by this head be based on 12 ft, 0 in. × 10 ft, 0 in. spacing, or should it be calculated on the actual square foot coverage of 7 ft, 0 in. × 8 ft, 0 in.? This is a difference of 64 square feet.

Answer: 120 sq ft.

(c) When sprinklers are installed above and below temporary obstructions such as overhead doors and such sprinklers are supplied from a common set of branch lines, the branch lines and the supply shall be calculated to supply the sprinklers both above and below the temporary obstruction.

Since it is difficult to predict whether an overhead door will be up or down at the time of fire and, regardless of the door position, whether sprinklers above or below will operate, it was considered more prudent to include the sprinklers below (in the calculated area) with the roof sprinklers.

7-4.3.1.3*   Each sprinkler in the design area and the remainder of the hydraulically designed system shall discharge at a flow rate at least equal to the stipulated minimum water application rate (density). Begin calculations at the hydraulically most remote sprinkler. Discharge at each sprinkler shall be based on the calculated pressure at that sprinkler.

When applying velocity pressure, sometimes the discharge from the second sprinkler from the end of the line may be lower than the end sprinkler. If this is a significant amount (over 3 percent of design flow), then either the pipe sizing should be changed or the end sprinkler discharge increased slightly to compensate.

The last sentence in the section was inserted to prevent discharge averaging over the design area in a gridded system. For example: If it is assumed that twelve sprinklers in the design area discharge 20.8 gpm each, then the grid piping may be simply sized to accommodate this flow. In actuality, however, every sprinkler will be discharging at a slightly different pressure so the rate would be 20.8 gpm plus the hydraulic increase. This increases the sophistication of the calculations.

A-7-4.3.1.3   When it is not obvious by comparison that the design area selected is the hydraulically most remote, additional calculations should be submitted. The most remote area, distance-wise, is not necessarily the hydraulically most remote area.

All systems are not as uniform as the one in Figure A-7-4.3(a), so it may be necessary to calculate several portions of a system in order to properly size the pipe and determine the required water supply.

7-4.3.1.4   Calculate pipe friction loss in accordance with the Hazen and Williams formula with "C" values from Table 7-4.3.1.4.

(a) Include pipe, fittings, and devices such as valves, meters, and strainers and calculate elevation changes which affect the sprinkler discharge.

(b) Calculate the loss for a tee or a cross where flow direction change occurs based on the equivalent pipe length of the piping segment in which the fitting is included. The tee at the top of a riser nipple shall be included in the branch line; the tee at the base of a riser nipple shall be included in the riser nipple; and the tee or cross at a cross main-feed main junction shall be included in the cross main. Do not include fitting loss for straight through flow in a tee or cross.

Formal Interpretation

Question: Regarding Section 7-4.3.1.4(b) in calculating the friction loss for a tee where the flow is turned 90°, is there any additional equivalent feet of pipe required for friction loss calculation because a bushing is used instead of a tee without a bushing? Referring to the sketch, would the equivalent feet of pipe be greater using a 1½ × 1½ × 2½ tee with 1½ × 1¼ bushing, as shown, than if a 1½ × 1¼ × 2½ tee were used?

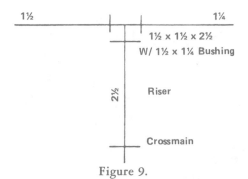

Figure 9.

Answer: The tee at the top of a riser nipple, where flow is turned 90°, should be considered part of the branch line segment. For example, if a 2½ in. riser supplies 1½ in. branch line piping on one side of the riser and 1¼ in. branch line piping on the other side, the loss through the tee should be represented by equivalent feet of 1½ in. pipe in the one direction and by 1¼ in. pipe in the other.

The use of bushings, as indicated by Paragraph 3-13.3, is discouraged by the standard. However, when used in crosses or tees, the method should be no different from that described in Paragraph 7-4.3.1.4(b). For a change in direction, the fitting friction loss equivalent length is sized according to the pipe supplied, not from any of the other outlet sizes.

(c) Calculate the loss of reducing elbows based on the equivalent feet value of the smallest outlet. Use the equivalent feet value for the "standard elbow" on any abrupt ninety-degree turn, such as the screw-type pattern. Use the equivalent feet value for the "long turn elbow" on any sweeping ninety-degree turn, such as a flanged, welded or mechanical joint-elbow type. (*See Table 7-4.2.*)

(d) Friction loss shall be excluded for the fitting directly connected to a sprinkler.

Table 7-4.3.1.4

Pipe or Tube	Hazen-William's "C" Value*
Unlined Cast or Ductile Iron	100
Black Steel (Dry Systems including Pre-action)	100
Black Steel (Wet Systems including Deluge)	120
Galvanized (all)	120
Plastic (listed) — Underground	150
Cement Lined Cast or Ductile Iron	140
Copper Tube	150

*The authority having jurisdiction may recommend other "C" values.

(a) Every component or elevation change that results in a change in pressure must be included in the calculations.

(b) Figure 7.1 may help in explaining this section.

Tee (A) is included with the 1½ in. starter piece.

Tee (B) is included with the 2 in. riser nipple.

Tees (C) and (D) are not included since the flow from the supply is straight through.

Tee (F) is included with the 3 in. piece of cross main.

(c) On a branch line with upright sprinklers screwed into the line tee, the fitting losses are excluded from the calculations because it is assumed that one fitting is included in the approval tests of a sprinkler.

In cases where sprinklers are installed on sprigs or drops, the reducing coupling is ignored but the tee on the branch line must be included.

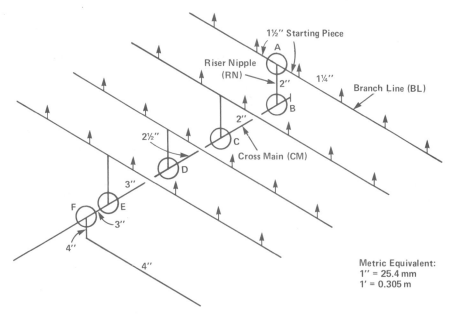

Figure 7.1.   Diagram of a typical sprinkler system used in making sample calculations.

7-4.3.1.5 Orifice plates or sprinklers of different orifice sizes shall not be used for balancing the system, except for special use such as exposure protection, small rooms or enclosures or directional discharge. (*See 4-4.20 for definition of small rooms.*)

Orifice plates are discouraged, because they may be removed while modifying a system and not reinstalled, which would nullify the hydraulic design. If installed in a horizontal run of pipe they can collect debris on the supply side and prevent adequate drainage on the system side.

7-4.3.1.6 Sprinkler discharge in closets, washrooms, and similar small compartments requiring only one sprinkler may be omitted from hydraulic calculations within the area of application. [Sprinklers in these small compartments shall, however, be capable of discharging minimum densities in accordance with Table 2-2.1(B).]

If the design area is 1500 sq ft or larger, sprinkler discharge in small rooms may be omitted. The philosophy here is that the sprinkler orifices will be small so the flow will be relatively insignificant. Also, in compartmented areas it is highly unlikely that all of the sprinklers in the design area will operate simultaneously.

This does not apply in cases where calculations are based on the area of the largest room. For example, if a hotel room is the largest room being considered, the sprinklers in the foyer, closet, and bathroom could operate from a fire in the bedroom. If these sprinklers have not been included in the calculations, they will rob water from the sprinkler(s) in the bedroom, possibly with adverse results.

Exception: This requirement shall not apply when areas of application are selected in accordance with 2-2.1.2.7.

Using sprinklers of different orifice size along the lines or on lines closer to the water supply is prohibited. It is feared that, during original installation or on replacement of sprinklers, the selection of the wrong sprinkler orifice size could adversely affect the hydraulics of the system.

Small rooms such as bathrooms, closets or foyers in hotels may be protected with small orifice sprinklers even through the sprinklers in the bedroom and hallway are ½-in. orifice or larger.

7-4.3.1.7*   Velocity pressure P_v may or may not be included in the calculations at the discretion of the designer. If velocity pressures are used, they shall be used on both branch lines and cross mains where applicable.

A-7-4.3.1.7   If the velocity pressure is included in the calculations, the following assumptions are to be used:

(a) At any flowing outlet along a pipe, except the end outlet, only the normal pressure (P_n) can act on the outlet. At the end outlet the total pressure (P_t) can act. The following are to be considered as end outlets.

1. The last flowing sprinkler on a dead-end branch line.

2. The last flowing branch line on a dead-end cross main.

3. Any sprinkler where a flow split occurs on a gridded branch line.

4. Any branch line where a flow split occurs on a loop system.

(b) At any flowing outlet along a pipe except the end outlet, the pressure acting to cause flow from the outlet is equal to the total pressure (P_t) minus the velocity presssure (P_v) on the upstream side.

(c) To find the normal pressure (P_n) at any flowing outlet except the end outlet, assume a flow from the outlet in question and determine the velocity pressure (P_v) for the total flow on the upstream side. Because normal pressure (P_n) equals total pressure (P_t) minus velocity pressure, the value of the normal pressure (P_n) so found should result in an outlet flow approximately equal to the assumed flow. If not, a new value should be assumed and the calculations repeated.

Velocity pressure is small compared to the total pressure, so normally will not have a major effect on the end result of the hydraulic calculations.

7-4.3.2   Minimum operating pressure of any sprinkler shall be 7 psi (0.5 bar).

When sprinklers are submitted to the testing laboratories for examination and listing, they are subjected to water distribution and fire tests. These tests are conducted at a minimum flow of 15 gpm per sprinkler. The pressure required to produce this flow through the ½-in. nominal orifice sprinkler is approximately 7 psi.

$$P = \frac{(Q)^2}{(K)^2} = \frac{(15)^2}{(5.6)} = 7.17 \text{ psi}$$

Therefore, 7 psi is specified as the minimum pressure to be used in calculations.

Exception: When higher minimum operating pressure for the desired application is specified in the listing of the sprinkler, it shall govern.

7-4.4 Dwelling Units.   When residential sprinklers are used, design shall comply with this section.

7-4.4.1 Design Discharge.   The system shall provide a discharge of not less than 18 gal/min (68 L/min) to any operating sprinkler and not less than 13 gal/min (49 L/min) per sprinkler to all operating sprinklers in the design area. Other discharge rates may be used in accordance with flow rates indicated in individual residential sprinkler listings.

These values of system discharge are based primarily on the Los Angeles and North Carolina fire tests. See 7, 8, 9, 10 of the NFPA 13D - References Cited in Commentary.

7-4.4.2* Number of Design Sprinklers.   The number of design sprinklers shall include all sprinklers within a compartment to a maximum of 4 sprinklers. When a compartment contains less than 4 sprinklers, the number of design sprinklers shall include all sprinklers in that compartment plus sprinklers in adjoining compartments to a total of 4 sprinklers. Adjoining corridors may be considered as compartments for purposes of these calculations. In all cases the design area shall include the 4 most hydraulically demanding sprinklers. See Figure A-7-4.4.2.

NFPA 13D, *Standard for the Installation of Sprinkler Systems in One- and Two-Family Dwellings and Mobile Homes* allows for a design to be based on 2 sprinklers operating. This is satisfactory for a one- or two-family dwelling but the 4 sprinklers required for an NFPA 13 system allows for a sufficient factor of safety to allow for variables that might occur in a larger dwelling occupancy such as an apartment or hotel, including obstructions, multiple ignition, etc.

7-4.4.3   The definition of compartment for use in 7-4.4.2 to determine the number of design sprinklers is a space which is completely enclosed by walls and a ceiling. The compartment enclosure may have openings to an adjoining space if the openings have a minimum lintel depth of 8 in. (203 mm) from the ceiling.

The 8-in. minimum depth is to ensure that heat will be "banked," enhancing operation over the fire.

7-4.4.4 Water Demand.   The water demand for the dwelling unit shall be determined by multiplying the design discharge of 7-4.4.1 by the number of design sprinklers specified in 7-4.4.2.

7-4.4.5   When areas such as attics, basements or other types of occupancies are outside of dwelling units but within the same structure, these areas shall be protected in accordance with all sections of this standard including appropriate water supply requirements of Table 2-2.1(B).

B-7 Sprinkler System Performance Criteria.

The following material is included in the standard in order to give guidance regarding the factors involved in using a conservative approach to the design of a sprinkler system.

B-7-1   Sprinkler system performance criteria have been based on test data. The factors of safety are generally small and are not definitive and can depend on expected (but not guaranteed) inherent characteristics of the sprinkler systems involved. These inherent factors of safety consist of the following:

(a) The flow-declining pressure characteristic of sprinkler systems whereby the initial operating sprinklers discharge at a higher flow than with all sprinklers operating within the designated area.

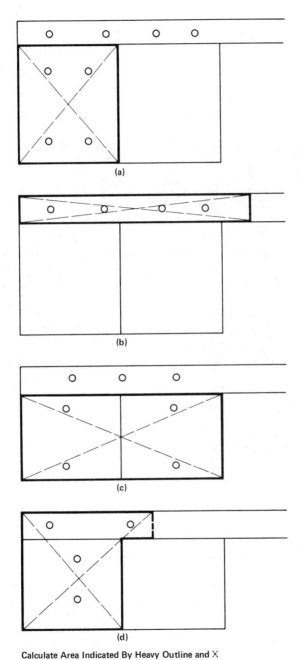

Calculate Area Indicated By Heavy Outline and X
○ Indicates Sprinklers

Figure A-7-4.4.2   Examples of Design Area for Dwelling Units.

(b) The flow-declining pressure characteristic of water supplies. This is particularly steep where fire pumps are the water source. This characteristic similarly produces higher than design discharge at the initially operating sprinklers.

The user of these standards may elect an additional factor of safety if the inherent factors are not considered adequate.

B-7-1.1   Performance specified sprinkler systems as opposed to scheduled systems can be designed to take advantage of multiple loops or gridded configurations. This results in minimum line losses at expanded sprinkler spacing, in contrast to the older "tree-type" configurations where advantage cannot be taken of two-way type flows.

Where the water supply characteristics are relatively flat with pressures being only slightly above the required sprinkler pressure at the spacing selected, gridded systems with piping designed for minimal economic line losses can all but eliminate the inherent flow-declining pressure characteristic generally assumed to exist in sprinkler systems. In contrast, economic design of a "tree-type" system would likely favor a system design with closer sprinkler spacing and greater line losses, demonstrating the inherent flow-declining pressure characteristic of the piping system.

Elements that enter into the design of sprinkler systems include:

1. Selection of density and area of application

2. Geometry of the area of application (remote area)

3. Permitted pressure range at sprinklers

4. Determination of the water supply available

5. Ability to predict expected performance from calculated performance

6. Future upgrading of system performance

7. Size of sprinkler systems.

In developing sprinkler specifications, each of these elements needs to be considered individually. The most conservative design will be based on the application of the most stringent conditions for each of the elements.

B-7-1.2 Selection of Density and Area of Application.   Specifications for density and area of application are developed from NFPA and other standards. It is desirable to specify densities rounded upward to 0.05 gpm.

Prudent design should consider reasonable-to-expect variations in occupancy. This would include not only variations in types of occupancy, but, in the case of warehousing the anticipated future range of materials to be stored, clearances, types of arrays, packaging, pile height and pile stability, as well as other factors.

Design also considers some degree of adversity at the time of a fire. To take this into account, the density and/or area of application may be increased. Another way is to use a dual performance specification where, in addition to the normal primary specifications, a secondary density and area of application is specified. The objective of such a selection is to control the declining pressure-flow characteristic of the sprinkler system beyond the primary design flow.

A case can be made for designing feed and cross mains to lower velocities than branch lines to achieve the same result as specifying a second density and area of application.

B-7-1.3 Geometry of the Area of Application (Remote Area).   It is expected that over any portion of the sprinkler system equivalent in size to the "area of application," the system will achieve the minimum specified density for each sprinkler within that area.

Where a system is computer-designed, ideally the program should verify the entire system by shifting the area of application the equivalent of one sprinkler at a time so as to cover all portions of the system. Such a complete computer verification of performance of the system is most desirable, but unfortunately not all available computer verification programs currently do this.

The selection of the proper Hazen-Williams coefficient is important. New unlined steel pipe has a Hazen-Williams coefficient close to 140. However, it quickly deteriorates to 130, and after a few years' usage, to 120. Hence, the basis for normal design is a Hazen-Williams coefficient of 120 for steel-piped wet systems. A Hazen-Williams coefficient of 100 is generally used for dry-pipe systems because of the increased tendency for deposits and corrosion in these systems. However, it should be realized that a new system will have fewer line losses than calculated, and the distribution pattern will be affected accordingly.

Conservatism can also be built into systems by intentionally designing to a lower Hazen-Williams coefficient than those indicated.

B-7-1.4 Ability to Predict Expected Performance from Calculated Performance.   Ability to accurately predict the performance of a complex array of sprinklers on piping is basically a function of the pipe line velocity. The greater the velocity the greater is the impact on difficult-to-assess pressure losses. These pressure losses are presently determined by empirical means which lose validity as velocities increase. This is especially true for fittings with unequal and more than two flowing ports.

The inclusion of velocity pressures in hydraulic calculations improves the predictability of the actual sprinkler system performance. Calculations should come as close as practicable to predicting actual performance. Conservatism in design should be arrived at intentionally by known and deliberate means. It should not be left to chance.

B-7-1.5 Future Upgrading of System Performance. It may be desirable in some cases to build into the system the capability to achieve a higher level of sprinkler performance than needed at present. If this is to be a consideration in conservatism, consideration needs to be given to maintaining sprinkler operating pressures on the lower side of the optimum operating range, and/or by designing for low pipe line velocities, particularly on feed and cross mains, to facilitate future reinforcement.

8

HIGH-RISE BUILDINGS

8-1 Application. This chapter deals with automatic sprinkler system design for life safety and fire protection in high-rise buildings of Type I or Type II construction, as defined in NFPA 220, *Standard on Types of Building Construction*, which is used predominantly for Light Hazard Occupancies. It is intended to cover totally sprinklered buildings only, and shall not apply to partially sprinklered buildings.

Where partial protection is installed, a fire occurring in an unsprinklered portion can result in overpowering of the system and resultant severe loss. Section 4-1.1.1 includes as a primary basic principle "sprinklers installed throughout the premises." This concept is reinforced here with regard to high-rise buildings.

8-2 Definition.

High-Rise Building. One in which fire must be fought internally because of height.

The definition of high-rise buildings contains only the element regarding the need for fighting the fire internally due to height. Difficulty in evacuation and stack effect are deliberately not included in the definition. Requirements contained under Section 8-5 regarding alarms and reliability reflect a concern for these installed fire protection systems, which become the first line of defense in buildings where external fire fighting is not possible.

8-3 Design Criteria.

8-3.1 The installation may be either a pipe schedule system or a hydraulically designed system. Pipe schedule systems shall comply with Chapters 1 through 6 of this standard. Hydraulically designed systems shall comply with Chapters 1 through 7 of this standard as modified by 8-3.2 and shall comply with Chapter 8.

8-3.2 In Light Hazard Occupancies, special sprinklers may be installed with larger protection areas than indicated in 4-2.2.1.1 or greater distances between sprinklers or branch lines than indicated in 4-2.1.1 when such installations are made in accordance with approvals or listings of a testing laboratory.

This section reinforces the encouragement for innovative technology and development of new devices capable of providing protection equivalent to that contemplated by the standard. See also Section 4-1.1.3.

8-4 Water Supplies.

8-4.1   Acceptable water supplies are as follows:

(a) Public water system where pressure and discharge capacity meet the design requirements of the system as calculated.

(b) Automatic fire pumps supplied under head from a water supply source adequate to meet hydraulically designed system requirements, including public mains, reservoirs, and wells.

(c) Pressure tanks.

(d) Gravity tanks.

8-4.2   Each water supply source shall be automatic and of adequate capacity and pressure to supply the sprinkler system calculated demand for a period of not less than 30 minutes.

8-4.3   Hose connections may be supplied from sprinkler risers. (*See 3-8.6 and 3-8.7.*)

8-5 Alarms.

8-5.1   A separate and distinct supervisory signal shall be provided to indicate a condition that will impair the satisfactory operation of the sprinkler system. This shall include, but need not be limited to, monitoring of control valves, fire pump power supplies and running conditions, water tank levels, and temperatures. Pressure supervision shall also be provided on pressure tanks.

To improve the reliability of the automatic sprinkler systems in high-rise buildings, supervision of any portion of the system that would impair its operation is required.

8-5.2   When each sprinkler system on each floor is equipped with a separate waterflow device, it shall be connected to an alarm system in such a manner that operation of one sprinkler will actuate the alarm system, and the location of the operated flow device shall be indicated on an annunciator and/or register. Annunciator or register shall be located at grade level at the normal point of fire department access, at a constantly attended building security control center, or both locations.

Sprinkler waterflow alarm annunciated by floor is not required as some systems may take on configurations which would not lend themselves to this type zoning. On the other hand, where sprinkler waterflow alarms are provided on each floor, it is the intent that they be annunciated at a point to allow rapid identification of the fire location by the fire department upon arrival.

8-5.3   When the location within the protected buildings where supervisory or alarm signals are received is not under constant supervision by qualified personnel in the employ of the owner, a connection shall be provided to transmit a signal to a remote monitoring station.

The overall reliability of the system is further improved by requiring remote monitoring of supervisory signals in the numerous cases where 24-hour-per-day surveillance of on-site supervisory equipment is not provided.

8-5.4   Alarm and supervisory systems in connection with the sprinkler system shall be installed in accordance with NFPA 71, *Standard for Central Station Signaling Systems*; NFPA 72C, *Standard for Remote Station Protective Signaling Systems*; or NFPA 72D, *Standard for Proprietary Protective Signaling Systems*.

9
LARGE-DROP SPRINKLERS

9-1 General.

9-1.1 Applications. This chapter provides requirements for the installation of large-drop sprinklers. Listed sprinklers other than large-drop sprinklers are not covered by this chapter.

Listed large-drop sprinklers are tested to evaluate mechanical and distribution properties under non-fire conditions, and are also tested to evaluate penetration and cooling, as well as resistance to skipping, under fire conditions. Sprinklers having a discharge coefficient (K-factor) similar to that for large-drop sprinklers, but which have not undergone the rigorous testing required to evaluate their performance against high challenge fires, may not be capable of achieving the degree of fire control performance contemplated by the protection designs contained here.

9-1.2* Definition — Large-Drop Sprinkler. A listed large-drop sprinkler is characterized by a K factor between 11.0 and 11.5, and proven ability to meet prescribed penetration, cooling and distribution criteria prescribed in the large-drop sprinkler examination requirements. The deflector/discharge characteristics of the large-drop sprinkler generate large drops of such size and velocity as to enable effective penetration of the high-velocity fire plume.

A-9-1.2 Large-drop sprinkler development began with the idea that a sprinkler which would produce a high proportion of large drops would be more effective against high challenge fires than standard nominal ½ in. and $^{17}\!/_{32}$ in. sprinklers. Large drops are better able to fight their way through the strong updraft generated by a severe fire (hence the name large-drop sprinkler). The general philosophy behind this idea was to create an offensive weapon for direct attack against the burning fuel, but with as little diminution of defensive capability as possible. As originally conceived, it was supposed that the large drops were the predominate reason for increased effectiveness. It later became evident, however, that other factors such as distribution and pressure play an equal (and possibly more important) role. A consideration of all characteristics indicated that the desired sprinkler would provide high penetration, adequate cooling, reasonable distribution, low skipping potential, and increased discharge capacity.

9-1.3* Applicability.

A-9-1.3   Large-drop sprinklers were developed for use against severe fire challenges requiring sprinkler discharges of 55 gal/min. (208 L/min) per sprinkler or more. The sprinklers were designed to deliver a heavy discharge through the strong updrafts generated by high challenge fires. They depend primarily on direct attack of the burning fuel to gain rapid control of the fire, and to a lesser extent on prewetting and cooling.

Large-drop sprinklers will provide excellent protection against high challenge fires, but careful design and close adherence to the rules given in this chapter are required. This is especially important because experimental findings indicate that the characteristics of large-drop sprinklers are not the same as those attributed to standard sprinklers.

9-1.3.1*   Large-drop sprinklers are suitable for use with the hazards listed in Table A-9-3 and may be used in other specific hazard classifications and configurations only when proven by large scale fire testing.

A-9-1.3.1   Fire tests have not been conducted with large-drop sprinklers over Ordinary and Extra Hazard Occupancies. Therefore, the protection requirements remain unknown.

At a given pressure, large-drop sprinklers will discharge approximately 100 percent and 40 percent more water than standard (½ in., 13 mm) and large orifice (¹⁷⁄₃₂ in., 14 mm) sprinklers, respectively.

Large-drop sprinklers were designed to cope with high challenge fires where high levels of penetration and cooling are required. Since both characteristics fall off sharply at low pressures, large-drop sprinklers cannot be used effectively to compensate for weak water supplies by taking advantage of the increased discharge capacity.

9-1.3.2   Requirements of this standard apply except those portions dealing with subjects specifically addressed in this chapter.

9-2 Installation.

9-2.1 Operating Pressure.

Large-drop sprinklers are effective against high challenge fires primarily because of quality of the sprinkler discharge and the high degree of water penetration through the fire plume. Both of these characteristics are closely dependent upon maintaining a minimum sprinkler discharge pressure at all times. Large-drop sprinkler protection designs specified in this chapter are determined on the basis of a minimum pressure and a minimum number of sprinklers. Traditional density/area design concepts do not apply to large-drop sprinklers.

Large-drop sprinkler development tests investigated the density/area concept. Tests conducted with varying discharge pressures and sprinkler spacings, but with an identical sprinkler discharge density, showed that when pressure was decreased below the minimum required in this chapter, both the number of sprinklers that opened and the measured roof steel temperatures were unacceptable, and were both materially higher than in the tests with higher sprinkler discharge pressures.

9-2.1.1 Large-drop sprinkler systems shall be designed such that the minimum operating pressure is not less than 25 psi (1.7 bar), unless large-scale testing proves a lesser pressure to be adequate for a particular hazard.

9-2.1.2 For design purposes, 95 psi (6.5 bar) shall be the maximum discharge pressure used at the starting point of the hydraulic calculations.

9-2.2 Type of System.

9-2.2.1 Large-drop sprinkler systems shall be limited to wet-pipe or pre-action systems.

Since fires in the occupancies contemplated by this chapter tend to grow in an exponential fashion, significant increases in heat release rates and fire plume velocities may occur in a relatively short period of time. The delay in water arrival inherent in dry-pipe sprinkler systems will allow a fire to gain sufficient headway to cause a diminution in penetrating ability of water discharging from large-drop sprinklers and, ultimately, loss of fire control due either to operation of too many sprinklers and degradation of the water supply to a point of noneffectiveness, or to excessive roof steel temperatures.

9-2.2.2 Galvanized steel or copper pipe and fittings shall be used in pre-action systems to avoid scale accumulation.

9-2.3 System Design.

9-2.3.1 Pipe shall be sized by hydraulic calculation.

9-2.3.2* The nominal diameter of branch line pipes (including riser nipples) shall be not less than 1¼ in. nor greater than 2 in., except starter pieces which may be 2½ in.

The intent of limiting maximum branch sizes or providing riser nipples is to minimize the effect of pipe shadow below the sprinkler, which can result in significant areas with materially reduced water application and loss of sprinkler water penetration.

Exception: *When branch lines are larger than 2 in., the sprinkler shall be supplied by a riser nipple to elevate the sprinkler 13 in. (330 mm) for 2½ in. pipe and 15 in. (38 mm) for 3 in. pipe. These dimensions are measured from the centerline of the pipe to the deflector. In lieu of this, sprinklers may be offset horizontally a minimum of 12 in. (305 mm).*

A-9-2.3.2 This test data was developed with 2 in. diameter pipe maximum. Riser nipples are required when branch lines exceed 2 in. diameter.

9-2.4 Temperature Rating. Sprinkler temperature ratings shall be the same as those used in large-scale fire testing to determine the protection requirements for the hazard involved.

Exception: *Sprinklers of intermediate and high temperature ratings shall be installed in specific locations as required by 3-16.6.3.*

9-2.5* Spacing.

Spacing requirements were determined by testing at various sprinkler spacings and discharge pressures, on the basis of the following parameters:
1) Maintain a minimum distance between sprinklers to ensure sufficient overlapping of sprinkler discharge patterns for adequate penetration of the fire plume.
2) Maintain sufficient separation of sprinklers to prevent sprinkler skipping.
3) Maintain sufficient cooling at ceiling level.
The importance of satisfying these three criteria cannot be overstated. They are so closely intertwined that it is difficult to evaluate one without considering the effect on the others, and vice versa. Certain combinations of spacing dimensions, when used with the discharge pressures and fire hazards contemplated here, will result in skipping, loss of sprinkler water penetration, and excessively high ceiling temperatures. If any of these adverse conditions occur, the adequacy of the sprinkler protection may be seriously compromised.

A-9-2.5 It is important that sprinklers in the immediate vicinity of the fire center not skip and this requirement imposes certain restrictions on the spacing.

9-2.5.1* The area of coverage shall be limited to a minimum of 80 ft^2 (7.4 m^2) and a maximum of 130 ft^2 (12.18 m^2).

A-9-2.5.1 Tests involving areas of coverage over 100 ft^2 (9.3 m^2) were limited in number and the use of areas of average over 100 ft^2 (9.3 m^2) should be carefully considered.

9-2.5.2 The distance between branch lines and between sprinklers on the branch lines shall be limited to not more than 12 ft (3.7 m) nor less than 8 ft (2.4 m).

Exception: Under open wood joist construction, the maximum distance shall be limited to 10 ft (3.0 m).

9-2.6 Clear Space Below Sprinklers. At least 36 in. (914 mm) shall be maintained between sprinkler deflectors and the top of storage.

At clearances less than 36 in., large-drop sprinklers installed within the allowable spacing guidelines will not provide sufficient overlapping of sprinkler discharge patterns between adjacent sprinklers.

9-2.7* Distance Below Ceiling. Sprinklers shall be positioned so that the tops of deflectors are in conformance with Table 9-2.7

Table 9-2.7   Minimum and Maximum Distance of Deflectors Below Ceiling for Various Construction Types

Construction Type[1]	Minimum Distance, In. (mm)	Maximum Distance, In. (mm)
Smooth ceiling and bar joist	6 (152)	8 (203)
Beam and girder	6 (152)	12 (305)
Panel up to 300 ft^2 (27.9m^2)	6 (152)	14 (358)
Open wood joist	1 (25) below bottom of joists	6 (152) below bottom of joists

[1]See Chapter 4 for definitions of construction types.

A-9-2.7 If all other factors are held constant, the operating time of the first sprinkler will vary exponentially with the distance between the ceiling and deflector. At distances greater than 7 in. (178 mm), for other than open wood joist construction, the delayed operating time will permit the fire to gain headway, with the result that substantially more sprinklers operate. At distances less than 7 in. (178 mm), other effects come into play. Changes in distribution, penetration, and cooling nullify the advantage gained by faster operation. The net result is again increased fire damage accompanied by an increase in the number of sprinklers operated. The optimum clearance between deflectors and ceiling is, therefore, 7 in. (178 mm). For open wood joist construction the optimum clearance between deflectors and the bottom of joists is 3½ in. (89 mm).

9-2.8 Location of Sprinklers in Beam and Girder and Panel Construction.

9-2.8.1   Under beam and girder construction and under panel construction, the branch lines may run across the beams, but sprinklers shall be located in the bays and not under the beams.

Large-drop sprinklers cannot be located directly below beams. The delay in sprinkler operating time may seriously degrade the protection capabilities of the sprinkler system, as further explained in A-9-2.7. Sprinklers should be located in the bays formed by the beams and also meet the positioning requirements in 9-2.8.2.

9-2.8.2   The maximum distance of deflector above the bottom of beams shall be limited to the values specified in Chapter 4.

9-2.9* Obstruction in Piping.

A-9-2.9   The plugging of a single sprinkler in the vicinity of the fire origin can cause a significant increase in the number of operating sprinklers as well as an increase in fire damage. Therefore, it is essential that the sprinkler piping be kept free of obstructing material.

Fouling or plugging of piping or heads may be prevented by several methods: sound installation practices, including the provision of screens or strainers when necessary; avoidance of dry-pipe systems with their history of plugging; and a high level of maintenance.

9-2.9.1   Screens located in the inlet piping directly connected to rivers, lakes, ponds, reservoirs, uncovered tanks and similar sources (*see NFPA 24, Installation of Private Fire Service Mains and Their Appurtenances*) shall be cleaned and serviced at least annually and after any work has been performed on nearby underground mains.

9-2.9.2*   Visual and/or flushing investigations shall be conducted of all systems for foreign material at intervals not exceeding five years.

A-9-2.9.2   Investigations should be conducted more frequently when justified by local conditions.

9-2.10* Obstructions to Distribution.

Because effective protection by large-drop sprinklers depends heavily on water penetration through the fire plume, physical obstructions which would reduce the amount of water penetration must be minimized. Since a given portion of the discharge is already sacrificed to vaporization or being carried away by the fire plume gases, further reductions in water available to reach the fire would be dangerous. The intent here is to provide guidance for location of sprinklers with respect to obstructions commonly found in areas where large-drop sprinklers may be expected to be installed.

A-9-2.10   To a great extent, large-drop sprinklers rely on direct attack to gain rapid control of both the burning fuel and ceiling temperatures. Therefore, interference with the discharge pattern and obstructions to the distribution should be avoided.

9-2.10.1 Obstruction Located at the Ceiling.   When sprinkler deflectors are located above the bottom of beams, girders, ducts, fluorescent lighting fixtures, or other obstructions located at the ceiling, the sprinklers shall be positioned so that the maximum distance from the bottom of the obstruction to the deflectors does not exceed the value specified in Chapter 4.

9-2.10.2 Obstructions Located Below the Sprinklers.

9-2.10.2.1   Sprinklers shall be positioned with respect to fluorescent lighting fixtures, ducts and obstructions more than 24 in. (610 mm) wide and located entirely below the sprinklers so that the minimum horizontal distance from the near side of the obstruction to the center of the sprinkler is not less than the value specified in Table 9-2.10.2.1. (*See Figure 9-2.10.2.1.*)

Table 9-2.10.2.1   Position of Sprinklers in Relation to
Obstructions Located Entirely Below the Sprinklers

Distance of Deflector Above Bottom of Obstruction	Minimum Distance to Side of Obstruction, ft (m)
Less than 6 in. (152 mm)	1½ (0.5)
6 in. (152 mm) to less than 12 in. (305 mm)	3 (0.9) 3 (0.9)
12 in. (305 mm) to less than 18 in. (457 mm)	4 (1.2)
18 in. (457 mm) to less than 24 in. (610 mm)	5 (1.5)

This section applies to obstructions no more than 24 in. below the sprinklers. The intent is to minimize obstruction to the lateral distribution from individual sprinklers.

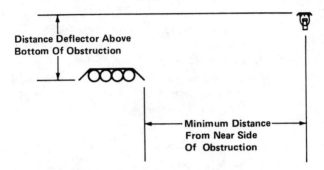

**Figure 9-2.10.2.1   Position of Sprinklers in Relation to Obstructions Located Entirely below the Sprinklers.
(To be used with Table 9-2.10.2.1.)**

9-2.10.2.2   When the bottom of the obstruction is located 24 in. (610 mm) or more below the sprinkler deflectors:

(a) Sprinklers shall be positioned so that the obstruction is centered between adjacent sprinklers. See Figure 9-2.10.2.2.

(b) The obstruction shall be limited to a maximum width of 24 in. (610 mm). See Figure 9-2.10.2.2.

Exception: When obstruction is greater than 24 in. (610 mm) wide, one or more lines of sprinklers shall be installed below the obstruction.

(c) The obstruction shall not extend more than 12 in. (305 mm) to either side of the midpoint between sprinklers. See Figure 9-2.10.2.2.

Exception: When extensions exceed 12 in. (305 mm), one or more lines of sprinklers shall be installed below the obstruction.

(d) At least 18 in. (457 mm) clearance shall be maintained between the top of storage and the bottom of the obstruction. See Figure 9-2.10.2.2.

At distances more than 24 in. below the sprinklers, obstructions will interfere with the distribution from more than one sprinkler, due to characteristics of the sprinkler's discharge pattern. The provisions here will ensure that sufficient overlap of discharge patterns from adjacent sprinklers is achieved, and thus ensure that adequate water distribution is available to attain fire plume penetration.

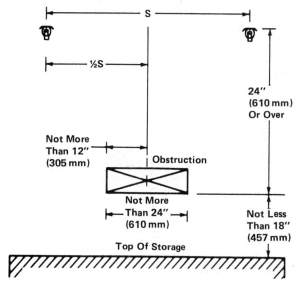

Figure 9-2.10.2.2   Position of Sprinklers in Relation to Obstructions Located 24 in. (610 mm) or More Below Deflectors.

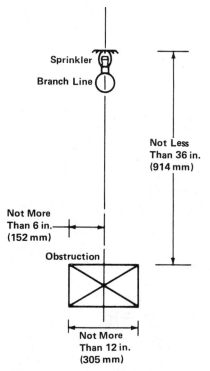

Figure 9-2.10.2.3   Position of Sprinklers in Relation to Obstructions Running Parallel to and Directly Below Branch Lines.

9-2.10.3 Obstructions Parallel to and Directly below Branch Lines. In the special case of an obstruction running parallel to and directly below a branch line:

(a) The sprinkler shall be located at least 36 in. (914 mm) above the top of the obstruction. See Figure 9-2.10.2.3.

(b) The obstruction shall be limited to a maximum width of 12 in. (305 mm). See Figure 9-2.10.2.3.

(c) The obstruction shall be limited to a maximum extension of 6 in. (152 mm) to either side of the centerline of the branch line. See Figure 9-2.10.2.3.

> The obstruction to distribution as expected under these conditions would not be significantly greater than that which would occur due to pipe shadow.

9-3* Protection Requirements.

A-9-3 Large scale fire testing has shown that large-drop sprinkler systems designed per the rules of Chapter 9 to provide the minimum operating pressures for the number of design sprinklers for the specific hazards of Table A-9-3 will adequately control the fire.

> Protection requirements for the various hazards listed in Table A-9-3 were, for the most part, developed from results of large-scale tests. In some cases, where large-scale tests were conducted on only one specific commodity for a particular storage arrangement, judgments were applied in determing protection requirements for lesser hazard commodities stored in the same manner.

9-3.1* Protection shall be provided as specified in Table A-9-3 or appropriate NFPA standards in terms of minimum operating pressure and the number of sprinklers to be included in the design area.

A-9-3.1 In testing where discharge density and all other conditions (except pressure and spacing) remained constant, sprinkler demand area, steel temperatures, and fire damage all exhibited a marked increase as the spacing was decreased. The importance of these findings cannot be overemphasized. They state clearly that the term "density" has no meaning when applied to large-drop sprinklers and it (density) cannot be used as a design parameter.

In testing the effect of discharge pressure, series of tests were run with two different fuel types. Although the relationship of decreased number of operated sprinklers with increased pressure held, the associated curves did not parallel each other. Therefore, it is not possible to extrapolate the protection requirements from one hazard to a higher hazard, nor is it possible to develop a curve from one test point. However, the protection requirements developed for a particular hazard (through large scale fire tests) may be applied safely to hazards known to be less severe, provided that the less severe hazards are arranged in the same configuration.

Table A-9-3   Pressure and Number of Design Sprinklers for Various Hazards (Note 6)

Minimum Operating Pressure (Note 1), psi (bar)	25 (1.7)	50 (3.4)	75 (5.2)
Hazard (Note 2)	Number Design Sprinklers		
Palletized Storage			
Class I, II and commodities up to 25 ft (7.6 m) with maximum 10 ft (3.0 m) clearance to ceiling	15	Note 3	Note 3
Class IV commodities up to 20 ft (6.1 m) with maximum 10 ft (3.0 m) clearance to ceiling	20	15	Note 3
Unexpanded plastics up to 20 ft (6.1 m) with maximum 10 ft (3.0 m) clearance to ceiling	25	15	Note 3
Idle wood pallets up to 20 ft (6.1 m) with maximum 10 ft (3.0 m) clearance to ceiling	15	Note 3	Note 3
Solid-Piled Storage			
Class I, II and III commodities up to 20 ft (6.1 m) with maximum 10 ft (3.0 m) clearance to ceiling	15	Note 3	Note 3
Class IV commodities and unexpanded plastics up to 20 ft (6.1 m) with maximum 10 ft (3.0 m) clearance to ceiling	Does Not Apply	15	Note 3
Double-Row Rack Storage with Minimum 5½ ft (1.7 m) Aisle Width (Note 4)			
Class I and II commodities up to 25 ft (7.6 m) with maximum 5 ft (1.5 m) clearance to ceiling	20	Note 3	Note 3
Class I, II, and III commodities up to 20 ft (6.1 m) with maxiumum 10 ft (3.0 m) clearance to ceiling	15	Note 3	Note 3
Class IV commodities up to 20 ft (6.1 m) with maximum 10 ft (3.0 m) clearance to ceiling	Does Not Apply	20	15
Unexpanded plastics up to 20 ft (6.1 m) with maximum 10 ft (3.0 m) clearance to ceiling	Does Not Apply	30	20
Unexpanded plastics up to 20 ft (6.1 m) with maximum 10 ft (3.0 m) clearance to ceiling (Note 7)	Does Not Apply	20	Note 3
Class IV commodities and unexpanded plastics up to 20 ft (6.1 m) with maximum 5 ft (1.5 m) clearance to ceiling	Does Not Apply	15	Note 3

**Table A-9-3   Pressure and Number of Design Sprinklers
for Various Hazards (Note 6) (continued)**

On-End Storage of Roll Paper (Note 5)			
Heavyweight paper, in closed array, banded in open array, or banded or unbanded in a standard array, up to 26 ft (7.9 m) with a maximum 34 ft (10.4 m) clearance to ceiling	Does Not Apply	15	Note 3
Any grade of paper, EXCEPT LIGHTWEIGHT paper, with stacks in closed array, or banded or unbanded standard array up to 20 ft (6.1 m) with maximum 10 ft (3.0 m) clearance to ceiling	Does Not Apply	15	Note 3
Medium weight paper completely wrapped (sides and ends) in one or more layers of heavyweight paper, or lightweight paper in two or more layers of heavyweight paper with stacks in closed array, banded in open array, or banded or unbanded in a standard array, up to 26 ft (7.9 m) with maximum 34 ft (10.4 m) clearance to ceiling	Does Not Apply	15	Note 3
Record Storage			
Paper records and/or computer tapes in multitier steel shelving up to 5 ft (1.5 m) in width and with aisles 30 in. (76 cm) or wider, without catwalks in the aisles, up to 15 ft (4.6 m) with maximum 5 ft (1.5 m) clearance to ceiling	15	Note 3	Note 3
Same as above, but with catwalks of expanded metal or metal grid with minimum 50 percent open area, in the aisles	Does Not Apply	15	Note 3

Notes:

1. Open Wood Joist Construction. Testing with open wood joist construction showed that each joist channel should be fully firestopped to its full depth at intervals not exceeding 20 ft (6.1 m). In unfirestopped open wood joist construction, or if firestops are installed at intervals exceeding 20 ft (6.1 m), the minimum operating pressures should be increased by 40 percent.

2. Building steel required no special protection for the occupancies listed.

3. The higher pressure will successfully control the fire, but the required number of design sprinklers should not be reduced from that required for the lower pressure.

4. In addition to the transverse flue spaces required by NFPA 231C, minimum 6 in. (152 mm) longitudinal flue spaces were maintained.

5. See NFPA 231F for definitions.

6. Unless otherwise specified the sprinklers used in the tests were high temperature rating.

7. Based on tests using sprinkler or ordinary temperature rating.

10

REFERENCED PUBLICATIONS

10-1   The following documents or portions thereof are referenced within this standard and shall be considered part of the requirements of this document. The edition indicated for each reference is current as of the date of the NFPA issuance of this document. These references are listed separately to facilitate updating to the latest edition by the user.

10-1.1 NFPA Publications.   National Fire Protection Association, Batterymarch Park, Quincy, MA 02269.

NFPA 13D-1984, *Installation of Sprinkler Systems for One- and Two-Family Dwellings and Mobile Homes*

NFPA 14-1983, *Installation of Standpipe and Hose Systems*

NFPA 20-1983, *Installation of Centrifugal Fire Pumps*

NFPA 22-1984, *Water Tanks for Private Fire Protection*

NFPA 24-1984, *Installation of Private Fire Service Mains and Their Appurtenances*

NFPA 51B-1984, *Fire Prevention in Use of Cutting and Welding Processes*

NFPA 71-1982, *Central Station Signaling Systems*

NFPA 72A-1985, *Local Protective Signaling Systems*

NFPA 72B-1979, *Auxiliary Protective Signaling Systems*

NFPA 72C-1982, *Remote Station Protective Signaling Systems*

NFPA 72D-1979, *Proprietary Protective Signaling Systems*

NFPA 72E-1984, *Automatic Fire Detectors*

NFPA 81-1981, *Fur Storage, Fumigation and Cleaning*

NFPA 96-1984, *Removal of Smoke and Grease-Laden Vapors from Commercial Cooking Equipment*

NFPA 220-1979, *Standard on Types of Building Construction*

NFPA 231-1985, *General Storage*

NFPA 231C-1980, *Rack Storage of Materials*

NFPA 1963-1985, *Screw Threads and Gaskets for Fire Hose Connections.*

10-1.2   The following additional NFPA codes and standards contain specific sprinkler design criteria on various objects (*see 2-2.1.3 and A-2-1*).

NFPA 15-1985, *Water Spray Fixed Systems*

NFPA 16-1980, *Deluge Foam-Water Sprinkler Systems and Foam-Water Spray Systems*

NFPA 43A-1980, *Storage of Liquid and Solid Oxidizing Materials*

NFPA 45-1982, *Fire Protection for Laboratories Using Chemicals*

NFPA 87-1980, *Construction and Protection of Piers and Wharves*

NFPA 214-1983, *Water Cooling Towers*

NFPA 231-1985, *General Storage*

NFPA 231C-1980, *Rack Storage of Materials*

NFPA 231D-1980, *Storage of Rubber Tires*

NFPA 231F-1984, *Storage of Roll Paper*

NFPA 409-1985, *Aircraft Hangars*

10-1.3 Other Codes and Standards.

10-1.3.1 ANSI Publications.   American National Standards Institute, Inc., 1450 Broadway, New York, New York 10018.

ANSI B1-20.1-1983, *Pipe Threads, General Purpose*

ANSI B16.1-1975, *Cast-Iron Pipe Flanges and Flanged Fittings, Class 25, 125, 250 and 800*

ANSI B16.3-1977, *Malleable-Iron Threaded Fittings, Class 150 and 300*

ANSI B16.4-1977, *Cast-Iron Threaded Fittings, Class 125 and 250*

ANSI B16.5-1981, *Pipe Flanges and Flanged Fittings, Steel, Nickel, Alloy, and Other Special Alloys*

ANSI B16.9-1978, *Factory-Made Wrought Steel Buttwelding Fittings*

ANSI B16.11-1980, *Forged Steel Fittings, Socket Welding and Threaded*

ANSI B16.18-1978, *Cast Copper Alloy Solder Joint Pressure Fittings*

ANSI B16.22-1980, *Wrought Copper and Copper Alloy Solder Joint Pressure Fittings*

ANSI B16.25-1979, *Buttwelding Ends*

ANSI B36.10-1979, *Welded and Seamless Wrought-Steel Pipe*

10-1.3.2 ASTM Publications.   American Society for Testing and Materials, 1916 Race Street, Philadelphia, PA 19105.

ASTM 88-1981, *Specifications for Seamless Copper Water Tube*

ASTM A53-1980, *Specifications for Welded and Seamless Steel Pipe*

ASTM A120-1980, *Welded and Seamless Steel Pipe for Ordinary Uses, Specifications for Black and Hot-Dipped Zinc Coated (Galvanized)*

ASTM A234-1980, *Specifications for Piping Fittings of Wrought-Carbon Steel and Alloy for Moderate and Elevated Temperatures*

ASTM A795-1982, *Specification for Black and Hot-Dipped Zinc-Coated (Galvanized) Welded and Seamless Steel Pipe for Fire Protection*

ASTM B32-1976, *Solder Metal, 95-5 (Tin-Antimony-Grade 95TA)*

ASTM B75-1981, *Specifications for Seamless Copper Tube*

ASTM B251-1976, *Specifications for General Requirements for Wrought Seamless Copper and Copper-Alloy Tube*

ASTM E380-1979, *Standard for Metric Practice*

10-1.3.3 AWS Publications.   American Welding Society, 2501 N.W. 7th Street, Miami, FL 33125.

AWS A5.8-1981, *Specification for Brasing Filler Metal*

AWS D10.9-1980, *Qualification of Welding Procedures and Welders for Piping and Tubing*

APPENDIX A

This Appendix is not a part of the requirements of this NFPA document . . . but is included for information purposes only.

The material contained in Appendix A of the 1985 edition of the *Standard for the Installation of Sprinkler Systems* is included in the text within this Handbook and therefore is not repeated here.

APPENDIX B

This Appendix is not a part of the requirements of this NFPA document . . . but is included for information purposes only.

The material contained in Appendix B of the 1985 edition of the *Standard for the Installation of Sprinkler Systems* is included within the text of this Handbook and therefore is not repeated here.

APPENDIX C

This Appendix is not a part of the requirements of this NFPA document . . . but is included for information purposes only.

In this Handbook, Appendix C follows 4-3.6 of the *Standard for the Installation of Sprinkler Systems.*

APPENDIX D
REFERENCED PUBLICATIONS

D-1   The following documents or portions thereof are referenced within this standard for informational purposes only and thus should not be considered part of the requirements of this document. The edition indicated for each reference is current as of the date of the NFPA issuance of this document. These references are listed separately to facilitate updating to the latest edition by the user.

D-1.1 NFPA Publications.   National Fire Protection Association, Batterymarch Park, Quincy, MA 02269.

NFPA 13A-1981, *Inspection, Testing and Maintenance of Sprinkler Systems*

D-2   The following NFPA Recommended Practices contain specific sprinkler design criteria on various subjects (*see 2-2.1.3 and A-2-1*).

NFPA 16A-1983, *Closed-Head Foam-Water Sprinkler Systems*

NFPA 231F-1984, *Storage of Baled Cotton*

Index

347

Reference

Reference

Reference   Reference

Reference   Reference

Reference   Reference

Reference

Reference

NFPA 13A

Recommended Practice for the

Inspection, Testing and Maintenance

of Sprinkler Systems

1981 Edition

NOTICE: Information on referenced publications can be found in the Appendix.

FOREWORD

An automatic sprinkler system provides for the extinguishment of fire in a building by the prompt and continuous discharge of water directly upon burning material. This is accomplished by means of an arrangement of pipes to which are attached outlet devices known as automatic sprinklers. These sprinklers are so constructed as to open automatically whenever the surrounding temperature reaches a predetermined point.

In general, there are two types of automatic sprinkler equipment, dry-pipe systems and wet-pipe systems. In locations which are not subject to freezing temperatures, wet-pipe systems, in which the pipe lines contain water under pressure, may be installed, but in buildings or portions thereof which are subject to freezing temperatures, the dry-pipe system is ordinarily used. In the latter type of system, water is admitted to the pipes automatically after elevated ceiling temperature has caused the automatic sprinkler to operate.

If not properly maintained, a sprinkler system may become inoperative. The following offers advice and suggestions relative to the inspection, testing and maintenance of sprinkler equipment upon which the safety of life and property may depend.

1

GENERAL INFORMATION

1-1 Scope. This recommended practice provides for the inspection, testing and maintenance of sprinkler systems.

Because this publication is a recommended practice, it is advisory only. Unlike a standard, which specifies regulations, this document provides guidance only and the word shall, which indicates a requirement, is not used. Regulations governing the installation of the components of sprinkler systems are found in the appropriate standards. It is the feeling of many that requirements for adequate maintenance of automatic sprinkler systems should be mandatory rather than advisory. Unfortunately, the multiplicity of conditions which actually exist in the field would make the application of a mandatory standard in all instances difficult, if not impossible, and thereby render it meaningless. It is, therefore, felt that the guidance provided by the recommended practice is more appropriate.

1-2 Purpose. The purpose of this recommended practice is to provide guidance for inspection, testing and maintenance of sprinkler systems.

The responsibility for properly maintaining a sprinkler system is the obligation of the owners of the property. Suggestions as to details for proper maintenance by the owner are expanded in Section 1-5.

1-3 Definitions.

Antifreeze System. Antifreeze system means a system employing automatic sprinklers attached to a piping system containing an antifreeze solution and connected to a water supply. The antifreeze solution, followed by water, discharges immediately from sprinklers opened by a fire.

See Section 5-5 of NFPA 13, *Standard for the Installation of Sprinkler Systems.*

Approved.   Means "acceptable to the authority having jurisdiction."

NOTE: The National Fire Protection Association does not approve, inspect or certify any installations, procedures, equipment or materials nor does it approve or evaluate testing laboratories. In determining the acceptability of installations or procedures, equipment or materials, the authority having jurisdiction may base acceptance on compliance with NFPA or other appropriate standards. In the absence of such standards, said authority may require evidence of proper installation, procedure or use. The authority having jurisdiction may also refer to the listings or labeling practices of an organization concerned with product evaluations which is in a position to determine compliance with appropriate standards for the current production of listed items.

Authority Having Jurisdiction.   The "authority having jurisdiction" is the organization, office, or individual responsible for "approving" equipment, an installation, or a procedure.

NOTE: The phrase "authority having jurisdiction" is used in NFPA documents in a broad manner since jurisdictions and "approval" agencies vary as do their responsibilities. Where public safety is primary, the "authority having jurisdiction" may be a federal, state, local, or other regional department or individual such as a fire chief, fire marshal, chief of a fire prevention bureau, labor department, health department, building official, electrical inspector, or others having statutory authority. For insurance purposes, an insurance inspection department, rating bureau, or other insurance company representative may be the "authority having jurisdiction." In many circumstances the property owner or his designated agent assumes the role of the "authority having jurisdiction"; at government installations, the commanding officer or departmental official may be the "authority having jurisdiction."

Butterfly Valve.   An indicating-type control valve incorporating wafer-type body with gear-operated, quarter-turn disc in the waterway.

Cold Weather Valve.   An indicating-type valve for the control of 10 sprinklers or less in a wet system protecting an area subject to freezing. The valve is normally closed and the system drained during freezing weather.

Control Valve.   A valve which may be opened or closed to regulate the flow of water to all or part of a sprinkler system.

See Section 3-14.2 of NFPA 13, *Standard for the Installation of Sprinkler Systems.*

Deluge System.   A system employing open sprinklers installed in a water supply through a valve which is opened by the operation of a fire detection system installed in the same areas as the sprinklers. When this valve opens, water flows into the piping system and discharges from all sprinklers attached thereto.

See Section 5-3 of NFPA 13, *Standard for the Installation of Sprinkler Systems.*

Dry-Pipe System.   A system employing automatic sprinklers installed in a piping system containing air or nitrogen under pressure, the release of which, as from the opening of a sprinkler, permits the water pressure to open a valve known as a dry-pipe valve. The water then flows into the piping system and out the opened sprinklers.

See Section 5-2 of NFPA 13, *Standard for the Installation of Sprinkler Systems.*

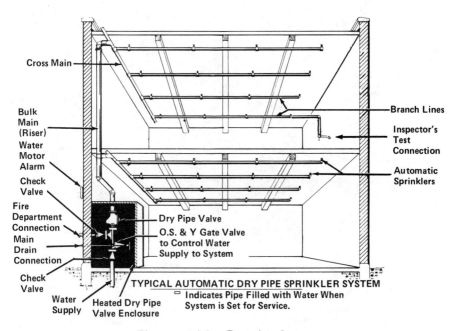

Figure 1-3(a)   Dry-pipe System.

Emergency Impairment.   A condition wherein a sprinkler system or a portion thereof is out of order due to an unexpected occurrence such as a ruptured pipe, operated sprinkler, interruption of water supply to the system, etc.

Indicator Post.   A control extending above ground or through a wall for operating sprinkler control valves. A target or indicator visible through an opening in the post shows whether the valve is open or shut.

Inspection.   A visual examination of a sprinkler system or portion thereof to verify that it appears to be in operating condition and is free from physical damage.

Listed.   Equipment or materials included in a list published by an organization acceptable to the "authority having jurisdiction" and concerned

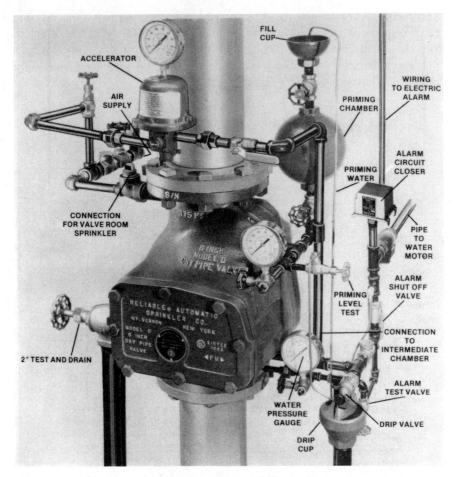

Figure 1-3(b)   Typical Dry-pipe Valve with Trimmings.

with product evaluation, that maintains periodic inspection of production of listed equipment or materials and whose listing states either that the equipment or material meets appropriate standards or has been tested and found suitable for use in a specified manner.

NOTE: The means for identifying listed equipment may vary for each organization concerned with product evaluation, some of which do not recognize equipment as listed unless it is also labeled. The "authority having jurisdiction" should utilize the system employed by the listing organization to identify a listed product.

Locks and Chains.   Heavy-duty locks and chains sufficiently durable to resist any cutting device less than heavy-duty bolt cutters.

Maintenance.   Work performed to keep equipment operable, or to make repairs.

Outside Screw and Yoke (O. S. & Y.) Valve.   A gate valve with a rising stem which indicates if the valve is open or closed.

Pre-Action System.   A closed sprinkler system containing air that may or may not be under pressure, with a supplemental fire detection system installed in the same areas as the sprinklers. Actuation of the fire detection system opens a valve which permits water to flow into the sprinkler piping and to be discharged from any sprinklers which may be open.

See Section 5-3.5 of NFPA 13, *Standard for the Installation of Sprinkler Systems.*

Pre-Planned Impairment.   A condition where a sprinkler system or a portion thereof is out of service due to work which has been planned in advance such as revisions to the water supply or sprinkler system piping.

Qualified Inspection Service.   A service program provided by a fire protection contractor and/or owner's representative in which all of the Chapter 7 provisions are included.

Quick-Opening Device.   A listed device such as an accelerator or an exhauster used to cause a dry-pipe valve to operate more rapidly.

See Section 5-2.4 of NFPA 13, *Standard for the Installation of Sprinkler Systems.*

Roadway Box.   A sleeve providing access to an underground control valve.

Seals.   The use of small-diameter wire which is threaded and knotted within a lead seal. Alternately, plastic devices are available for the same purpose.

The closing of the seal makes it impossible to turn a valve unless the seal is broken. The broken seal indicates that the valve may have been moved from its normal position.

Should.   Indicates a recommendation or that which is advised but not required.

Sprinkler System.   A sprinkler system, for fire protection purposes, is an integrated system of underground and overhead piping designed in accordance with fire protection engineering standards. The installation includes a water supply such as a gravity tank, fire pump, reservoir or pressure tank and/or connection by underground piping to a city main. The portion of the sprinkler system aboveground is a network of specially sized or hydraulically designated piping installed in a building, structure or area, generally overhead, and to which sprinklers are connected in a systematic pattern. The system includes a controlling valve and a device for actuating an alarm when the system is in operation. The system is usually activated by heat from a fire and discharges water over the fire area.

Tamper Switch.   An electrical device for control-valve supervision which initiates an alarm when the control valve is moved from the normal position.

Testing.   Conducting periodic physical checks on the sprinkler system such as water flow tests, alarm tests, or dry-pipe valve trip tests.

Waterflow Alarm.   A listed device so constructed and installed that any flow of water from a sprinkler system equal to or greater than that from a single automatic sprinkler will result in an alarm signal.

See Section 3-17 of NFPA 13, *Standard for the Installation of Sprinkler Systems.*

Wet-Pipe System.   A system employing automatic sprinklers installed in a piping system containing water and connected to a water supply. Water discharges immediately from sprinklers opened by a fire.

See Section 5-1 of NFPA 13, *Standard for the Installation of Sprinkler Systems.*

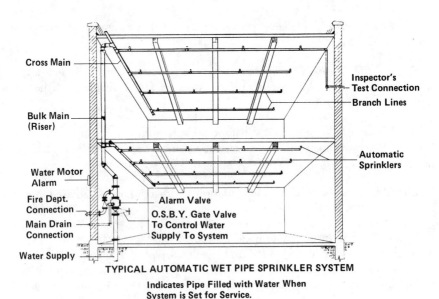

TYPICAL AUTOMATIC WET PIPE SPRINKLER SYSTEM

Indicates Pipe Filled with Water When
System is Set for Service.

Figure 1-3(c)   Wet-pipe System.

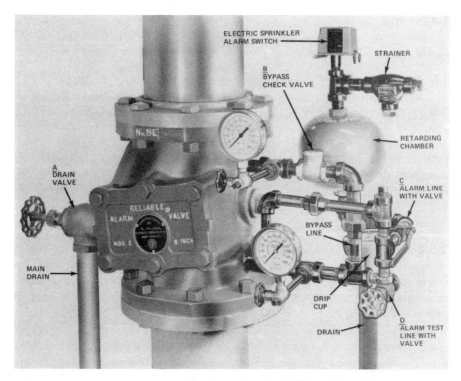

Figure 1-3(d)   Typical Alarm Valve and Trimmings.

1-4 Units.   Metric units of measurement in this recommended practice are in accordance with the modernized metric system known as the International System of Units (SI). Two units (liter and bar), outside of but recognized by SI, are commonly used in international fire protection. These units are listed in Table 1-4 with conversion factors.

Table 1-4

Name of Unit	Unit Symbol	Conversion Factor
liter	L	1 gal = 3.785 L
cubic decimeter	dm^3	1 gal = 3.785 dm^3
pascal	Pa	1 psi = 6894.757 Pa
bar	bar	1 psi = 0.0689 bar
bar	bar	1 bar = 10^5 Pa

For additional conversions and information see ASTM E380, *Standard for Metric Practice*.

1-4.1   If a value for measurement as given in this recommended practice is followed by an equivalent value in other units, the first stated is to be regarded as the requirement. A given equivalent value may be approximate.

1-4.2   The conversion procedure for the SI units has been to multiply the quantity by the conversion factor and then round the result to the appropriate number of significant digits.

1-5 Responsibility of the Owner or Occupant.

1-5.1   The responsibility for properly maintaining a sprinkler system is the obligation of the owners of the property.

By means of periodic tests, the equipment is shown to be in good operating condition or any defects or impairments are revealed. Such tests are made, however, at the owner's responsibility and risk. Intelligent cooperation in the performance of these tests shows evidence of the owner's interest in property conservation.

1-5.2   Automatic sprinkler systems installed in accordance with NFPA standards require a minimum of inspection, testing and maintenance; however, deterioration or impairment may result from neglect. Definite provision for periodic competent attention is a prime requirement if the system is to serve its purpose effectively.

1-5.3   Arrangements should be made to keep all stock piles, racks and other possible obstructions the proper distance below sprinklers. [The minimum recommended distance below sprinkler deflectors at the ceiling is 18 in. (457 mm). The minimum recommended distance below sprinkler deflectors in racks can be found in NFPA 231, *Standard for General Storage*; NFPA 231C, *Standard for Rack Storage*; and NFPA 231D, *Standard on Storage of Rubber Tires*.]

1-5.4   A competent and reliable employee should be given the responsibility of regularly inspecting, testing and maintaining the system and reporting any troubles or defects to his employer. This employee should have proper instruction and training and a general understanding of the mechanical requirements of operation.

1-5.5   Support personnel should be trained in inspection, testing and maintenance and be fully capable of taking over the functions at any time when the authorized individual is unavailable.

1-5.6 Public Fire Department.

1-5.6.1   It is advisable to notify the fire department of the installation of automatic sprinkler equipment so that it may become familiar with the system. The fire department should know the extent of the protection and the location and arrangement of the control valves and the connections for fire department use.

The fire department should also be notified if the system or a major portion of it is temporarily taken out of service. This notification allows the fire department to preplan in the event of any emergency and also provides it an opportunity for making suggestions for provision of emergency or temporary water supplies during the impairment period.

1-5.7 Security Personnel.

1-5.7.1   Instruct security personnel in the following:

(a) Location and use of control valves, drain valves and alarm devices.

(b) Prompt transmittal of a fire alarm to a fire department or brigade, before attempting to extinguish the fire.

(c) Proper notification in case of fire or impairment of sprinkler equipment.

(d) Daily visual inspection of all sprinkler control valves on the guard's first round to ascertain that they are open.

(e) Proper notification immediately of any valve found closed.

(f) Proper notification when sprinkler alarms operate, to determine the cause of water flow.

(g) Do not close sprinkler control valves until it has definitely been established that there is no fire.

(h) During cold weather, verify that windows or other openings are closed and that proper temperature is being maintained to prevent freezing.

The importance of inspection to ascertain that all sprinkler control valves are open cannot be overemphasized. Closed valves are by far the greatest cause of sprinkler system failure. It is also important that security personnel not close valves until it has been established that a fire has been completely extinguished. There have been a significant number of large losses where the sprinkler valve was prematurely closed, either because of concern regarding water damage or because the party closing the valve incorrectly thought the fire had been extinguished. When closed valves are found with no detectable reason for their being closed, security personnel should be especially alert for possible arson attempts.

1-6 Sprinkler Inspection Service.

1-6.1   The level of reliability of the protection offered by an automatic sprinkler system is promoted when there is a qualified inspection service. Qualified inspection service should include:

(a) Four visits per year, at regular intervals.

(b) All services indicated in summary Table 7-3.

(c) The completion of a report form with copies furnished to the property owner. (*See Chapter 7, Report of Inspection, Exhibit I.*)

1-6.2   The outside inspection services are an adjunct to, and are not intended to replace, the owners' obligations.

2

WATER SUPPLIES

2-1 General. The source and quantity of water is of fundamental importance. To ensure the continued existence of proper flow, it is necessary that periodic inspections and tests be conducted by qualified personnel.

2-2 Gravity Tanks and Suction Tanks. (*See NFPA 22, Standard for Water Tanks for Private Fire Protection.*)

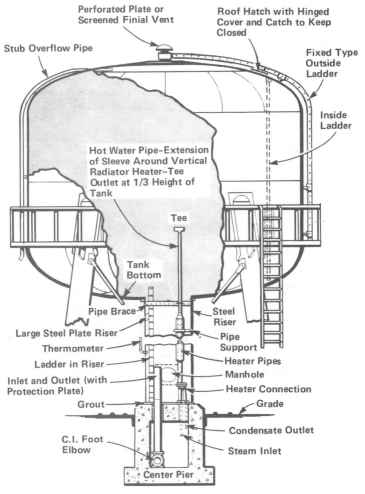

Figure 2.1. Typical gravity tank installation.

2-2.1   Monthly inspections should be made to check the maintenance of water at proper level in the tank.

Constant maintenance of a full supply of water in gravity tanks is necessary not only to ensure proper performance of the sprinkler system in the event of a fire, but to prevent shrinkage of wooden tanks and minimize corrosion of steel tanks.

2-2.2   Heating devices should be kept in order and the water temperature in the tank should be checked daily during freezing weather to maintain a minimum temperature of 40°F (4°C).

2-2.3   The tank roof should be kept tight and in good repair, with the hatches fastened closed and the frostproof casing of the tank riser in good repair.

2-2.4   Ice should not be allowed to form on any part of the tank structure. The prevention of freezing in the riser or the formation of ice in the tank itself is extremely important. Freezing in the riser of an elevated tank may obstruct the flow of water from the tank. The formation of a layer of ice on the water of elevated or suction tanks also may impede or prevent the flow from the tank. The formation of heavy icicles through leaking of the tank is dangerous as tank collapse may ensue or people may be endangered by falling icicles.

2-2.5   The bases of the tower columns should be kept free from dirt and rubbish which would otherwise permit the accumulation of moisture with consequent corrosion. The tops of foundation piers should always be at least 6 in. (152 mm) above the ground level.

Coal or ashes or combustible material of any kind should not be piled near the columns as this may cause failure of the steel work due to fire, heating or corrosion. The tank site should be kept cleared of weeds, brush, and grass.

2-2.6   Before repainting, the surface should be thoroughly dried and all loose paint, rust, scale and other surface contamination should be removed. After proper surface preparation, the original paint system should be restored. It may be necessary or economical to repaint the entire inside surface. On the exterior, normal maintenance will involve local patching and periodic application of one complete finish coat when the preceding has weathered thin or for improved appearance after patching.

The painters should not allow any scrapings or other foreign material to fall down the riser or outlet. If the opening is covered for protection, only a few sheets of paper tied over the end of the settling-basin-stub should be used. The paper should be removed upon the completion of the job.

For detailed information refer to NFPA 22, *Standard for Water Tanks for Private Fire Protection*, Care and Maintenance Section.

2-2.7 Necessary periodic emptying of steel tanks for repainting can be minimized by use of a cathodic corrosion prevention system that counteracts the natural electrolytic action that is the basis for most corrosion. Such a system needs periodic attention to the condition of suspended electrodes. If chemical water additives are used to inhibit corrosion, semi-annual chemical analysis of the water should be made. (*See also NFPA 22, Standard for Water Tanks for Private Fire Protection, Section A-2-7.13.*)

If cathodic protection is maintained in a steel tank, the tank should be cleaned out sufficiently often to prevent sediment and scale entering the discharge pipe.

2-2.8 The authority having jurisdiction should always be notified in advance when and for how long the tank is to be out of service.

2-3 Pressure Tanks. (*See NFPA 22, Standard for Water Tanks for Private Fire Protection.*)

2-3.1 Pressure tanks should be inspected regularly, checking the water level and air pressure monthly.

2-3.2 The interior of pressure tanks should be inspected carefully at three-year intervals to determine if corrosion is taking place and if repainting or repairing is needed. When necessary, they should be thoroughly scraped and wire brushed and repainted with an approved metal-protective paint.

2-3.3 Applicable safety codes should be consulted with respect to the maintenance and testing of pressure tanks.

2-3.4 The tank should be pressure tested at intervals as required by the ASME, *Non-Fired Pressure Vessel Code.*

2-3.5 Sight gage valves should be kept closed except when test for water level is being made.

By keeping these valves closed, the water and air in the tank are isolated from the sight glass, so that breaking of the glass will not affect the volume of water nor the pressure available for fire protection.

2-3.6 The tank and its supports should be examined and painted as recommended for gravity tanks.

2-3.7 The heat within the tank enclosure should be checked daily during cold weather to maintain a 40°F (4°C) room temperature.

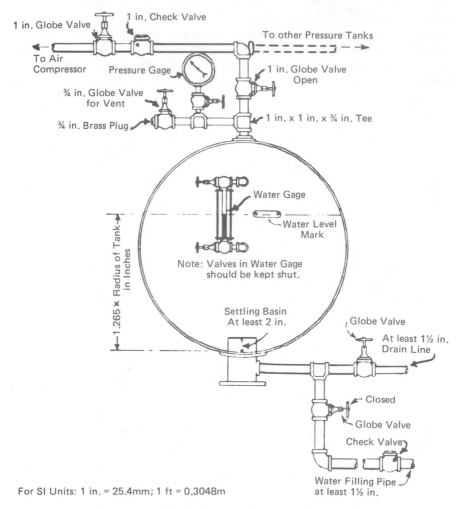

For SI Units: 1 in. = 25.4mm; 1 ft = 0.3048m

Figure 2.2.   Typical connection to pressure tanks.

2-4 Fire Pumps.   *(See NFPA 20, Standard for Installation of Centrifugal Fire Pumps and NFPA 21, Standard for the Operation and Maintenance of National Standard Steam Fire Pumps.)*

2-4.1 General.

2-4.1.1   The pump room should be kept clean and accessible at all times. The fire pump, driver and controller should be protected against possible interruption of service through damage caused by explosion, fire, flood, earthquake, rodents, insects, windstorm, freezing, vandalism and other adverse conditions.

2-4.1.2   The suction pipes, intakes, foot valves, and screens of fire pumps should be examined frequently to make sure that they are free from any obstruction. Mud, gravel, leaves and other foreign material entering the suction pipe may cause damage to the pump or obstruction of the piping of the sprinkler system. The formation of ice may also impair the operation of the pump.

NOTE: Horizontal pumps should be provided with water under a positive head.

2-4.1.3   Suitable means should be provided for maintaining the temperature of a pump room or pump house, where required, above 40°F (4°C). Where pumps are driven by internal combustion engines the temperature of the pump room, pump house, or area where engines are installed should never be less than the minimum recommended by the engine manufacturer.

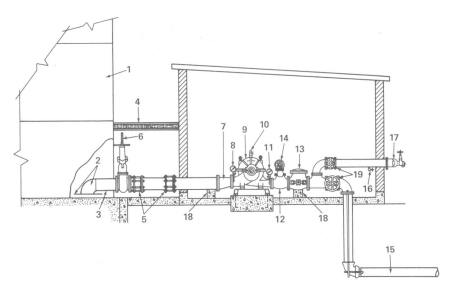

Figure 2.3.   Horizontal split case fire pump installation with the water supply under a positive head: 1) aboveground suction tank; 2) entrance elbow; 3) suction pipe; 4) frostproof casing; 5) flexible couplings for strain relief; 6) O.S. & Y. gate valve; 7) eccentric reducer; 8) suction gage; 9) horizontal split case fire pump; 10) automatic air release; 11) discharge gage; 12) reducing discharge tee; 13) discharge check valve; 14) relief valve if required; 15) discharge pipe; 16) drain valve or ball drip; 17) hose valve manifold with hose valves; 18) pipe supports; 19) indicating gate or indicating butterfly valve.

2-4.1.4   Pump rooms and pump houses should be dry and free of condensate. Accumulation of water in the steam pump supply line or drainage equipment may be dangerous and should be avoided. Where condensate is a problem some heat should be provided.

2-4.1.5   Fire pumps should be operated only in connection with fire protection service and not for plant use.

2-4.1.6   Oil in internal combustion engine pumps should be changed in accordance with manufacturer's instructions, but not less than annually.

2-4.1.7   Storage batteries should be tested frequently to determine the condition of battery cells and the amount of charge in the battery. Only distilled water should be used in battery cells. The plates should be kept submerged at all times.

2-4.1.8   Fuel storage tanks should be kept full at all times.

2-4.2   Periodic Operation and Testing.

2-4.2.1   The pump should be operated every week at rated speed. Inspect the condition of the pump, bearings, stuffing boxes, suction pipe strainers and the various other details pertaining to the driver and control equipment. The examination should be extended to include the condition and reliability of the electric power supply and, if the pump is engine driven, the storage batteries, lubricated system and oil and fuel supplies.

Exception: Electric motor driven fire pumps should be tested monthly.

The packing glands of horizontal shaft centrifugal pumps are part of the pump's lubrication system and should drip slowly when the pump is in operation.

2-4.2.2   When automatically controlled pumping units are to be tested weekly by manual means, at least one start should be accomplished by reducing the water pressure either with the test drain on the pressure sensing line or with a larger flow from the system.

The pressure switch in the control panel that activates the pump automatically is connected to the system through small diameter, noncorrosive piping. The test drain valve will be found in that piping.

2-4.2.3   If the driver has an internal combustion engine, it should be run for at least 30 minutes to bring it up to normal running temperature and to make sure it is running smoothly at rated speed. Automatically controlled equipment should be arranged to automatically start the engine with the initiating means being a solenoid valve drain on the pressure control line.

2-4.2.4   Steam pumps should be operated until water is discharged freely from the relief valve. Regular inspections should be made: checking the maintenance

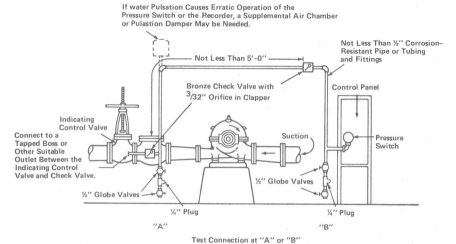

If water Pulsation Causes Erratic Operation of the
Pressure Switch or the Recorder, a Supplemental Air Chamber
or Pulastion Damper May be Needed.

Not Less Than 5'-0''

Not Less Than ½'' Corrosion-
Resistant Pipe or Tubing
and Fittings

Bronze Check Valve with
$3/32''$ Orifice in Clapper

Control Panel

Indicating
Control Valve

Connect to a
Tapped Boss or
Other Suitable
Outlet Between the
Indicating Control
Valve and Check Valve.

Suction

Pressure
Switch

½'' Globe Valves

½'' Globe Valves

¼'' Plug

¼'' Plug

"A"

"B"

Test Connection at "A" or "B"

If Water is Clean, Ground–Face Unions with Noncorrosive Diaphrams
Drilled for $3/32''$ Orifices May be Used in Place of the Check Valves.

For SI Units: 1 in. = 25.4 mm; 1 ft = 0.3048 m

Figure 2.4. Piping connection for each automatic pressure switch for fire pumps or jockey pumps. Solenoid drain valve used for engine-driven pumps may be at "A," "B," or inside of controller enclosure.

of ample pressure; proper supply of lubricating oil; operative condition of relief valve and level of water in the priming tank.

2-4.2.5   A yearly flow test should be made to make sure that neither pump nor suction pipe is obstructed and the pump is operating properly. When the water supply is from a public service main, pump operation should not reduce the suction head at the pump below the pressure allowed by the local authority. At this time both the static and pumping water level of vertical shaft pumps should be determined.

Arrangements must be made for the safe discharge and disposal of the large volume of water.

2-5 Hydrants.   (*See NFPA 24, Standard for Private Fire Service Mains and Their Appurtenances.*)

2-5.1 Inspection of Hydrants.

(a) Public hydrants near the building should be observed for any signs of damage or vandalism.

(b) Private hydrants should be inspected monthly to verify that they are visible and readily accessible with caps in place.

2-5.2 Maintenance.

(a) Lubricate private hydrants twice yearly.

(b) Private hydrants should be serviced as recommended by manufacturers.

2-5.3 Testing.   At least annually, private hydrants should be opened and closed to ensure proper water flow and drainage.

2-6 Riser Flow Tests.

2-6.1   Water flow tests should be made quarterly from water supply test pipes (main drain valves).

Test at the main drain valves includes noting of pressure gage readings with unrestricted flow of water with the drain valve wide open, as compared with the reading with the drain valve closed. If the readings vary materially from those previously established or from normal readings, the condition should be investigated. These tests are intended to show whether or not the normal water supply is available on the system and to indicate the possible presence of closed valves or other obstructions in the supply pipe.

NOTE: Water flow test of a system having a direct connnection to central station or fire department should be made only after proper notice is given to the signal receiving station.

See Figure 2-9.1 of NFPA 13, *Standard for the Installation of Sprinkler Systems.*

2-7 Control Valves.   *(See NFPA 26, Recommended Practices for the Supervision of Valves Controlling Water Supplies for Fire Protection.)*

2-7.1 General.

2-7.1.1   Valves should be numbered and each should have a sign indicating the portion of the system it controls.

2-7.1.2   A valve seal and tag system should be used in connection with the supervision and maintenance of a sprinkler system.

2-7.1.3   Each control valve in the sprinkler system should be secured in its normal or open position by means of a seal, lock, or tamper switch.

2-7.1.4   All control valves of the sprinkler system should be inspected at regular intervals.

(a) Sealed valves — weekly

(b) Locked valves and valves with tamper switches — monthly.

Sections 2-7.1.3 and 2-7.1.4 are the most important in the entire recommended practice. Closed valves are the greatest cause of sprinkler system failure. A conscientiously applied program of securing control valves in the open position in combination with a regular inspection procedure will minimize the likelihood of such an occurrence.

2-7.1.5   If a normally open sprinkler valve is closed, shutting off any part of the system, the owner or manager of the property should be notified immediately so that the owner may follow his normal valve supervision procedure, including notifying the authority having jurisdiction. (*See Chapter 6, Impairments.*)

2-7.1.6   Valves should be kept in normal position and the sprinkler system in service to the greatest extent possible during alterations and repairs.

When alterations or repairs would otherwise take a system out of service for more than a few hours, the portion of the system being modified should be blanked off and the remainder of the system retained in service. Whenever possible, the portion of the system being modified should be returned to service at the completion of each day's work.

When the alteration or repair isolates the system from its supply, a temporary supply, such as connecting the opened main drain valve to an opened hydrant with fire hose, should be provided when possible.

2-7.1.7   After any alterations or repairs, an inspection should be made to ensure that the valves are in the fully open position, properly sealed, locked or equipped with a tamper switch, and the system is in commission.

2-7.1.8   Valve stems should be oiled or greased at least once a year. At this time, completely close and reopen the valve to test its operation and distribute the lubricant.

2-7.2 Valve Inspection Report.

2-7.2.1   A valve inspection report should show that the valves are:

(a) In normal open or closed position

(b) Properly sealed, locked, or equipped with a tamper switch

(c) In good operating condition

(d) Readily accessible

(e) Provided with wrenches where required.

2-7.3 Indicator Post.

2-7.3.1 Quarterly, each post indicator valve should be opened until spring or torsion is felt in the rod assuring that the rod has not become detached from the valve. Valves should be backed one-quarter turn from the wide open position to prevent jamming.

2-7.4 Underground Gate Valves with Roadway Boxes.

2-7.4.1 Quarterly, each valve should be operated with a T-handle wrench to verify that it is in the open position.

2-7.4.2 The location of each such valve should be clearly indicated by a sign on a nearby wall or by a marker. The sign should also indicate direction of valve opening, clockwise or counterclockwise.

2-7.4.3 The roadway box for the valve should always be readily accessible, and the cover should be kept in place.

2-8 Fire Department Connections.

See Section 2-7 of NFPA 13, *Standard for the Installation of Sprinkler Systems.*

2-8.1 Fire department connections should be visible and accessible at all times. They should be inspected monthly.

2-8.2 Caps or plugs should be in place, threads in good condition, ball drip or drain in working order, and check valve not leaking. Prior to replacing caps or plugs, ensure that waterway is clear of foreign material.

2-9 Hose and Hose Stations. *(See NFPA 1962, Standard for the Care, Use and Maintenance of Fire Hose Including Connections and Nozzles.)*

See NFPA 14, *Standard for the Installation of Standpipe and Hose Systems.*

2-9.1 Hose stations should be inspected monthly to ensure that all equipment is in place and in good condition. Hose racks or reels and nozzles should be checked for obvious signs of mechanical damage. Hose station control valves should be checked for signs of leakage.

2-9.2 Hose including gaskets should be removed and re-racked at least annually.

3

AUTOMATIC SPRINKLERS

See Section 3-16 of NFPA 13, *Standard for the Installation of Sprinkler Systems.*

3-1 General.

3-1.1 Sprinklers should be visually checked regularly. Sprinklers should be free from corrosion, foreign material, and paint, and not bent or damaged.

3-1.2 The standard sprinkler is the type manufactured since 1953, incorporating a uniform, hemispherical discharge pattern. Water is discharged in all directions below the plane of the deflector. Little or no water is discharged upward to wet the ceiling. Sprinkler deflectors are stamped as follows:

Upright Sprinkler Marked SSU

Pendent Sprinkler Marked SSP

3-1.3 The old style sprinkler is the type manufactured before 1953. It discharged approximately 40 percent of the water upward to the ceiling. It can be installed in either the upright or pendent position.

3-1.4 Only listed sprinklers may be used. Sprinklers may not be altered in any respect nor have any type of ornamentation, paint or coatings applied after shipment from the place of manufacture.

3-1.5 Corrosion resistant or specially coated sprinklers are installed in locations where chemicals, moisture or other corrosive vapors exist.

3-1.6 Temperature Ratings.

3-1.6.1 The standard temperature rating of automatic sprinklers is shown in Table 3-1.6.1. Automatic sprinklers are manufactured with their frame arms colored in accordance with color code designated in Table 3-1.6.1.

3-1.6.2 When higher temperature sprinklers are necessary to meet extraordinary conditions, special sprinklers as high as 650°F (343°C) are obtainable and may be used.

Table 3-1.6.1   Temperature Ratings, Classifications and Color Codings

Maximum Ceiling Temperature		Temperature Rating		Temperature Classification	Color Code
°F	°C	°F	°C		
100	38	135 to 170	57 to 77	Ordinary	Uncolored
150	66	175 to 225	79 to 107	Intermediate	White
225	107	250 to 300	121 to 149	High	Blue
300	149	325 to 375	163 to 191	Extra High	Red
375	191	400 to 475	204 to 246	Very Extra High	Green
475	246	500 to 575	260 to 302	Ultra High	Orange

3-1.6.3   Information regarding the highest temperature that may be encountered in any location in a particular installation should be obtained by use of a thermometer that will register the highest temperature encountered, which should be hung for several days in the questionable location under the normal ambient temperature condition.

3-2 Replacement Sprinklers.

3-2.1   Care should be taken to ensure that replacement sprinklers have the proper characteristics for the location:

(a) Style

(b) Orifice size

(c) Temperature rating

(d) Coating, if any

(e) Deflector type (upright, pendent, sidewall, etc.).

Standard upright or pendent sprinklers having the characteristics indicated in 3-2.1 in common may be used interchangeably regardless of model or manufacturer. The listing specifications of other types of sprinklers should be checked if the use of a replacement sprinkler of a different model or manufacturer is contemplated. The five characteristics listed materially affect the operation of the sprinkler and, therefore, its ability to extinguish a particular fire. Orifice size is of particular importance, especially when replacing other than ½-in. orifices (when replacing either large or small orifice sprinklers) due to the considerably different hydraulic and discharge characteristics. Replacement of sprinklers with improper temperature rating could result in premature opening where the higher temperature sprinkler was originally specified or installed, due to exposure to some particular heat source such as a unit heater, boiler, etc. This is also important when replacing sprinklers in storage occupancies where higher temperature sprinklers may have been specified to reduce the number of sprinklers opening in the event of fire in this type of high heat release exposure.

In recent years special sprinklers having protection areas or distances between sprinklers different from those specified for standard sprinklers in NFPA 13, *Standard for the Installation of Sprinkler Systems*, have been listed. Extreme care must be taken to ensure that such sprinklers are replaced with sprinklers having comparable characteristics.

3-2.2 Standard sprinklers manufactured after 1952 may be used to replace old style sprinklers manufactured prior to 1953.

Exception: Piers and wharves. See 3-16.2.8 of NFPA 13, Installation of Sprinkler Systems.

Old style sprinklers are used for the protection of fur vaults. See Section 4-4.17 of NFPA 13, *Standard for the Installation of Sprinkler Systems*.

3-2.3 Old style sprinklers may be used to replace existing old style sprinklers.

3-2.4 Old style sprinklers should not be used to replace standard sprinklers without a complete engineering review of the system.

Because of their improved discharge characteristics, standard sprinklers are installed to protect a greater coverage area per sprinkler than was permitted for the old style sprinklers. Old style sprinklers, therefore, installed in systems designed for standard sprinklers will not provide adequate coverage.

3-2.5 Secondhand sprinklers should not be used.

3-3 Automatic Sprinkler Replacement and Testing Program.

3-3.1 Representative samples of solder-type sprinklers with temperature classification of Extra High (325°F)(163°C) or greater which are exposed on a semicontinuous to continuous maximum allowable ambient temperature condition should be tested at 5-year intervals for operation by a testing laboratory acceptable to the authority having jurisdiction.

The fusing element in 360°F solder-type sprinklers will, if exposed to temperatures approaching the fusing element's rating, gradually change its melting point.

3-3.1.1 A representative sample of sprinklers should normally consist of a minimum of two per floor or individual riser, and in any case not less than four, or 1 percent of the number of sprinklers per individual sprinkler system, whichever is greater.

3-3.2 All automatic sprinklers should be replaced when painted, corroded, damaged or loaded with foreign materials or when representative samples fail to pass test.

3-3.3 When sprinklers have been in service for 50 years, representative samples should be submitted to a testing laboratory acceptable to the authority having jurisdiction for operational testing. Test procedure should be repeated at 10-year intervals.

3-3.3.1 Sprinklers made previous to 1920 should be replaced.

3-4 Sprinkler Guards. Sprinklers which are so located as to be subject to mechanical injury should be protected with approved sprinkler guards.

Generally, sprinklers which are located closer than 7 ft (2.1 m) from the floor are considered to be subject to mechanical injury and guards should be considered in these cases. Also, sprinklers located under the rakes of stairwells at lower levels should be provided with guards. Guards are also important in storage locations and for sprinklers installed in storage racks.

3-5 Stock of Spare Sprinklers.

3-5.1 A supply of spare sprinklers (never less than six) should be stored in a cabinet on the premises for replacement purposes. The cabinet should be so located that it will not be exposed to moisture, dust, corrosion, or a temperature exceeding 100°F (38°C).

3-5.1.1 The stock of spare sprinklers should be as follows:

(a) For buildings having not over 300 sprinklers — not less than 6 sprinklers

(b) For buildings having 300 to 1,000 sprinklers — not less than 12 sprinklers

(c) For buildings having over 1,000 sprinklers — not less than 24 sprinklers

(d) Stock of spare sprinklers should include all types and ratings installed.

The stock of spare sprinklers required is a minimum. Spare sprinklers of all types and ratings installed should be available. For an occupancy with a variety of types and ratings of sprinklers installed, the stock of spare sprinklers should be increased above the minimum.

3-5.1.2 A special sprinkler wrench should be provided and kept in the cabinet, to be used in the removal and installation of sprinklers. Other types of wrenches may damage the sprinklers.

3-5.1.3 Automatic sprinklers and fusible links protecting commercial-type cooking equipment and their associated ventilation systems should be inspected twice yearly and replaced annually.

Bulb-type sprinklers and bulb-type spray nozzles showing no buildup of grease or other material need not be replaced.

3-5.1.4 Sprinklers protecting spraying areas should be clean and protected against overspray residue so that they will operate quickly in event of fire. If covered, polyethylene or cellophane bags having a thickness of 0.003 in. (0.076 mm) or less, or thin paper bags should be used. Coverings should be replaced or heads cleaned frequently so that heavy deposits or residue do not accumulate. If not covered, the sprinklers should be replaced annually.

4
SPRINKLER SYSTEM COMPONENTS

4-1 General.

4-1.1 The sprinkler contractor provides instructional literature describing operation and proper maintenance of fire protection devices. This instructional literature should be posted near the system riser.

4-2 Piping.

See Section 3-1 of NFPA 13, *Standard for the Installation of Sprinkler Systems.*

4-2.1 General Provisions.
Piping should be kept in good condition and free from mechanical injury. Sprinkler piping should not be used for support of ladders, stock or other material.

4-2.2 When the piping is subject to corrosive atmosphere, a protective coating that resists corrosion should be provided and maintained in proper condition.

4-2.3 When the age or service conditions of the sprinkler equipment warrant, an internal examination of the piping should be made. When it is necessary to flush a part or all of the piping system, this work should be done by sprinkler contractors or other qualified workers.

4-3 Hangers.

See Section 3-15 of NFPA 13, *Standard for the Installation of Sprinkler Systems.*

4-3.1 Hangers should be kept in good repair. Broken or loose hangers should be replaced or refastened.

4-3.2 Broken or loose hangers may put undue strain on piping and fittings, cause breaks, and interfere with proper drainage.

4-4 Gages.

See Section 2-9.2 of NFPA 13, *Standard for the Installation of Sprinkler Systems.*

4-4.1   Gages on wet-pipe sprinkler systems should be checked monthly to ensure that normal water supply pressure is being maintained. Gages on dry, pre-action and deluge systems should be inspected weekly to ensure that normal air and water pressures are being maintained.

A pressure reading on the gage on the system side of an alarm valve in excess of the pressure recorded on the gage on the supply side of the valve is normal as the highest pressure from the supply will get locked into the system. Equal gage readings could indicate a leak in the system. If there are no visible leaks, it is a good possibility that the alarm valve itself is leaking. See Figure 1-3(d).

4-4.2   Gages should be checked with an inspector's gage every five years.

4-5 Water Flow Alarm Devices.

See Section 3-17.3 of NFPA 13, *Standard for the Installation of Sprinkler Systems.*

4-5.1   Water-flow alarm devices include mechanical water motor gongs, vane-type water flow devices and pressure switches which provide audible and/or visual signals.

4-5.2   Valves controlling water supply to alarm devices should be sealed or locked in the normally open position.

4-5.3   Water-flow alarm devices should be tested at least quarterly, weather permitting.

4-6 Notification to Supervisory Service.

4-6.1   To avoid false alarms where supervisory service is provided, including proprietary, remote alarm receiving facility or fire department, the central station should always be notified before operating any valve or otherwise disturbing the sprinkler system.

4-7 Wet Systems — Alarm Valves.

4-7.1   Test alarms quarterly by opening the inspector's test connection.

Exception: Where weather conditions or other circumstances prohibit using the inspector's test connection, the by-pass test connection may be used.

4-7.2   Cold weather valves should be closed at the approach of freezing weather. Drain the piping in the area subject to freezing. The drain valves on the exposed piping should be left slightly open. (Automatic protection should be restored when danger of freezing is past.)

NOTE: To provide year-round protection, it is recommended that cold weather valves be replaced with dry-pipe valves or antifreeze systems.

One manufacturer refers to its 2-in. dry-pipe valve as being a cold weather valve. This section refers only to manually operated control valves.

4-7.3   The freezing point of solutions in antifreeze systems should be checked annually by measuring the specific gravity with a hydrometer, and adjusting the solutions if necessary. The use of antifreeze solutions should be in conformity with any state or local health regulations.

See Figures 5-5.4 and A-5-5.3.3 of NFPA 13, *Standard for the Installation of Sprinkler Systems.*

4-7.4   Buildings should be inspected to verify that windows, skylights, doors, ventilators, and other openings and closures will not unduly expose sprinkler piping to freezing. Blind spaces, unused attics, stair towers, low spaces under buildings and roof houses are often subject to freezing.

4-8 Dry Systems — Dry Valves, Accelerators, Exhausters.

See Section 5-2 of NFPA 13, *Standard for the Installation of Sprinkler Systems.*

4-8.1   Dry-pipe systems should not be converted to wet-pipe during warm weather. This will cause corrosion and accumulation of foreign matter in the pipe system and loss of alarm service.

4-8.2 Inspection and Maintenance.

4-8.2.1   The priming water should be inspected quarterly and maintained at the proper level as recommended by the dry valve manufacturer.

4-8.2.2   Grease or other sealing material must not be used on seats of dry, pipe valves. Force should not be used in attempting to make dry valves tight.

4-8.2.3   Test water flow and low air pressure alarms and perform a water-flow test through the main drain connection quarterly.

A valved bypass is provided in the dry-pipe valve trim to facilitate water flow alarm tests. It should be utilized when the alarms are to be tested without a trip test of the dry-pipe valve. [See Figure 1-3(b)].

4-8.2.4   The air or nitrogen pressure on each dry-pipe system should be checked at least once a week and maintained as per manufacturer's instructions. All leakage resulting in pressure loss greater than 10 psi (0.7 bar) per week should be repaired.

4-8.2.5   The dry-pipe valve enclosure should be maintained at a minimum temperature of 40°F (4°C).

4-8.2.6   Before and during freezing weather, all low-point drains on dry-pipe systems should be drained as frequently as required to remove all moisture. This process should be repeated daily until all condensate has been removed. The freezing of a small amount of water in the system piping may cause rupture of the sprinkler system resulting in extensive damage to the sprinkler system and water damage to the building and contents. Drum drip assemblies should be in a warm area or in a heated enclosure, when practical.

See Figure 3-11.3.3 of NFPA 13, *Standard for the Installation of Sprinkler Systems.*

The dry-pipe valve need not be tripped for moisture to enter a dry-pipe system. It will condense out of the air pressurizing the piping. When draining drum drips, the normally open top valve is closed to isolate the drum drip from the system.

4-8.3 Trip Tests.

4-8.3.1   Trip tests of each dry-pipe valve, including quick-opening devices, if any, should be done in the spring to allow all condensate to drain from the system piping. At this time, thoroughly clean the dry-pipe valve, renew parts as required and reset the valve.

4-8.3.2   Each dry-pipe valve should be trip tested with control valve partially open, cleaned and reset at least once each year during warm weather. The shut-off valve should be kept open at least far enough to permit full flow of water at good pressure through the main drain when it is fully opened.

4-8.3.3   Before any dry-pipe valve is tripped or tested, the water supply line to it should be thoroughly flushed. The main drain below the valve should be opened wide, and water at full pressure should be discharged long enough to clear the pipe of any accumulation of scale or foreign material. If there is a hydrant on the supply line, this hydrant should be flushed before the main drain is opened.

4-8.3.4 Caution: The tripping of dry-pipe valves with throttled water supplies will not completely operate some models that require a high rate of flow to complete movement of the clapper assemblies.

4-8.3.5   All dry-pipe valves should have a tag or card attached showing the date on which the valve was last tripped and showing the name of the person and the organization making the test. Separate records of initial air and water pressures, tripping time and tripping air pressure, and dry-pipe valve operating condition should be kept for comparison with previous test records.

4-8.4 Trip Test Full Flow.   Each dry-pipe valve should be trip tested with control valve wide open at least once every three years or when the system is altered. This test should be conducted by opening the inspector's test pipe. Test should be terminated when the dry-pipe valve has tripped and clean water is flowing at the inspector's test pipe.

A full flow trip test is recommended only once each three years with a restricted flow trip test the other two years because, while full flow tests must be periodically conducted, they have some undesirable side effects. The high velocity flow will tend to draw foreign material into the system, and the wetting of the pipe wall will result in the development of scale. There is also the problem of draining all the water from the system.

4-9 Air Compressor.

4-9.1   An air compressor should be lubricated only if recommended by the manufacturer and in accordance with his instructions. The motor unit should be kept dirt free. Filters and strainers should be cleaned as required. Crystals in air dryers should be replaced when color changes indicate they have absorbed moisture.

4-10 Air Maintenance Device.

4-10.1   Strainers, filters and restriction orifices should be cleaned as required. If regulator is provided with a drain cock, periodically remove condensation.

4-11 Quick-Opening Devices (Accelerator or Exhauster).

See Section 5-2.4 of NFPA 13, *Standard for the Installation of Sprinkler Systems.*

4-11.1   The quick-opening device should be tested at least twice a year.

The manufacturer's instructions for testing and reseting the device should be carefully followed. When the device does not operate properly on test, the dry-pipe system should be kept in service and the device repaired or replaced immediately. Repair parts or replacement device should be obtained from the original manufacturer.

4-12 Deluge, Preaction and Automatic On-Off Preaction Systems.

See Section 5-3 of NFPA 13, *Standard for the Installation of Sprinkler Systems.*

4-12.1   Complete charts are furnished by the installing company, showing the proper method of operating and testing these systems. Only competent mechanics fully instructed with respect to the details and operation of such systems should be employed in their repair and adjustment. It is highly advisable for the owner to arrange with the installing company for at least annual inspection and testing of the equipment.

4-12.2   In preaction systems when it is necessary to repair the actuating system, as distinguished from the piping system itself, the water may be turned into the sprinkler piping, and automatic sprinkler protection thus maintained without alarm service, provided there is no danger of freezing.

4-12.3   Test detection systems semi-annually and alarms quarterly according to the procedures suggested by the manufacturer.

5

FLUSHING

5-1 Flushing.

5-1.1   For effective control and extinguishment of fire, automatic sprinklers should receive an unobstructed flow of water. Although the overall performance record of automatic sprinklers has been very satisfactory, there have been numerous instances of impaired efficiency because sprinkler piping or sprinklers were plugged with pipe scale, mud, stones, or other foreign material. If the first sprinklers to open in a fire are plugged, the fire in that area will not be extinguished, an excessive number of sprinklers will operate causing increased water damage, and possibly the fire will spread out of control.

5-2 Types of Obstruction Material.

5-2.1   Obstructions may consist of compacted fine materials, such as rust, mud, or sand. Pipe scale is found more frequently in dry-pipe than in wet systems. Dry-pipe systems that have been maintained wet or dry alternately over a period of years are particularly susceptible to the accumulation of scale. Also, in systems continuously dry, condensation of moisture in the air supply may result in the formation of a hard scale along the bottom of the piping. When sprinklers open, the scale is broken loose and carried along the pipe, plugging some of the sprinklers or forming obstructions at the fittings.

5-2.2   Stones of various sizes, cinders, cast-iron pipe tubercles, chips of wood, or other coarse materials may be found. Sprinkler piping is sometimes partially obstructed by such objects as pieces of wood, paint brushes, broken pump valves or springs, or excess materials from improperly poured pipe joints. Materials may be sucked from the bottom of streams or reservoirs by fire pumps with poorly arranged or inadequately screened intakes and forced into the system. Sometimes floods damage intakes. Other materials may be permitted to enter by careless workers during installation or extensions of mains.

5-3 Preventing Entrance of Obstructive Material.   The following measures should be taken to assure, as far as possible, that sprinkler systems are clear of obstructive foreign matter and will remain unobstructed.

5-3.1   Take care when installing underground mains, both public and private, to prevent entrance of stones, soil, or other foreign material. As assurance that such material has not entered newly installed sprinkler systems from underground mains, installers are required as a condition of acceptance to flush all newly installed mains before connecting the inside piping. Private fire service mains should also be flushed after repairs or when breaks have occurred in public mains.

5-3.2   Screen pump suction supplies and maintain screens in good condition. Equip connections from penstocks with strainers or grids, unless the penstock inlets themselves are so equipped.

5-3.3   Keep dry-pipe systems on air the year round, instead of alternately on air or water, to inhibit formation of rust and scale.

5-3.4   Use extreme care when cleaning tanks and open reservoirs to prevent material from entering piping. Materials removed from the interior of gravity tanks during cleaning should not be permitted to enter the discharge pipe.

5-4 Conditions Showing Need for Investigation.   Although precautions for preventing entrance of obstructive materials are generally followed at well-maintained premises, evidence based on fire experience and hundreds of flushing investigations shows that some sprinkler systems are obstructed to an extent that would seriously impair their effectiveness during a fire.

5-4.1   Conditions that may indicate the need of investigation include the following:

(a) Defective intake screens for fire pumps taking suction from streams and reservoirs.

(b) Discharge of obstructive material during routine water tests.

(c) Foreign material in fire pumps, in dry-pipe valves, or in check valves.

(d) Heavy discoloration of water during drain tests or plugging of inspector's test connections.

(e) Plugging of sprinklers.

(f) Plugged piping in sprinkler systems dismantled during building alterations.

(g) Failure to flush underground mains following installations or repairs.

(h) A record of broken public mains in the vicinity.

(i) Abnormally frequent tripping of dry-pipe valve.

5-4.2 Sprinkler systems should be examined internally at periodic intervals for obstructions. Where unfavorable conditions such as those itemized above are found, the system should be examined at five-year intervals after installation or possibly sooner. Where conditions are favorable, dry-pipe systems should be examined at ten-year intervals after installation.

5-4.3 Dry-pipe systems found obstructed should be flushed and re-examined at intervals of not more than five years.

5-5 Precautions. When sprinkler systems are to be shut off for investigation or for flushing, take all the precautions outlined earlier. To prevent accidental water damage, control valves should be shut tight and the system completely drained before sprinkler fittings are removed or pipes disconnected. Cover stock and machinery susceptible to water damage, and provide equipment for mopping up any accidental discharge of water.

5-5.1 Large quantities of water are required for effective flushing by the hydraulic method, and it is important to plan in advance the most convenient methods of disposal.

5-6 Investigation Procedure.

5-6.1 From the plan of the fire protection system, determine the sources of water supply, age of mains and sprinkler systems, types of systems, and general piping arrangement.

5-6.2 Examine the fire pump suction supply and screening arrangements. If needed, have the suction cleaned before using the pump in tests and flushing operations. Gravity tanks should be inspected internally, except steel tanks recently cleaned and painted. If possible, have the tank drained and determine whether there is loose scale on the shell or sludge or other obstruction on the tank bottom. Cleaning and repainting may be required, particularly if not done within five years.

5-7 Test—Flushing Mains.

5-7.1 Use hydrants near the ends of mains for flow tests to determine whether mains contain obstructive material. If such material is found, mains should be thoroughly flushed before investigating sprinkler systems. Connect two lengths of 2½-in. hose to the hydrant. Attach burlap bags to free ends of the hose from which the nozzles have been removed to collect any material flushed out, and flow water long enough to determine condition of the main being investigated. If there are several sources of water supply, investigate each independently avoiding any unnecessary interruptions to sprinkler protection. On extensive layouts repeat the tests at several hydrants to determine general conditions.

5-8 Testing Sprinkler Systems.

5-8.1   Investigate the dry systems first. Tests on several carefully selected, representative systems usually are sufficient to indicate general conditions throughout the premises. When preliminary investigations indicate considerable obstructive material, this would justify investigating all systems (both wet and dry) before outlining needed flushing operations.

5-8.2   In selecting specific systems or pipes for investigation, consider:

(a) Pipes found obstructed during a fire or during maintenance work

(b) Systems adjacent to points of recent repair to yard mains, particularly if hydrant flow shows material in the main

(c) Pipes involving long horizontal runs of feed and cross mains. Obstructions are most likely to be found in the most remote branch lines at the end of the longest cross main from the longest feed main, particularly if the branch lines are lower than part of the feed main, as under a deck or platform.

5-8.3   Tests should include flows through 2½-in. fire hose directly from cross mains [*see Figures 5-8.3(a) and 5-8.3(b)*] and flows through 1½-in. hose from representative branch lines.

5-8.4   The fire pump should be operated for the large volume flows, as maximum pressure is desirable. Burlap bags should be used to collect dislodged material as is done in the investigation of yard mains, and each flow should be continued long enough to show the condition of the piping interior. After a test, leave all valves open and locked or sealed.

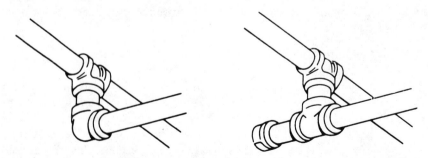

Figure 5-8.3(a)   Replacement of Elbow at End of Cross Main with a Flushing Connection Consisting of a 2-in. Nipple and Cap.

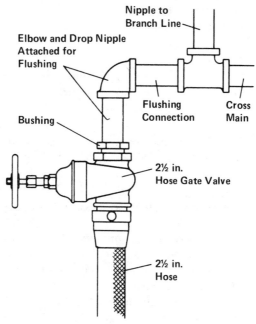

Figure 5-8.3(b)   Connection of 2½-in. Hose Gate Valve with 2-in. Bushing, and Nipple and Elbow to 2-in. Cross Main.

5-9 Dry-Pipe Systems.

5-9.1 Having selected the test points of a dry-pipe system, close the main control valve and release air from the system. Check the piping visually with a flashlight while it is being dismantled. Attach hose valves and 1½-in. hose to ends of branch lines to be tested, shut these valves, and have air pressure restored on the system and the control valve reopened. Open the hose valve on the end branch line allowing the system to trip in simulation of normal action. If this test plugs the hose or piping, the extent of plugging should be noted and cleared from the branch line before proceeding with further tests.

5-9.2 After flowing the small end line, shut its hose valve and test the feed or cross main by discharging water through a 2½-in. fire hose, collecting any foreign material in a burlap bag.

5-9.3 After the test, the dry-pipe valve should be internally cleaned and reset in the normal manner. Its control valve should be opened, sealed and a drain test made.

5-10 Wet Systems.

5-10.1 Testing of wet systems is similar to that of dry systems except that the system must be drained after closing the control valve to permit installation of hose valves for the test. Slowly reopen the control valve and make a small hose flow as prescribed for the branch line, followed by the 2½-in. hose flow for the cross main.

5-10.2 In any case, if lines become plugged during the tests, piping must be dismantled and cleaned, the extent of plugging noted, and a clear flow obtained from the branch line before proceeding further.

5-10.3 Make similar tests on representative systems to indicate the general condition of the wet systems throughout the installation, keeping a detailed record of what is done.

5-11 Outside Sprinklers for Protection Against Exposure Fires.

5-11.1 Outside or open sprinkler equipment should be flow tested once each year during warm weather. Before making flow tests, proper precautions should be taken to prevent damage from water discharge. Flow tests will determine that the sprinklers and the system piping are in good condition and free of obstructions. Obstructed sprinklers or piping should be cleared immediately.

5-12 Flushing Procedure.

5-12.1 If investigation indicates the presence of sufficient material to obstruct sprinklers, a complete flushing program should be carried out. The work should be done by qualified competent personnel. The source of the obstructing material should be determined and steps taken to prevent further entrance of such material. This entails such work as inspection and cleaning of pump suction screening facilities or cleaning of private reservoirs. If recently laid public mains appear to be the source of the obstructing material, waterwork authorities should be requested to flush their system. For recommendations and procedures for cleaning pump suctions see NFPA 20, *Standard for the Installation of Centrifugal Fire Pumps.*

5-13 Private Fire Service Mains.

5-13.1 Mains should be thoroughly flushed before flushing any interior piping. Flush piping through hydrants at dead ends of the system or through blowoff valves, allowing the water to run until clear. If the water is supplied from more than one direction or from a looped system, close divisional valves to produce a high velocity flow through each single line. A velocity of at least 6 ft per second (1.8 m/s) is necessary for cleaning the pipe and for lifting foreign material to an aboveground flushing outlet. Use the flow specified in Table 5-13.1 for the size of the main under investigation.

Table 5-13.1   Waterflow Recommended for Flushing Piping

Size of Pipe	Flow	
In.	gpm	(L/min.)
4	400	(1514)
6	750	(2839)
8	1000	(3785)
10	1500	(5678)
12	2000	(7570)

5-13.2   Connections from main to sprinkler riser should be flushed. Although flow through a short open-ended 2-in. drain may create sufficient velocity in a 6-in. main to move small obstructing material, the restricted waterway of the globe valve usually found on a sprinkler drain may not allow stones and other large objects to pass. If presence of large size material is suspected, a larger outlet will be needed to pass such material and to create the 750-gpm (2839-L/min) flow necessary to move it. Fire department connections on sprinkler risers can be used as flushing outlets by removing the clappers. Mains can also be flushed through a temporary siamese fitting attached to the riser connection before the sprinkler system is installed (*see Figure 5-13.2*).

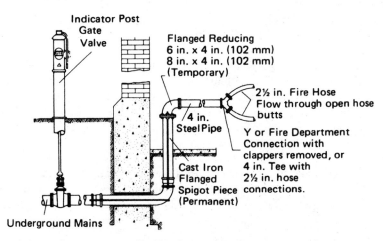

Figure 5-13.2   Arrangement for Flushing Branches from Underground Mains to Sprinkler Risers.

5-13.3 Sprinkler Piping.

5-13.3.1   Two methods are commonly used for flushing sprinkler piping: (a) the hydraulic method and (b) the hydropneumatic method.

(a) The hydraulic method consists of flowing water progressively from the mains, sprinkler risers, feed mains, cross mains, and finally the branch lines in the same direction in which it would flow during a fire.

(b) The hydropneumatic method utilizes special equipment and compressed air to blow a charge of approximately 30 gal (114 L) of water from the ends of branch lines back into feed mains and down the riser, washing the foreign material out of an opening at the base of the riser.

The hydraulic method of flushing gridded systems is explained in 5-13.4.8. This recommended practice does not address hydropneumatic flushing of gridded systems, as there has not yet been enough experience to establish a preferred procedure.

5-13.3.2 The choice of method depends on conditions at the individual premises. If examination indicates the presence of loose sand, mud, or moderate amounts of pipe scale, the piping can generally be satisfactorily flushed by the hydraulic method. Where the material is more difficult to remove and available water pressures are too low for effective scouring action, the hydropneumatic method is generally more satisfactory.

5-13.3.3 In some cases, where obstructive material is solidly packed or adheres tightly to the walls of the piping, the pipe will have to be dismantled and cleaned by rodding or other positive means.

5-13.3.4 Successful flushing by either the hydraulic or hydropneumatic method is dependent on establishing sufficient velocity of flow in the respective pipes to remove silt, scale, and other obstructive material. With the hydropneumatic method, this is accomplished by the air pressure behind the charge of water.

5-13.3.5 When flushing a branch line through the end pipe, sufficient water should be discharged to scour the largest pipe in the branch line. Lower rates of flow may reduce the efficiency of the flushing operation. To establish the recommended flow, remove small end piping and connect hose to larger section, if necessary.

5-13.3.6 To determine that the piping is clear after it has been flushed, representative branch lines and cross mains should be investigated, using both visual examination and sample flushings.

5-13.3.7 Whenever any section of piping is found severely or completely obstructed with packed material, such as hard scale, cinders, or gravel, the piping will usually have to be disassembled to remove the material.

5-13.3.8   Where pipe scale indicates internal or external corrosion, a section of the pipe affected should be thoroughly cleaned to determine if the walls of the pipe have seriously weakened.

5-13.4 Hydraulic Method.   After the mains have been thoroughly cleared, flush risers, feed mains, cross mains, and finally the branch lines. Following this sequence will prevent drawing obstructing material into the interior piping.

5-13.4.1   Water should be turned into dry-pipe systems for one to two days before flushing, if possible, to soften pipe scale and deposits. When alarm is turned off due to this procedure, consideration should be given to providing watch service during the unattended hours. To flush risers, feed mains, and cross mains, attach 2½-in. hose gate valves to the extreme ends of these lines [*see Figure 5-8.3(b).*] Such valves usually can be procured from the manifold of fire pumps or hose standpipes. As an alternative, an adapter with 2½-in. hose thread and standard pipe thread can be used with a regular gate valve. A length of fire hose without a nozzle should be attached to the flushing connection. To prevent kinking of the hose and to obtain maximum flow, an elbow should usually be installed between the end of the sprinkler pipe and the hose gate valve. Attach the valve and hose so that no excessive strain will be placed on the threaded pipe and fittings. Support hose lines properly.

5-13.4.2   Where feed and cross mains and risers contain pipe 4, 5, and 6 in. in size, it may be necessary to use a fire department connection with two hose connections to obtain sufficient flow to scour this larger pipe.

5-13.4.3   In multistory buildings, systems should be flushed by starting at the lowest story and working up. Branch line flushing in any story may follow immediately the flushing of feed and cross mains in that story, allowing one story to be completed at a time.

5-13.4.4   Where a repetition of the trouble is probable, leave a 2-in. capped nipple at the ends of the cross mains for flushing piping. Sprinkler installation rules require that a flushing connection be provided at the end of each cross main terminating in 1¼-in. or larger pipe.

5-13.4.5   Flush branch lines after feed and cross mains have been thoroughly cleared. This will avoid drawing obstructing material from these pipes into the branches. Equip the ends of several branch lines with gate valves, and flush individual lines of the group consecutively. This will eliminate the need for shutting off and draining the sprinkler system to change a single hose line. The hose should be 1½ in. and as short as practicable. Branch lines may be flushed in any order that will expedite the work.

5-13.4.6   Branch lines also may be flushed through pipe 1½ in. or larger extending through a convenient window. If pipe is used, 45° elbows should be provided at the ends of branch lines. When flushing branch lines, hammering the pipes is an effective method of moving obstructions.

5-13.4.7   All pendent sprinklers should be removed and cleaned of obstructions.

5-13.4.8 Flushing Gridded Sprinkler Systems.   All new gridded sprinkler systems should be arranged so that they can be thoroughly flushed. Figure 5-13.4.8 should be used as a guide. Other arrangements accomplishing the same results are acceptable.

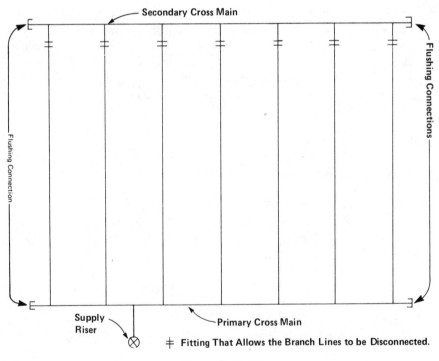

Figure 5-13.4.8

In the case of a system such as in Figure 5-13.4.8, the flushing procedure is as follows:

(a) Disconnect all branch lines close to the secondary cross main, and cap or valve all open ends supplied by the primary cross main. Visually examine the interior of each branch line connected to the secondary cross main and plug or cap.

(b) Flush the primary cross main, first one end, then the other.

(c) Flush each branch line independently.

(d) Flush the secondary cross main from an auxiliary water source.

(e) Reconnect branch lines.

(f) Flush the secondary cross main, first one end, then the other.

5-13.5 Hydropneumatic Method. The apparatus used for hydropneumatic flushing consists of a hydropneumatic machine, a source of compressed air, 1-in. (25-mm) air supply hose, 1½-in. hose for connecting to the sprinkler system, and 2½-in. hose.

5-13.5.1 The hydropneumatic machine (*see Figure 5-13.5.1*) consists of a 30-gal (114-L) water tank mounted over a 25-cu ft [(185-gal) (700-L)] compressed air tank. The compressed air tank is connected to the top of the water tank through a 2-in. lubricated plug cock. The bottom of the water tank is connected through hose to a suitable water supply. The compressed air tank is connected through suitable air hose to either the plant air system or a separate air compressor.

5-13.5.2 To flush the sprinkler piping, the water tank is filled with water, the pressure raised to 100 psi (6.90 bars) in the compressed air tank, and the plug cock between tanks opened to put air pressure on the water. The water tank is connected by hose to the sprinkler pipe to be flushed. Then the lubricated plug cock on the discharge outlet at the bottom of the water tank is snapped open, permitting the water to be "blown" through the hose and sprinkler pipe by the compressed air. The water tank and air tank should be recharged after each blow.

5-13.5.3 Outlets for discharging water and obstructing material from the sprinkler system must be arranged. With the clappers of dry-pipe valves and alarm check valves on their seats and cover plates removed, sheet metal fittings can be used for connection to 2½-in. hose lines or for discharge into a drum. [Maximum capacity per blow is about 30 gal (114 L).] If the main riser drain is to be used, the drain valve should be removed and a direct hose connection made. For wet-pipe systems with no alarm check valves, the riser should be taken apart just below the drain opening and a plate inserted to prevent foreign material from dropping to the base of the riser. Where dismantling of a section of the riser for this purpose is impractical, the hydropneumatic method should not be used.

5-13.5.4 Before starting a flushing job, each sprinkler system to be cleaned should be studied and a schematic plan prepared showing the order of the blows.

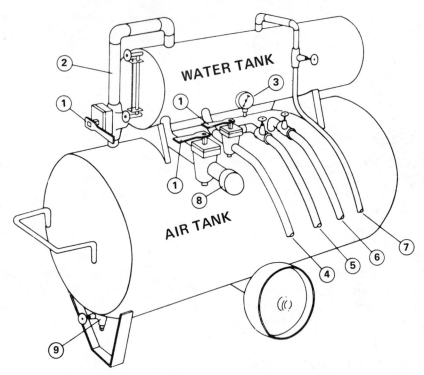

Figure 5-13.5.1   Hydropneumatic Machine.

1. Lubricated plug cocks.

2. Pipe connection between air and water tanks. This connection is open when flushing sprinkler system.

3. Air pressure gage.

4. 1-in. (25-mm) rubber hose (air type). Used to flush sprinkler branch lines.

5. Hose connected to source of water. Used to fill water tank.

6. Hose connected to ample source of compressed air. Used to supply air tank.

7. Water tank overflow hose.

8. 2½-in. pipe connection. When flushing large interior piping, connect woven jacket fire hose here and close 1-in. (25-mm) plug cock hose connection (4) used for flushing sprinkler branch lines.

9. Air tank drain valve.

5-13.5.5 Hydropneumatic Method of Flushing Branch Lines.   With the mains already flushed or known to be clear, the branch sprinkler lines should next be flushed. The order of cleaning individual branch lines must be carefully laid out if an effective job is to be done. In general, flush the branch lines starting with the line closest to the riser and work toward the dead end of the cross main (*see Figure 5-13.5.5*).

The order of flushing the branch lines is shown by the circled numerals. In this example the southeast quadrant is flushed first, then the southwest, next the northeast, and last the northwest.

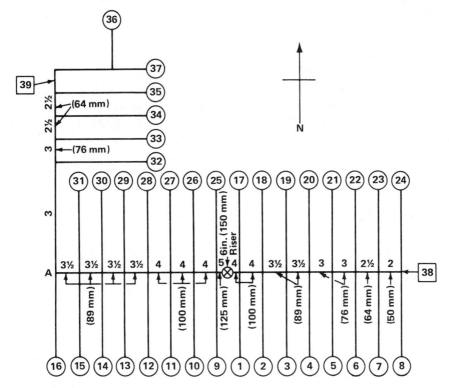

Figure 5-13.5.5   Schematic Diagram of Sprinkler System Showing Sequence to Be Followed when Hydropneumatic Method Is to Be Used.

5-13.5.5.1   Air hose 1 in. (25 mm) in diameter is used to connect the machine with the end of the branch line being flushed. This hose should be as short as practicable. When the blow is made, the air pressure should be allowed to drop to 85 psi (5.9 bars) before the valve is closed. The resulting short slug of water will have less friction loss and a high velocity and hence do a more effective cleaning job than if the full 30 gal (114 L) of water is used. One blow is made for each branch line.

5-13.5.6 Hydropneumatic Method of Flushing Large Interior Piping.

When flushing cross mains, completely fill the water tank and raise the pressure in the air receiver to 100 psi (6.9 bars). Connect the machine to the end of the cross main to be flushed with not over 50 ft (15.2 m) of 2½-in. hose. After opening the valve, allow air pressure in the machine to drop to zero. Two to six blows are necessary at each location, depending on the size and length of the main.

5-13.5.6.1   In Figure 5-13.5.5, the numerals in squares indicate the location and order of the cross main blows. Since the last branch line blows were west of the riser, clean the cross main east of the riser first. Where large cross mains are to be cleaned, it is suggested, if practical, to make one blow at 38, one at 39, the next at 38, then at 39, alternating in this manner until the required number of blows has been made at each location.

5-13.5.6.2   Cross mains are best flushed by introducing the blow at a point where water moving through the piping will make the least number of right-angle bends. In Figure 5-13.5.5, blows at 39 should be adequate to flush the cross mains back to the riser. Do not attempt to clean the cross mains from A to the riser by backing out branch line 16 and connecting the hose to the open side of the tee. If this were done, a considerable portion of the blow would pass northward up the 3-in. pipe list supplying branches 32 to 37, and the portion passing eastward to the riser could be ineffective. When the size, length, and condition of cross mains require blowing from a location corresponding to A, the connection should be made directly to the cross main corresponding to the 3½-in. pipe so that the entire flow would travel to the riser.

5-13.5.6.3   When flushing through a tee, always flush the run of tee after flushing the branch. Note the location of blows 35, 36, and 37 in Figure 5-13.5.5.

5-13.5.6.4   When flushing feed mains, arrange the work so that the water will pass through a minimum of right-angle bends.

5-13.5.6.5   The importance of doing a thorough flushing job should be strongly emphasized to those in charge of the work. In a number of instances, sprinklers in systems that had supposedly been flushed clear became plugged during a subsequent fire, permitting it to get out of control and cause serious damage.

6
IMPAIRMENTS

6-1 General. Valves controlling the water supply to all or part of a sprinkler system should be electrically supervised, or sealed, locked, or equipped with tamper switches and should be inspected frequently because of their importance to fire protection. The closing of control valves without proper authorization or preparation can seriously jeopardize operations. If sprinkler systems, fire hydrants, ground storage tanks, gravity tanks, fire pumps, etc. are impaired, the consequences may result in loss of life and damage to property. It is essential that adequate measures are taken during a fire protection impairment to ensure that the increased risks are minimized.

6-2 Impairment Coordinator. A representative of the building owner, manager or tenant should be assigned to coordinate all impairments and restoration of protection.

6-3 Preplanned Impairment Programs.

6-3.1 All preplanned impairments should be authorized by the Coordinator. Before authorization is given, he should be responsible for verifying that the following is accomplished:

(a) Determine the exact extent of the intended impairment.

(b) Inspect the area or buildings to be involved and determine the increased risks.

(c) Submit recommendations to management. Consideration should be given to the need for temporary protection, termination of all hazardous operations and frequent inspections of the area involved with 24-hour per day watchman service.

(d) Notify the fire department.

(e) Notify the insurance carrier, the alarm company, and other appropriate authority and implement Tag Impairment System (if such system is in use).

(f) Notify the supervisors in the areas to be affected.

6-3.2   When all impaired equipment is restored to normal working order, the following should be accomplished:

(a) Verify that all control valves are fully opened and locked, sealed, or equipped with a tamper switch.

(b) Conduct a main drain and alarm test on each sprinkler riser affected.

(c) Maintain as large a portion of the system in service as possible.

(d) Advise supervisors that protection has been restored.

(e) Advise the fire department that protection has been restored.

(f) Advise the insurance carrier, the alarm company, and other appropriate authority that protection has been restored.

6-4 Emergency Impairments.   Emergency impairments include sprinkler leakage, frozen or ruptured piping, equipment failure, etc. When this occurs, appropriate emergency action should be taken. The Coordinator should be contacted, and he should proceed to the extent possible to implement the preplanned impairment program including the restoration of sprinkler protection.

6-5 Restoring Systems to Service after Disuse.

6-5.1   Occasionally, automatic sprinkler systems in idle or vacant properties are shut off and drained. When the equipment in such properties is restored to service, it is recommended that such work be performed by a responsible and experienced sprinkler contractor. In such cases, the following procedures are recommended:

6-5.1.1   All lines of sprinkler piping should be traced from the extremities of the system to the main connections with a careful check for blank gaskets in flanges, closed valves, corroded or damaged sprinklers or piping, insecure or missing hangers and insufficient support. Proper repairs or adjustments should be made and needed extensions or alterations of the equipment should be completed.

6-5.1.2   Air should be used to test the system for leaks before turning on the water. Water should be admitted slowly to the system, with proper precautions against damage by escape of water from previously undiscovered defects. When the system has been filled under normal service pressure, drain valve tests should be made to detect any closed valve that possibly could have been overlooked. All available test pipes then should be flushed, and where such pipes are not provided in accordance with the present standards, the proper equipment should be installed.

6-5.1.3 Where the sprinkler system has been long out of service, damaged by freezing or subject to extensive repairs or alterations, the entire system should be hydrostatically tested in accordance with NFPA 13, *Standard for the Installation of Sprinkler Systems*. Special care should be taken to detect any sprinklers showing leaks and to make replacements where necessary.

6-5.1.4 Dry-pipe valves, quick-opening devices, alarm valves and all alarm connections should be examined, put in proper condition and tested. Fire pumps, pressure and gravity tanks, reservoirs and other water supply equipment should receive proper attention before being placed in service. Each supply should be tested separately.

6-5.1.5 An investigation for obstruction or stoppage in the sprinkler system piping should be made. (*See Chapter 5.*)

6-5.1.6 All control valves should be operated from closed to fully open position and should be left sealed, locked or equipped with a tamper switch.

6-6 Sprinkler System Alterations.

6-6.1 Alterations will usually involve an impairment to all or part of the sprinkler system. Any alteration to a sprinkler system should be done in accordance with NFPA 13, *Standard for the Installation of Sprinkler Systems*, or other applicable NFPA standards.

7

FIRE RECORDS

7-1 Protection Records.

7-1.1   In all businesses, it is desirable to keep records of inspection, testing and maintenance of protection equipment. The exact program for any building or set of buildings should be tailored to a specific plan for the particular building, considering occupancy, types of protection, alarms provided, etc. In the development of a fire protection record plan, it is advisable to consider advice from various sources including the:

Authority Having Jurisdiction (Rating Bureaus, Fire Prevention Bureaus, Fire Marshals, etc.).

Manufacturers of Various Devices

Fire Insurance Companies

Independent Fire Protection Consultants

Sprinkler Contractors

Other Applicable NFPA Codes as outlined in the Appendix.

7-1.2   An individual within the organization should be designated (Protection Record Administrator) to implement inspection, testing and maintenance programs. Some firms may elect to do the basic functions themselves and contract the more technically involved operations.

7-1.3   The person in charge should consider the use of master control records, including a copy of the manufacturer's instructions covering all devices.

7-2 Valve Tag Systems.

7-2.1   Closed valves have caused over 30 percent of all sprinkler system failures. Adoption of the valve tag systems should visually highlight and minimize this significant cause of unsatisfactory sprinkler system performance. (*See Figure 7-2.1.*)

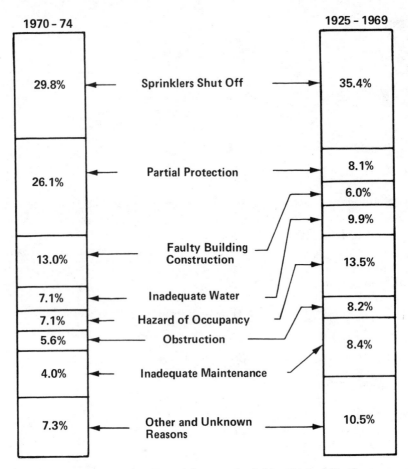

Figure 7-2.1 Reasons for Unsatisfactory Sprinkler System Performance.

7-2.2 Tag Impairment System. This system usually includes a three-part tag which is easily identifiable. One part is tied to a valve to be closed temporarily. The second part is sent to the authority having jurisdiction. The third part is displayed in the protection record administrator's office and sent to the authority having jurisdiction when protection is restored.

7-2.3 Valve Record Tag System. A tag on each valve indicating date sealed or locked, date and results of maintenance procedures should be provided. This provides a chronological record of valve maintenance.

7-3 Inspection Records. Inspection and maintenance records should be kept by the protection record administrator of the following activities:

Table 7-3

Summary: Minimum Inspection – Testing – Maintenance

Records – Inspection	= Visual Observation
Testing	= Handling Equipment, etc.
Maintenance	= Periodic Servicing and Repair

For Guidance on Specific Valves, Pumps, Hydrants, etc. Refer to the Manufacturer's Instructions.

Parts	Activity	Frequency	Section Number
Flushing Piping	Test	5 Years 10 Years	5-4.2
Fire Department Connections	Inspection	Monthly	2-8
Indicator Post Valve or Tamper Switch	Inspection Test	Weekly-Sealed Monthly-Locked Quarterly	2-7.1.4 2-7.3.1 2-7.4.1
Main Drain	Flow Test	Quarterly	2-6.1
Open Sprinklers	Test	Annual	5-11.1
Pressure Gage	Calibration Test	5 Years	4-4.2
Sprinklers	Test	50 Years	3-3.3
Sprinklers High Temp.	Test	5 Years	3-3.1
Valves in Roadway Boxes	Inspection Test	Weekly-Sealed Monthly-Locked Quarterly	2-7.1.4 2-7.3.1 2-7.4.1
Valves on Systems	Inspection Inspection Maintenance	Weekly-Sealed Monthly-Locked Yearly	2-7.1.4 2-7.1.4 2-7.1.8
Water Flow Alarms	Inspection	Monthly	4-12.3
Hydrants	Inspection Test (Open and Close) Maintenance	Monthly Semi-Annually	2-5.1 2-5.3 2-5.2
Antifreeze Solution	Test	Annually	4-7.3
Cold Weather Valves	Open and Close Valves	Fall, Close; Spring, Open	4-7.2
Dry-Pipe Valve Air Pressure, Water Pressure Enclosure Priming Water Level	Inspection Inspection Inspection	Weekly Daily-Cold Weather Quarterly	Chapter 4 4-8.2.5 4-8.2.8
Dry-Pipe Valves	Trip Test	Annual-Spring	1-6.1 4-8.3.1

Dry-Pipe Valves	Full Flow Trip Test	3 Years-Spring	4-8.4
Low Point Drains	Test	Fall	4-8.2.8
Quick Opening Devices	Test	Semi-Annually	4-11.1
Gravity Tank			
Water Level	Inspection	Monthly	2-2.1
Heat	Inspection	Daily-Cold Weather	2-2.2
Condition	Inspection	Bi-Annual	2-2.6
Pressure Tank			
Water Level & Pressure	Inspection	Monthly	2-3.1
Heat Enclosures	Inspection	Daily-Cold Weather	2-3.7
Condition	Inspection	3 Years	2-3.2
Pump, Centrifugal	Test-Operate	Weekly	2-4.2.2
	Test-Reduced Water	Weekly	2-4.2.2
	Pressure Start	Weekly	2-4.2.2
	Test-Capacity	Yearly	2-4.2.5
Steam Pump	Test-Operate		2-4.2.4
	Test-Capacity	Yearly	2-4.2.5

Exhibit I   Report of Inspection

Owner's Section (To be answered by Owner or Occupant)

A. Explain any occupancy hazard changes since the previous inspection.

B. Describe fire protection modifications made since last inspection.

C. Describe any fires since last inspection.

D. When was the system piping last checked for stoppage, corrosion or foreign material?

E. When was the dry-piping system last checked for proper pitch?

F. Are dry valves adequately protected from freezing?

Inspector's Section (All responses reference current inspection)

  1. General

    a. Is the building occupied?

    b. Are all systems in service?

    c. Is there a minimum of 18 in. (457 mm) clearance between the top of the storage and the sprinkler deflector?

d. In areas protected by wet system, does the building appear to be properly heated in all areas, including blind attics and perimeter areas, where accessible? Do all exterior openings appear to be protected against freezing?

e. Does the hand hose on the sprinker system appear to be satisfactory?

2. Control Valves (*See Item 14.*)

a. Are all sprinkler system control valves and all other valves in the appropriate open or closed position?

b. Are all control valves in the open position and locked, sealed or equipped with a tamper switch?

3. Water Supplies (*See Item 15.*)

a. Was a water flow test of main drain made at the sprinkler riser?

4. Tanks, Pumps, Fire Department Connections

a. Are fire pumps, gravity tanks, reservoirs and pressure tanks in good condition and properly maintained?

b. Are fire department connections in satisfactory condition, couplings free, caps in place, and check valves tight? Are they accessible and visible?

5. Wet Systems (*See Item 13.*)

a. Are cold weather valves (O.S.&Y.) in the appropriate open or closed position?

b. Have antifreeze system solutions been tested?

c. Were the antifreeze test results satisfactory?

6. Dry Systems (*See Items 10 to 14.*)

a. Is the dry valve in service?

b. Are the air pressure and priming water level in accordance with the manufacturer's instructions?

c. Has the operation of the air or nitrogen supply been tested? Is it in service?

d. Were low points drained during this inspection?

e. Did quick-opening devices operate satisfactorily?

f. Did the dry valve trip properly during the trip pressure test?

g. Did the heating equipment in the dry-pipe valve room operate at the time of inspection?

7. Special Systems — as defined in Section 1-3 (*See Item 16.*)

a. Did the deluge or pre-action valves operate properly during testing?

b. Did the heat-responsive devices operate properly during testing?

c. Did the supervisory devices operate during testing?

8. Alarms

a. Did water motor and gong test satisfactorily?

b. Did electric alarm test satisfactorily?

c. Did supervisory alarm service test satisfactorily?

9. Sprinklers

a. Are all sprinklers free from corrosion, loading or obstruction to spray discharge?

b. Are sprinklers over 50 years old, thus requiring sample testing?

c. Is stock of spare sprinklers available?

d. Does the exterior condition of sprinkler system appear to be satisfactory?

e. Temperature. Are sprinklers of proper temperature ratings for their locations?

10. Date dry-pipe valve trip tested (control valve partially open)

_____

(See Trip Test Table which follows.)

11. Date dry-pipe valve trip tested (control valve fully open)

_____

(See Trip Test Table which follows.)

12. Date quick-opening device tested

_____

(See Trip Test Table which follows.)

13. Date deluge or preaction valve tested

_____

(See Trip Test Table which follows.)

Trip Test Table

DRY PIPE OPERATING TEST		DRY VALVE			Q.O.D.			
		MAKE	MODEL	SERIAL NO.	MAKE	MODEL	SERIAL NO.	

DRY PIPE OPERATING TEST		TIME TO TRIP THRU TEST PIPE		WATER PRESSURE	AIR PRESSURE	TRIP POINT AIR PRESSURE	TIME WATER REACHED TEST OUTLET		ALARM OPERATED PROPERLY	
		MIN.	SEC.	PSI	PSI	PSI	MIN.	SEC.	YES	NO
	Without Q.O.D.									
	With Q.O.D.									
	IF NO, EXPLAIN									

DELUGE & PREACTION VALVES	OPERATION		☐ PNEUMATIC	☐ ELECTRIC	☐ HYDRAULIC			
	PIPING SUPERVISED		☐ YES ☐ NO	DETECTING MEDIA SUPERVISED		☐ YES ☐ NO		
	DOES VALVE OPERATE FROM THE MANUAL TRIP AND/OR REMOTE CONTROL STATIONS					☐ YES ☐ NO		
	IS THERE AN ACCESSIBLE FACILITY IN EACH CIRCUIT FOR TESTING ☐ YES ☐ NO		IF NO, EXPLAIN					
	MAKE	MODEL	DOES EACH CIRCUIT OPERATE SUPERVISION LOSS ALARM		DOES EACH CIRCUIT OPERATE VALVE RELEASE		MAXIMUM TIME TO OPERATE RELEASE	
			YES	NO	YES	NO	MIN.	SEC.

14. See Control Valve Maintenance Table

Control Valve Maintenance Table

Control Valves	Number	Type	Open	Secured	Closed	Signs	Explain Abnormal Condition
City Connection Control Valve							
Tank Control Valves							
Pump Control Valves							
Sectional Control Valves							
System Control Valves							
Other Control Valves							

15. Water Flow Test at Sprinkler Riser

Water Supply Source		City	Tank	Pump
Date	Test Pipe Location	Size Test Pipe	Static Pressure	Residual (Flow) Pressure
Last Water Flow Test				
This Water Flow Test				

16. Heat Responsive Devices

Test Method _____

_____

Type of Equipment _____

Manufacturer _____

Test Results:

Valve No. _____A   B   C   D   E   F_____   Valve No. _____A   B   C   D   E   F_____

Valve No. _____A   B   C   D   E   F_____   Valve No. _____A   B   C   D   E   F_____

Valve No. _____A   B   C   D   E   F_____   Valve No. _____A   B   C   D   E   F_____

Valve No. _____A   B   C   D   E   F_____   Valve No. _____A   B   C   D   E   F_____

Auxiliary Equipment: No.? _____ Type? _____ Location? _____ Test Result? _____

17. Explain any "No" answers and comments. _____

_____

_____

_____

_____

18. Adjustments or corrections made during this inspection: _____

_____

_____

_____

_____

19. Although these comments are not the results of an engineering review, the following desirable improvements are recommended:

_____

_____

_____

_____

Signature: _____

Date: _____

APPENDIX

This Appendix is not a part of this NFPA document, but is included for information purposes only.

A-1 Referenced Publications.

A-1-1 NFPA Publications.   This publication make reference to the following NFPA documents and the year dates shown indicate the latest edition. They are available from the National Fire Protection Association, Batterymarch Park, Quincy, MA 02269.

NFPA 13–1980, *Standard for the Installation of Sprinkler Systems*

NFPA 13E–1978, *Recommendations for Fire Department Operations in Properties Protected by Sprinkler and Standpipe Systems*

NFPA 20–1980, *Standard for the Installation of Centrifugal Fire Pumps*

NFPA 21–1975, *Standard for the Operation and Maintenance of National Standard Steam Fire Pumps*

NFPA 22–1981, *Standard for Water Tanks for Private Fire Protection*

NFPA 24–1981, *Standard for Private Fire Service Mains and Their Appurtenances*

NFPA 26–1976, *Recommended Practices for the Supervision of Valves Controlling Water Supplies for Fire Protection*

NFPA 33–1977, *Standard for Spray Application Using Flammable and Combustible Materials*

NFPA 231D–1980, *Standard for Storage of Rubber Tires*

NFPA 1962–1979, *Standard for the Care, Maintenance and Use of Fire Hose Including Connections and Nozzles*

A-2 Other Publications.

A-2-1   A selected list of other NFPA publications related to the inspection, testing and maintenance of sprinkler systems is as follows:

NFPA 14–1980, *Standard for the Installation of Standpipe and Hose Systems*

NFPA 71–1977, *Standard for the Installation, Maintenance and Use of Central Station Signaling Systems*

NFPA 72A–1979, *Standard for the Installation, Maintenance, and Use of Local Protective Signaling Systems*

NFPA 72B–1979, *Standard for the Installation, Maintenance and Use of Auxiliary Protective Signaling Systems*

NFPA 87–1980, *Standard for the Construction and Protection of Piers and Wharves*

NFPA 231–1980, *Standard for Indoor General Storage*

NFPA 231C–1980, *Standard for Rack Storage of Materials*

NFPA 231D–1980, *Standard for the Storage of Rubber Tires*

A-2-2 Other Codes and Standards.   This publication makes reference to the following codes and standards.

No. 13-C, *Recommended Method of Reporting Dry Pipe Valve Tests*, available from the American Insurance Association Engineering and Safety Department, 85 John Street, New York, NY 10038.

ASME, *Non-Fired Pressure Vessel Code*, available from the American Society of Mechanical Engineers, East 47th Street, New York, NY 10017.

ASTM E380-1976, *Standard for Metric Practice*, available from the American Society for Testing and Materials, 1916 Race Street, Philadelphia, PA 19103.

NFPA 13D

Standard for the

Installation of Sprinkler Systems

In One- and Two-Family Dwellings

and Mobile Homes

1984 Edition

NOTICE: An asterisk (*) following the number or letter designating a paragraph indicates explanatory material on that paragraph in Appendix A. Material from Appendix A is integrated with the text and is identified by the letter A preceding the paragraph number to which it relates.

Information on referenced publications can be found in Chapter 5.

PREFACE

It is intended that this standard provide a method for those individuals wishing to install a sprinkler system for additional life safety and property protection. It is not the purpose of this standard to require the installation of an automatic sprinkler system. This standard assumes that one or more smoke detectors will be installed in accordance with NFPA 74, *Standard for the Installation, Maintenance, and Use of Household Fire Warning Equipment.*

National attention focused on the residential fire problem — primarily the result of a report published by the Presidential Commission on Fire Prevention & Control entitled "America Burning" (1973) — caused the NFPA Committee on Automatic Sprinklers to direct its attention to the residential fire problem. Thus, in the summer of 1973, the NFPA Sprinkler Committee established a Subcommittee on Residential and Light Hazard Occupancies. The Subcommittee was charged with developing a standard that would produce an adequately reliable but inexpensive sprinkler system for these occupancies where the majority of fire deaths were and are occurring. In its first meeting, basic philosophies for residential systems were established which carry through to the current edition of NFPA 13D.

- Cost was of major importance. A system having slightly less reliability or operational features than that described in NFPA 13 but which could be installed at a substantially lower cost was necessary, if acceptance of a residential system was to be achieved.
- Life safety would be the primary goal of NFPA 13D with property protection as a secondary goal.
- System design should be such that a fire could be controlled for sufficient time to enable people to escape; i.e., a 10-minute stored water supply with an adequate local audible alarm.
- Piping arrangements, components and hangers must be compatible with residential construction techniques and combined sprinkler-plumbing systems were acceptable from a fire protection standpoint.
- The fire record could reasonably serve as a baseline to permit sprinklers to be omitted in areas of low incidents causing fire deaths — thus saving cost.

423

The first draft document produced by the Subcommittee actually encompassed residential systems for one- and two-family dwellings, mobile homes, and multifamily housing up to four stories in height. However, when finally adopted in 1975, the multifamily housing portion had been eliminated because of strong feelings that such systems needed to be designed in accordance with NFPA 13. This first standard was put together largely without formalized test data and adopted by NFPA at its May 1975 meeting.

Beginning in 1976, the National Fire Prevention & Control Administration (later renamed the U.S. Fire Administration) funded a number of research programs to evaluate the residential fire problem in depth and determine the "what" and "how" of accomplishing a satisfactory solution. The NFPCA/USFA programs included studies to assess the probable impact of using sprinklers to reduce the incidence of deaths and injuries in residential fires.[1] Other studies evaluated the design, installation, practical usage, and user acceptance factors that would impact on accomplishing reliable and acceptable installed systems;[2] studies to evaluate minimum water discharge rates, automatic sprinkler flow, and response sensitivity design criteria;[3, 4] and full-scale tests to test prototype systems.[6, 7, 8, 9]

Collectively these research efforts provided an extensive data base from which a complete revision of the standard was developed and published in 1980. This portion of this handbook deals with the 1984 edition of the standard and is intended to give the reader some insight as to the rationale of the committee in arriving at the standard's requirements as well as the data base which supports this rationale.

The serious student is encouraged to study the various publications listed in the bibliography which appears at the end of this section. The student is also cautioned that extensive research continues in this rapidly changing area of residential fire protection and that current data may necessitate the updating of the information contained herein.

1

GENERAL INFORMATION

1-1* Scope. This standard deals with the design and installation of automatic sprinkler systems for one- and two-family dwellings and mobile homes.

> This standard applies only to one- and two-family dwellings and mobile homes as defined under Section 1-3. Sprinkler systems for residential occupancies that do not meet these definitions should be designed in accordance with NFPA 13.

Formal Interpretation

Question: Is NFPA 13D appropriate for use in multiple (three or more) attached dwellings under any condition?

Answer: No. NFPA 13D is appropriate for use only in one- and two-family dwellings and mobile homes. Buildings which contain more than two dwelling units shall be protected in accordance with NFPA 13. Section 3-16.2.9 of NFPA 13 permits residential sprinklers to be used in residential portions of other buildings provided all other requirements of NFPA 13, including water supplies, are satisfied.

Note: Building codes may contain requirements such as 2-hour fire separations which would permit adjacent dwellings to be considered unattached.

A-1-1 NFPA 13D is appropriate for use only in one- and two-family dwellings and mobile homes. Residential portions of any other building may be protected with residential sprinklers in accordance with 3-16.2.9 of NFPA 13-1983, *Standard for the Installation of Sprinkler Systems.* Other portions of such sections should be protected in accordance with NFPA 13.

The criteria in this standard are based on full-scale fire tests of rooms containing typical furnishings found in residential living rooms, kitchens, and bedrooms. The furnishings were arranged as typically found in dwelling units in a manner similar to that shown in Figures A-1-1(a), A-1-1(b) and A-1-1(c). Sixty full-scale fire tests were conducted in a two story dwelling in Los Angeles, California and 16 tests were conducted in a 14-ft (4.3-mm) wide mobile home in Charlotte, North Carolina. Sprinkler systems designed and installed according to this standard are expected to prevent flashover within the compartment of origin if sprinklers are installed in the compartment. A sprinkler system designed and installed according to this standard cannot, however, be completely expected to control a fire involving unusually higher average fuel loads than typical for dwelling units [10 lbs/ft^2 (49 kg/m^2)] and where the interior finish has an unusually high flame spread rating (greater than 225).

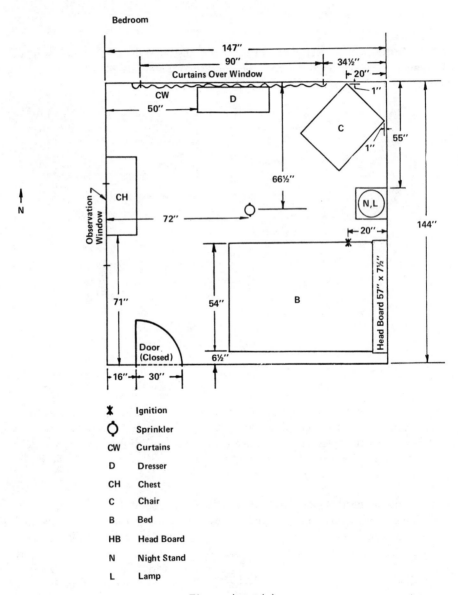

Figure A-1-1(a).

Mobile Home Bedroom

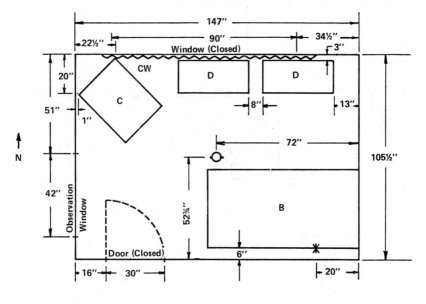

X Ignition
O Sprinkler
CW Curtains
D Dresser
C Chair
B Bed

Figure A-1-1(b).

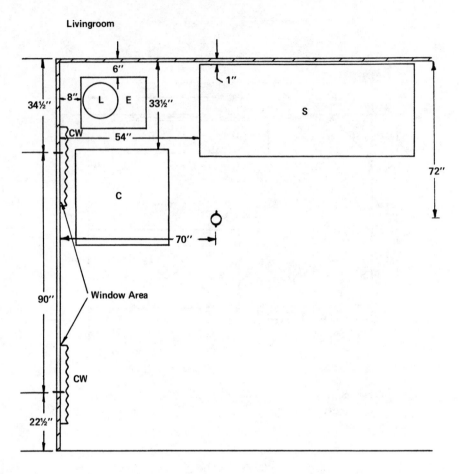

S Sofa
E End Table
L Lamp
C Chair
CW Curtains
○ Sprinkler

Figure A-1-1(c).

While the requirements of this standard are based on the Los Angeles and North Carolina fire tests,[7, 8, 9, 10] earlier exploratory tests by Factory Mutual Research Corporation and Battelle Columbus Laboratories[3, 4] indicated that there is a potential for systems to be developed that will control residential fires at application rates as low as 0.025 gpm per sq ft. These tests usually involved smoldering fires in typical residential room configurations and flaming fires using a prototype (i.e., gas burner) rather than a real-world fire array. Subsequent tests at Factory Mutual Research Corporation[6] and in test facilities involving room configurations at Los Angeles and in North Carolina[7, 8, 9, 10] established that design had to be based on the density and area of application found in Sections 4-1.1 and 4-1.2 of this standard.

1-2* Purpose.   The purpose of this standard is to provide a sprinkler system that will aid in the detection and control of residential fires and thus provide improved protection against injury, life loss, and property damage. A sprinkler system installed in accordance with this standard is expected to prevent flashover (total involvement) in the room of fire origin, when sprinklered, and to improve the chance for occupants to escape or be evacuated.

Guidelines are established for the design and installation of sprinkler systems for one- and two-family dwellings and mobile homes. Nothing in this standard is intended to restrict new technologies or alternate arrangements providing the level of safety prescribed by the standard is not lowered.

A-1-2 Levels of Protection.   Various levels of firesafety are available to dwelling occupants to provide life safety and property protection.

This standard recommends, but does not require, sprinklering of all areas in a dwelling; it permits sprinklers to be omitted in certain areas. These areas are the ones shown by NFPA statistics (*see Table A-1-2*) to be the ones where the incidence of life loss from fires in dwellings is low. Such an approach produces a reasonable degree of fire safety. Greater protection to both life and property will be achieved by sprinklering all areas.

Guidance for installation of smoke detectors and fire detection systems may be found in NFPA 74, *Standard for the Installation, Maintenance and Use of Household Fire Warning Equipment.*

Table A-1-2   Causal Factors in One- and Two-Family Dwelling Fires Which Caused One or More Deaths

Area of Origin		
Living Room	41%	
Bedroom	27%	
Kitchen	15%	Based on 6066 incidents where area of origin was reported
Storage Area	4%	
Heating Equipment Room	3%	
Structural Area	2%	
Other Areas	8%	

Form of Material		
Furniture	27%	
Bedding	18%	
Combustible Liquid or Gas	13%	
Interior Finish	9%	
Structural Member	9%	Based on 5080 incidents where form of material ignited was reported
Waste, Rubbish	4%	
Clothing, on a Person	3%	
Cooking Materials	3%	
Electrical Insulation	2%	
Curtains, Drapery	2%	
Other	10%	

Form of Heat of Ignition		
Smoking Materials	36%	
Heat from Fuel — Fire or Powered Object	25%	
Heat from Miscellaneous Open Flame (Including Match)	15%	
Heat from Electrical Equipment Arcing or Overload	14%	Based on 5016 incidents where form of heat of ignition was reported
Hot Objects Including Properly Operating Electrical Equipment	7%	
Other	3%	

Total number of incidents reported

10,194

Source: FIDO Data Base 1973 to 1982. NFPA Fire Analysis Department.

Both Underwriters Laboratories Inc. and Factory Mutual Research Corporation have developed test standards for the evaluation of residential sprinklers.[11, 12] These standards have significantly different fire test requirements from those found in the UL and FM standards for commercial sprinklers. The residential sprinkler criteria are:

• maximum ceiling air temperature — approximately 600°F. (i.e. prevent flashover)

- maximum temperature at 5 ft 3 in. above floor — 200°F.
- maximum of two sprinklers in test area to operate during test.

The laboratory test configuration is a 12-ft × 24-ft room with a combustible array simulating residential furnishings.

The committee adopted the concept of "levels of protection" in order to achieve a low cost system and to avoid the necessity of installing dry-pipe sprinkler systems in cold climates. The areas where omission of sprinklers are permitted per Section 4-6 of this standard are those where the record justifies such omission. (See Table A-1-2.) The committee felt that dry-pipe systems were a less desirable form of protection because of the apparent delay in application of water. (This was noted in early Factory Mutual tests and substantiated in the Los Angeles and North Carolina fire tests[7, 8, 9, 10]. The key to achieving control of the residential fire is the rapid application of water. Further, residential systems must be simple and easy to maintain — factors not associated with dry-pipe systems.

1-3* Definitions.

Approved.   Acceptable to the "authority having jurisdiction."

NOTE: The National Fire Protection Association does not approve, inspect or certify any installations, procedures, equipment, or materials nor does it approve or evaluate testing laboratories. In determining the acceptability of installations or procedures, equipment or materials, the authority having jurisdiction may base acceptance on compliance with NFPA or other appropriate standards. In the absence of such standards, said authority may require evidence of proper installation, procedure or use. The authority having jurisdiction may also refer to the listings or labeling practices of an organization concerned with product evaluations which is in a position to determine compliance with appropriate standards for the current production of listed items.

Authority Having Jurisdiction.   The "authority having jurisdiction" is the organization, office or individual responsible for "approving" equipment, an installation or a procedure.

NOTE: The phrase "authority having jurisdiction" is used in NFPA documents in a broad manner since jurisdictions and "approval" agencies vary as do their responsibilities. Where public safety is primary, the "authority having jurisdiction" may be a federal, state, local or other regional department or individual such as a fire chief, fire marshal, chief of a fire prevention bureau, labor department, health department, building official, electrical inspector, or others having statutory authority. For insurance purposes, an insurance inspection department, rating bureau, or other insurance company representative may be the "authority having jurisdiction." In many circumstances the property owner or his designated agent assumes the role of the "authority having jurisdiction"; at government installations, the commanding officer or departmental official may be the "authority having jurisdiction."

Check Valve.   A valve which allows flow in one direction only.

Control Valve.*   A valve employed to control (shut) a supply of water to a sprinkler system.

A-1-3   System control valves should be of the indicating type, such as plug valves, ball valves, butterfly valves, or OS & Y gate valves.

The type of control valve used should be an indicating type, i.e., one that has some external means that will indicate to the occupant of a home that the valve is in the open position.

Design Discharge.   Rate of water discharged by an automatic sprinkler expressed in gallons-per-minute.

Dry System.   A system employing automatic sprinklers attached to a piping system containing air under atmospheric or higher pressures. Loss of pressure from the opening of a sprinkler or detection of a fire condition causes the release of water into the piping system and out the opened sprinkler.

Because they deliver water to the fire less rapidly and entail the need for greater maintenance, dry systems are a less desirable form of protection. However, they must be used when any part of the sprinkler piping is located in an area of the residence which may freeze in cold weather.

Dwelling.   Any building which contains not more than one or two "dwelling units" intended to be used, rented, leased, let or hired out to be occupied, or which are occupied for habitation purposes.

Dwelling Unit.   One or more rooms arranged for the use of one or more individuals living together as in a single housekeeping unit, normally having cooking, living, sanitary and sleeping facilities.

Labeled.   Equipment or materials to which has been attached a label, symbol or other identifying mark of an organization acceptable to the "authority having jurisdiction" and concerned with product evaluation, that maintains periodic inspection of production of labeled equipment or materials and by whose labeling the manufacturer indicates compliance with appropriate standards or performance in a specified manner.

Listed.   Equipment or materials included in a list published by an organization acceptable to the "authority having jurisdiction" and concerned with product evaluation, that maintains periodic inspection of production of listed equipment or materials and whose listing states either that the equipment or material meets appropriate standards or has been tested and found suitable for use in a specified manner.

NOTE: The means for identifying listed equipment may vary for each organization concerned with product evaluation, some of which do not recognize equipment as listed unless it is also labeled. The "authority having jurisdiction" should utilize the system employed by the listing organization to identify a listed product.

Mobile Home. A factory-assembled structure equipped with service connections and made so as to be readily movable as a unit on its running gear and designed to be used as a dwelling unit with or without a foundation.

Multipurpose Piping Systems. Piping systems within dwellings and mobile homes intended to serve both domestic and fire protection needs.

Preengineered System. A packaged sprinkler system including all components connected to the water supply designed to be installed according to pretested limitations.

Pump. A mechanical device that transfers and/or raises the pressure of a fluid (water).

Residential Sprinkler. An automatic sprinkler which has been specifically listed for use in residential occupancies.

Residential sprinklers are specifically listed by UL and FM for residential service and have been tested for compliance with the residential sprinkler standard.[11, 12] Typically, they are fast responding sprinklers, and they may be either of the upright, pendent, or sidewall configuration.

Shall. Indicates a mandatory requirement.

Should. Indicates a recommendation or tnat which is advised but not required.

Sprinkler-Automatic. A fire suppression device which operates automatically when its heat-actuated element is heated to or above its thermal rating allowing water to discharge over a specific area.

Sprinkler System. An integrated system of piping connected to a water supply, with listed sprinklers which will automatically initiate water discharge over a fire area. When required, the sprinkler system also includes a control valve and a device for actuating an alarm when the system operates.

Standard. A document containing only mandatory provisions using the word "shall" to indicate requirements. Explanatory material may be included only in the form of "fine print" notes, in footnotes, or in an Appendix.

Supply Pressure. Pressure within the supply (i.e., city or private supply water source).

Supply pressure is the pressure that one can expect to attain from the city or other water supply as explained in A-4-4.3.

System Pressure.   A pressure within the system (i.e., above the control valve).

System pressure is determined by substracting friction losses between the street connection and the water supply valve in the residence from the water pressure in the street after making any necessary adjustments for changes in elevation as explained in A-4-4.3 of this standard.

Water Flow Alarm.   A sounding device activated by a water flow detector or alarm check valve and arranged to sound an alarm which will be audible in all living areas over background noise levels with all intervening doors closed.

Water Flow Detector.   An electric signaling indicator or alarm check valve actuated by water flow in one direction only.

Wet System.   A system employing automatic sprinklers attached to a piping system containing water and connected to a water supply so that water discharges immediately from sprinklers opened by a fire.

1-4* Maintenance.   The owner is responsible for the condition of a sprinkler system and shall keep the system in normal operating condition.

A-1-4   The responsibility for properly maintaining a sprinkler system is the obligation of the owner or manager who should understand the sprinkler system operation. A minimum monthly maintenance program should include the following:

(a) Visually inspect all sprinklers to ensure against obstruction of spray.

(b) Inspect all valves to assure that they are open.

(c) Test all water flow devices.

(d) The alarm system, if installed, should be tested.

NOTE: When it appears likely that the test will result in a response of the fire department, notification to the fire department should be made prior to the test.

(e) Pumps, where employed, should be operated. See NFPA 20, *Standard for the Installation of Centrifugal Fire Pumps.*

(f) The pressure of air used with dry systems should be checked.

(g) Water level in tanks should be checked.

(h) Care should be taken to see that sprinklers are not painted either at the time of installation or during subsequent redecoration. When painting sprinkler piping or painting in areas next to sprinklers, the sprinklers may be protected by covering with a bag which should be removed immediately after painting has been finished.

(i) For further information see NFPA 13A, *Recommended Practice for the Inspection, Testing and Maintenance of Sprinkler Systems.*

A good guide for the maintenance of sprinkler systems is found in NFPA 13A. The property owner may obtain further guidance on maintenance of the sprinkler system from his local sprinkler contractor. Generally, because of its simplicity, maintenance of a residential sprinkler system does not require any greater care than that which one normally exercises in maintaining a residential plumbing system.

1-5 Design and Installation.

1-5.1 Devices and Materials.

1-5.1.1*   Only new residential sprinklers shall be employed in the installation of sprinkler systems.

A-1-5.1.1   At least 3 spare sprinklers of each type, temperature rating and orifice size used in the system should be kept on the premises. When fused sprinklers are replaced by the owner, fire department or others, care should be taken to assure that the replacement sprinkler has the same operating characteristics.

The sprinklers used in residential systems are special types which have been tested for residential use.[12] They are essentially one-time operating valves which must be treated carefully in the process of installation. When the system is originally installed, only new sprinklers should be used to ensure that they are of the proper type and in the best condition.

On those occasions when it is necessary to replace a sprinkler because it has operated (fused), or has been damaged, one should carefully check the orifice size, the temperature rating, and the deflector configuration to be certain that the exact same type of sprinkler replaces the one that has operated. Many sprinklers have unique operating characteristics, i.e., upright, pendent, or sidewall, and some have special limitations on their area of coverage.

1-5.1.2   Only listed and approved devices and approved materials shall be used in sprinkler systems.

Exception: Listing may be waived for tanks, pumps, hangers, water flow detection devices, and water control valves.

At the time the 1984 edition of the standard was prepared, tanks pumps, hangers, water flow devices, and water control valves in the size range suitable for residential use had not been tested or listed by testing laboratories. It was the feeling of the committee that listing was desirable but not mandatory for these devices. The committee anticipated that the type of equipment used would be similar to that found in common use in residential plumbing systems.

1-5.1.3 Preengineered systems shall be installed within the limitations which have been established by the testing laboratories where listed.

Approval laboratories have sometimes tested and listed complete systems which include all of the components necessary for an application. In this case, the system might include sprinklers, water supply tanks, valves, pumps, special piping limitations, etc. The section permits a manufacturer to develop such a system for application as a residential sprinkler system but does not require the use of such a package system.

1-5.1.4* All systems shall be tested for leakage at normal system operating water pressure.

A-1-5.1.4 Testing of a system can be accomplished by filling the system with water and checking visually for leakage at each joint or coupling.

Fire department connections are not required for systems covered by this standard, but may be installed at the discretion of the owner. In these cases hydrostatic tests in accordance with NFPA 13, *Standard for the Installation of Sprinkler Systems* are required.

Dry systems should also be tested by placing the system under air pressure. Any leak which results in a drop in system pressure greater than 2 lb/sq in. (0.14 bars) in 24 hours should be corrected. Check for leaks using soapy water brushed on each joint or coupling. Leaks will be shown by the presence of bubbles. This test should be made prior to concealing of piping.

Exception: When a fire department pumper connection is provided, hydrostatic pressure tests shall be provided in accordance with NFPA 13, Standard for the Installation of Sprinkler Systems.

Unlike NFPA 13, residential systems do not require a hydrostatic test at 50 psi above the operating pressure. If such had been required, it would have added significantly to the installed cost of the system and would have made "do it yourself" installation difficult. Plumbing systems have been successfully installed in homes without special hydrostatic testing for years, and since it is intended that the residential sprinkler system be similar to, if not a part of, the plumbing system, the committee concluded that a special hydrostatic test was not generally required.

While a fire department connection to a residential system is not required, it can be installed. When this occurs, the system pressure can be increased using a fire department pumper, and in such instances it should be verified that the system has the integrity to withstand such pressures. Thus, a hydrostatic test should be done in accordance with NFPA 13 when a fire department pumper connection is included as a part of the system. Normally, one would expect this to be at a pressure of 200 psi.

1-6 Units.   Metric units of measurement in this standard are in accordance with the modernized metric system known as the International System of Units (SI). Two units (liter and bar), outside of but recognized by SI, are commonly used in international fire protection. These units are listed in Table 1-6 with conversion factors.

Table 1-6

Name of Unit	Unit Symbol	Conversion Factor
liter	L	1 gal = 3.785 L
pascal	Pa	1 psi = 6894.757 Pa
bar	bar	1 psi = 0.0689 bar
bar	bar	1 bar = 105 Pa

For additional conversions and information see ASTM E380, *Standard for Metric Practice.*

1-6.1   If a value for measurement as given in this standard is followed by an equivalent value in other units, the first stated is to be regarded as the requirement. A given equivalent value may be approximate.

1-6.2   The conversion procedure for the SI units has been to multiply the quantity by the conversion factor and then round the result to the appropriate number of significant digits.

2

WATER SUPPLY

2-1 General Provisions. Every automatic sprinkler system shall have at least one automatic water supply. When stored water is used as the sole source of supply, the minimum quantity shall equal the water demand rate times 10 minutes. *(See 4-1.3.)*

Since any sprinkler system is only as good as its water supply, this supply must be automatic and reliable. The adequacy of a public water supply is determined in the manner described in 4-4.3 and A-4-4.3 of this standard. When stored water is used, the amount should provide at least 10 minutes duration [at least 260 gal]. A fuel oil tank of 275 gal capacity would make a satisfactory storage tank. This provides sufficient water to hold a fire in control while giving occupants sufficient time to evacuate the residence.

2-2* Water Supply Sources. The following water supply sources are acceptable:

(a) A connection to a reliable water works system.

(b) An elevated tank.

(c) A pressure tank installed in accordance with NFPA 13, *Standard for the Installation of Sprinkler Systems*, and NFPA 22, *Standard for Water Tanks for Private Fire Protection*.

(d) A stored water source with an automatically operated pump.

A-2-2 Connection for fire protection to city mains is often subject to local regulation concerning metering and backflow prevention requirements. Preferred and acceptable water supply arrangements are shown in Figures A-2-2 (a), (b) and (c). When a meter must be used between the city water main and the sprinkler system supply, an acceptable arrangement is shown in Figure A-2-2(c). Under these circumstances, the flow characteristics of the meter must be included in the hydraulic calculation of the system. [*See Table 4-4.3(d).*] When a tank is used for both domestic and fire protection purposes, a low water alarm actuated when the water level falls below 110 percent of the minimum quantity specified in Section 2-1 should be provided.

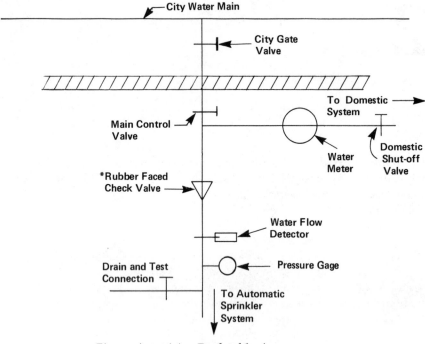

Figure A-2-2(a)   Preferable Arrangement.

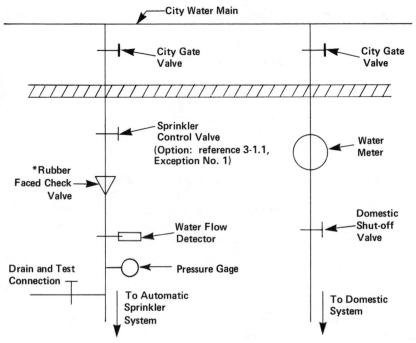

Figure A-2-2(b)   Acceptable Arrangement.

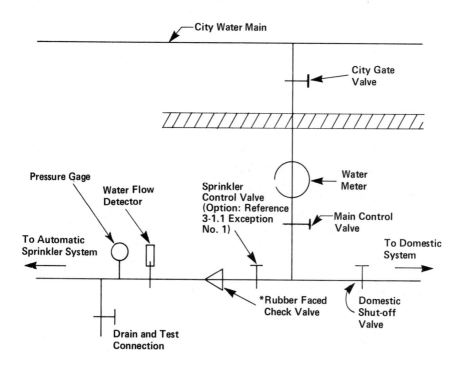

—City Water Main

City Gate
Valve

Pressure Gage

Water Flow
Detector

Sprinkler
Control Valve
(Option: Reference
3-1.1 Exception
No. 1)

Water
Meter

Main Control
Valve

To Automatic
Sprinkler System

To Domestic
System

*Rubber Faced
Check Valve

Domestic
Shut-off
Valve

Drain and Test
Connection

*Rubber faced check valves optional.

Figure A-2-2(c)   Acceptable Arrangement.

Figure A-2-2 shows preferred and acceptable arrangements of connections to city mains and includes control valves, meters, domestic take-offs, water flow detectors, pressure gages, and check valves appropriately located. When backflow preventors are required by local water company requirements, they will normally be located in the position where the rubber faced valve is shown on the diagram. The piping arrangement, including valves and fittings, must be taken into account when performing the calculations for system adequacy described in Section 4-4.3.

From a fire protection standpoint, meters are undesirable because of their high friction loss characteristic which must be included in the hydraulic calculation of the system in accordance with the values shown in Table 4-4.3(d). Unfortunately, many water authorities require metering of all water connections to a residence.

Occasionally, a system may be supplied from an elevated tank, and when this is done, the tank must have an elevation sufficient to provide adequate pressure to supply the system. This is determined by the methods described in 4-4.3 and 4-4.4. When a pressure tank is used for water supply, the amount of water and air in the elevation of the tank is determined by following Section 2-6.3 and A-2-6.3 of NFPA Standard 13 and 4-4.4 of this standard.

A stored water source with an automatically operated pump must have a pump with a capacity sufficient to supply the system's demand as described in Section 4-1.1, 4-1.3, and 4-1.2 and must have adequate pressure determined in accordance with 4-4.4.

2-3* Multipurpose Piping System. A piping system serving both sprinkler and domestic needs shall be acceptable when:

(a)* In common water supply connections serving more than one dwelling unit, 5 gallons per minute (19 L/min) is added to the sprinkler system demand to determine the size of common piping and the size of the total water supply requirements.

(b) Smoke detectors are provided in accordance with NFPA 74, *Standard for the Installation, Maintenance, and Use of Household Fire Warning Equipment.*

(c) All piping in the system conforms to the piping specifications of this standard.

(d) Permitted by the local plumbing or health authority.

A-2-3(a) In dwellings where long term use of lawn sprinklers is common, provision should be made for such usage.

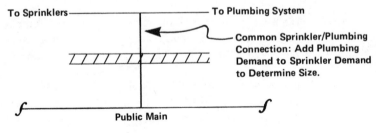

Figure A-2-3(a).

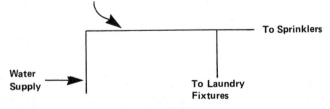

Figure A-2-3(b).

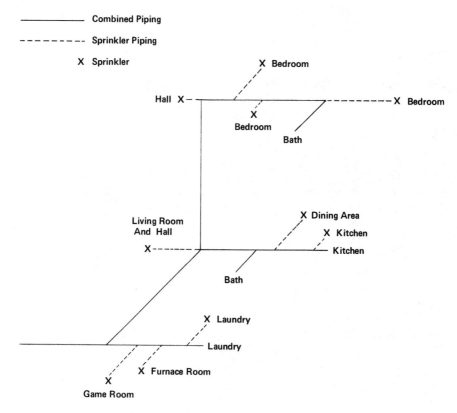

Figure A-2-3(c).

In setting this requirement, the committee recognized that typical domestic water usage does not span a great length of time (except for an automatic long-term demand for a domestic use such as a lawn sprinkling system or laundry) and is not likely to occur simultaneously with a fire. When one can anticipate an automatic domestic demand that might take place concurrent with operation of sprinklers to combat a fire, such demand should be taken into account in the system design.

In permitting combined piping systems, i.e., systems serving both domestic and sprinkler needs concurrently, the committee recognized there is some value in ensuring that the sprinkler system will be in operation when common piping serves both sprinklers and domestic needs. In a residence, if one does not have water for normal usage, it is likely that something will be done about it and promptly. The committee also recognized that restrictions may be placed on such combined systems by local plumbing or health regulations because there are typically trapped piping sections at the extremities of a sprinkler system which are not permitted in such codes. Further, some plumbing regulations permit piping methods of a lesser quality level than are prescribed by this standard.

2-4 Mobile Home Water Supply.   A water supply for a sprinklered dwelling manufactured off-site shall not be less than that specified on the manufacturer's nameplate [*see 4-4.3 (k) Exception*].

The piping for a sprinkler system installed in a mobile home will typically be installed by the mobile home manufacturer. This standard anticipates that, when this is done, the mobile home manufacturer will specify the capacity and pressure needed to supply the system from a water supply to which the system must ultimately be connected.

Figures A-2-3(a) and A-2-3(b) show a variety of arrangements for combining piping systems which include the location of control valves, meters, and alarms. Figures A-2-3(c), A-2-3(d) and A-2-3(e) show examples of how domestic demands should be accounted for in the design of a system. These point out where special demands should be accounted for in the system; the actual calculations are done in accordance with Section 4-4.3 of this standard.

3

SYSTEM COMPONENTS

3-1 Valves and Drains.

3-1.1   Each system shall have a single control valve arranged to shut off both the domestic and sprinkler systems, and a separate shutoff valve for the domestic system only.

> A control valve permits the occupant to shut off the system in order to replace a sprinkler which has operated to extinguish a fire. The control valve must be in the open position at all other times.
>
> With only one control valve, there is little possibility of inadvertently shutting off the sprinkler system because, with the domestic water shut off, something would be done rather quickly to get it back in operation. The separate control valve for the sprinkler system is acceptable when the valve is supervized.

Exception No. 1:   The sprinkler system piping may have a separate control valve where supervised by one of the following methods:

(a) Central station, proprietary or remote station alarm service,

(b) Local alarm service which will cause the sounding of an audible signal at a constantly attended point, or

(c) Locking the valves open.

Exception No. 2:   A separate shutoff valve is not required for the domestic water supply in multipurpose piping systems.

3-1.2   Each sprinkler system shall have a ½ in. or larger drain and test connection with valve on the system side of the control valve.

> This arrangement permits the occupant to open this valve to determine that water is actually being supplied to the system and to check operation of the alarm. It also permits the system piping to be drained when maintenance or repairs are being done.

3-1.3   Additional drains shall be installed for each trapped portion of a dry system which is subject to freezing temperatures.

Figure 3.1 illustrates a trapped section of piping and how a drain is arranged to comply with 3-1.3. It is important that such portions of the system be carefully drained before the onset of freezing weather each year. If water from the operation of the system or condensation collects in a trapped section of piping, it my freeze and break the piping with the possibility of extensive water damage.

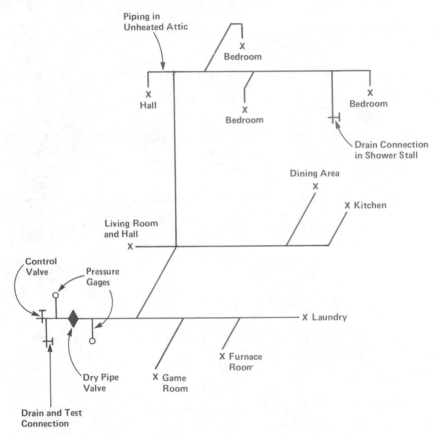

Figure 3.1.   Drain connection for a trapped section of piping.

3-2 Pressure Gages.

3-2.1   A pressure gage shall be installed to indicate air pressure on dry systems and on water supply pressure tanks.

Pressure gages enable the user to observe the presence of the proper pressure level in dry systems and on water supply pressure tanks.

3-3 Piping.

3-3.1 Pipe or tube used in sprinkler systems shall be of the materials in Table 3-3.1 or in accordance with 3-3.2 through 3-3.5. The chemical properties, physical properties and dimensions of the materials listed in Table 3-3.1 shall be at least equivalent to the standards cited in the table and designed to withstand a working pressure of not less than 175 psi (12.1 bars).

Table 3-3.1

Materials and Dimensions	Standard
Ferrous Piping (Welded and Seamless), Welded and Seamless Steel Pipe for Ordinary Uses, Specification for Black and Hot-Dipped Zinc Coated (Galvanized)	ASTM A120
Specification for Welded and Seamless Steel Pipe	ASTM A53
Wrought Steel Pipe	ANSI B36.10
Specification for Electric-Resistance Welded Steel Pipe	ASTM A135
Copper Tube (Drawn, Seamless) Specification for Seamless Copper Tube	ASTM B75
Specification for Seamless Copper Water Tube	ASTM B88
Specification for General Requirements for Wrought Seamless Copper and Copper-Alloy Tube	ASTM B251
Brazing Filler Metal (Classification BCuP-3 or BCuP-4)	AWS A5.8
Specification for Solder Metal, 95-5 (Tin-Antimony-Grade 95TA)	ASTM B32
Specification for Solder Metal, 50-50	ASTM B32

3-3.2 Other types of pipe or tube may be used, but only those investigated and listed for this service by a testing and inspection agency laboratory.

3-3.3 Whenever the word pipe is used in this standard, it shall be understood to also mean tube.

3-3.4 Schedule 10 steel pipe may be joined with mechanical groove couplings approved for service with grooves rolled on the pipe by an approved groove rolling machine.

3-3.5 Fittings used in sprinkler systems shall be of the materials listed in Table 3-3.5 or in accordance with 3-3.7. The chemical properties, physical properties and dimensions of the materials listed in Table 3-3.5 shall be at least equivalent to the standards cited in the table. Fittings used in sprinkler systems shall be designed to withstand the working pressures involved, but not less than 175 psi (12.1 bars) cold water pressure.

Table 3-3.5

Material and Dimensions	Standard
Cast Iron	
Cast Iron Threaded Fittings, Class 125 and 250	ANSI B16.4
Cast Iron Pipe Flanges and Flanged Fittings	ANSI B16.1
Malleable Iron	
Malleable Iron Threaded Fittings, Class 150 and 300	ANSI B16.3
Steel	
Factory-made Wrought Steel Buttweld Fittings	ANSI B16.9
Buttwelding Ends for Pipe, Valves, Flanges and Fittings	ANSI B16.25
Spec. for Piping Fittings of Wrought Carbon Steel and Alloy Steel for Moderate and Elevated Temperatures	ASTM A234
Pipe Flanges and Flanged Fittings, Steel Nickel Alloy and Other Special Alloys	ANSI B16.5
Forged Steel Fittings, Socket Welded and Threaded	ANSI B16.11
Copper	
Wrought Copper and Copper Alloy Solder-Joint Pressure Fittings .	ANSI B16.22
Cast Copper Alloy Solder-Joint Pressure Fittings	ANSI B16.18

3-3.6   Joints for the connection of copper tube shall be brazed.

Exception: Soldered joints (50-50 or 95-5 solder metal) may be used for wet-pipe copper tube systems.

The 50-50 solder is the type normally used in domestic plumbing systems. The 95-5 solder, which has a higher melting point, would be more desirable if available.

3-3.7   Other types of fittings may be used, but only those investigated and listed for this service by a testing and inspection agency laboratory.

The types of pipe and fittings that are described in this section of the standard are those which have been found satisfactory either due to their service record or based on their having been tested and listed by a testing laboratory. Currently, the most commonly used materials are ordinary and lightweight steel pipe and types K and L copper and copper-alloy tube. This does not preclude the use of plastic pipe, if it has been tested and listed by a testing and inspection agency laboratory. See Sections 3-1 and 3-12 of NFPA 13 for additional information on pipe and fittings.

3-4 Piping Support.

3-4.1 Piping shall be supported from structural members. This standard contemplates support methods comparable to those used in local plumbing codes. Piping layed on open joist or rafter shall be secured to minimize lateral movement.

Unlike the hanging requirements for sprinkler systems installed under NFPA 13, this standard simply requires adequate support by methods recognized in local plumbing codes. Generally, residential systems deal with pipe sizes of ¾-inch to 1¼-inch and will never be expected to exceed 2-inch. The loads which must be supported are therefore small, and most hangers that will be used will give adequate support.

3-4.2 Piping laid on open joists or rafters shall be secured to prevent lateral movement.

3-5 Sprinklers.

3-5.1 Listed residential sprinklers shall be used. The basis of such a listing shall be tests to establish the ability of the sprinklers to control residential fires under standardized fire test conditions. The standardized room fires shall be based on a residential array of furnishings and finishes.

Figure 3.2.   Grinnell Model F954 Residential Sprinkler.

Both Underwriters Laboratories Inc. and Factory Mutual Research Corporation published standards for testing and evaluation of residential sprinklers.[11, 12] These standards call for the sprinklers to pass room fire tests and be of a fast response characteristic. Standard commercial sprinklers of the type used in NFPA 13 sprinkler systems do not meet these criteria and should not be used. The residential sprinkler test requirements were verified in testing at Factory Mutual Research Corporation,[6] and in the Los Angeles and North Carolina residential sprinkler fire tests referenced in the bibliography.[7, 8, 9, 10]

Exception: Residential sprinklers shall not be used in dry systems unless specifically listed for that purpose.

3-5.2   The sprinklers shall have fusing temperatures not less than 35°F (19°C) above maximum expected ambient temperature.

As a practical matter, this section requires that the minimum operating (fusing) temperature rating of the sprinkler should be 135°F. The temperature rating range of residential sprinklers may vary from 135-170°F. The maximum temperature (ambient) in the residential environment should never exceed a value of 35°F below the temperature rating of the residential sprinkler which is used.

3-5.3   Operated or damaged sprinklers shall be replaced with sprinklers having the same performance characteristics as original equipment.

Underwriters Laboratories Inc. has conducted a program in which field samples of automatic sprinklers have been tested to determine their operating characteristics. Experience gained from this program has shown that, when a sprinkler has been damaged or painted after leaving the factory, it is unlikely to operate properly at the time of a fire and thus should be replaced. Obviously, a sprinkler which has been operated must also be replaced.

3-5.4 Painting and Ornamental Finishes.

3-5.4.1*   Sprinkler frames may be factory painted or enameled as ornamental finish in accordance with 3-5.4.2, otherwise sprinklers shall not be painted and any sprinklers which have been painted, except for factory applied coatings, shall be replaced with new listed sprinklers.

A-3-5.4.1   Decorative painting of a residential sprinkler is not to be confused with the temperature identification colors as referenced in 3-16.6 of NFPA 13-1983, *Standard for the Installation of Sprinkler Systems.*

3-5.4.2   Ornamental finishes shall not be applied to sprinklers by anyone other than the sprinkler manufacturer, and only sprinklers listed with such finishes shall be used.

It is extremely important that only manufacturers who have had their procedures listed apply painting or ornamental finishes to residential sprinklers. The homeowner does not have the facilities available to properly duplicate the manufacturers procedures. "Do it yourself" applications could seriously impair the sprinkler's operation — or could render it inoperable.

3-6* Alarms.   Local water flow alarms shall be provided on all sprinkler systems.

Exception: Dwellings or mobile homes having smoke detectors in accordance with NFPA 74, Standard for the Installation, Maintenance, and Use of Household Fire Warning Equipment.

A-3-6   Alarms should be of sufficient intensity to be clearly audible in all bedrooms over background noise levels with all intervening doors closed. The tests of audibility level should be conducted with all household equipment which may be in operation at night in full operation. Examples of such equipment are window air conditioners and room humidifiers. When off-premises alarms are provided, at least water flow and control valve position should be monitored.

Since the philosophy of this standard is to provide a system that will control a fire for 10 minutes while permitting occupants an opportunity to evacuate, an alarm is needed to warn occupants and initiate that evacuation. This standard suggests two methods by which this can be done. If smoke detectors are installed in accordance with NFPA 74, this standard anticipates that the smoke detectors will give an adequate warning for purposes of evacuation. However, if smoke detectors are not installed, then the sprinkler system must be equipped with a water flow device, which will sound an audible alarm throughout the residence to initiate evacuation.

To conduct audibility tests, have members of the family position themselves on each level of the house inside a room (such as a bedroom) with the door closed. Operate the alarm and have the family members report if it can be heard. Bear in mind that the alarm must be of sufficient level to awaken a person who is asleep.

4
SYSTEM DESIGN

4-1 Design Criteria.

4-1.1 Design Discharge.   The system shall provide a discharge of not less than 18 gallons per minute (68 L/min) to any single operating sprinkler and not less than 13 gallons per minute (49 L/min) per sprinkler to the number of design sprinklers.

This section of this standard prescribes a discharge of 18 gpm for a single operating sprinkler and 13 gpm per sprinkler for two sprinklers operating in the design area in a room. This translates to a density of 0.125 gpm per sq ft for one sprinkler and 0.09 gpm per sq ft for two sprinklers at the maximum spacing of 144 sq ft per sprinkler. Test data indicates that the primary criterion should be flow rate rather than density for a residential system.[7, 8, 9, 10]

These values of system discharge are based primarily on the Los Angeles and North Carolina fire tests.[7, 8, 9, 10]

4-1.2* Number of Design Sprinklers.   The number of design sprinklers shall include all sprinklers within a compartment to a maximum of 2 sprinklers.

A-4-1.2   It is intended that the design area is to include the 2 adjacent sprinklers producing the greatest water demand within the compartment.

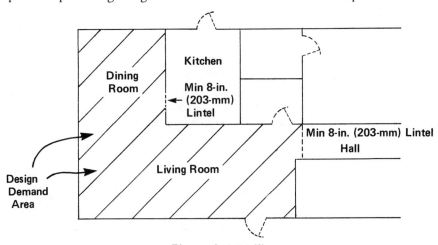

Figure A-4-1.2(i).

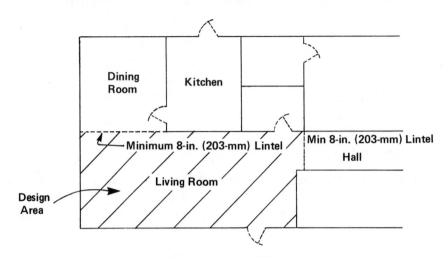

Figure A-4-1.2(ii).

In the Los Angeles and North Carolina fire tests,[7, 8, 9, 10] all of the test fires used typical residential room configurations. These were extinguished with one or two sprinklers when the water flow rate, spacing, and the discharge characteristics of the sprinklers were adequate. When one of these features was not adequate, the test failed because one of the environmental criteria was exceeded or because the water supply was overtaxed when an excessive number of sprinklers opened. Accordingly, design of the residential system should be based on two sprinklers operating (unless no compartment has more than one sprinkler). With two sprinkler designs, each sprinkler must produce a minimum flow of 13 gpm when calculations are done in accordance with 4-4.3. With only one sprinkler in any room, design is based on an 18-gpm flow rate.

A room compartment is determined by the enclosing walls or the enclosing geometrics of a soffet, lintel, or beam which forms a definite interruption in a smooth ceiling configuration. [See Figures A-4-1.2(i) and A-4-1.2(ii)]. In open plan configurations compartment boundaries may be determined by beams, soffets, or lintels over doorways as well as by walls or partitions. In any event, the designer selects no more than two sprinklers to define the design area. When only one sprinkler is within all compartment boundaries, it defines the design area.

4-1.2.1 The definition of compartment for use in 4-1.2 to determine the number of design sprinklers is a space which is completely enclosed by walls and a ceiling. The compartment enclosure may have openings to an adjoining space if the openings have a minimun lintel depth of 8 in. (203 mm) from the ceiling.

As previously described under Section 4-1.2, the design area may be determined either by walls and partitions or by beams, soffets or lintels provided they have a minimum depth of 8 in. This depth is to ensure that heat will be "banked," enhancing sprinkler operation over the fire.

4-1.3 Water Demand. The water demand for the system shall be determined by multiplying the design discharge of 4-1.1 by the number of design sprinklers of 4-1.2.

This calculation will produce a sprinkler demand of 18 gpm for a single sprinkler and 26 gpm for two sprinklers. These values are used in performing the calculations to determine pipe sizing or adequacy of public water supply pressure in Section 4-4.3.

4-1.4 Sprinkler Coverage.

4-1.4.1   Residential sprinklers shall be spaced so that the maximum area protected by a single sprinkler does not exceed 144 sq ft (13.4 m^2).

This section indicates that the design area for a single sprinkler should not exceed 144 sq ft and normally will be the actual area covered by the sprinklers included in the design and specified under 4-1.2. When test and listing indicate greater spacing, this is permitted by 4-1.6.

4-1.4.2   The maximum distance between sprinklers shall not exceed 12 ft (3.7 m) on or between pipelines and the maximum distance to a wall or partition shall not exceed 6 ft (1.8 m). The minimum distance between sprinklers within a compartment shall be 8 ft (2.4 m).

4-1.5   The minimum operating pressure of any sprinkler shall be in accordance with the listing information of the sprinkler and provide the minimum flow rates specified in 4-1.1.

4-1.6   Application rates, design areas, areas of coverage, and minimum design pressures other than those specified in 4-1.1, 4-1.2, 4-1.4, and 4-1.5 may be used with special sprinklers which have been listed for such specific residential installation conditions.

These limitations on spacing were determined from the Los Angeles and North Carolina fire tests.[7, 8, 9, 10] Since these tests were conducted with prototype fast response residential sprinklers, the committee has recognized that other fast response residential sprinklers may be developed which will operate properly with different spacing limitations. Such sprinklers may be used provided that spacings which differ from those described in this standard are justified, based on testing conducted by a testing and inspection agency/laboratory. (See definitions of Labeled and Listed in Section 1-3.)

4-2 Position of Sprinklers.

4-2.1   Pendent and upright sprinklers shall be positioned so that the deflectors are within 1 to 4 in. (25.4 to 102 mm) from the ceiling.

Exception: Special residential sprinklers shall be installed in accordance with the listing limitations.

This limitation is based on the results of the Los Angeles and North Carolina fire tests.[7, 8, 9, 10]

4-2.2   Sidewall sprinklers shall be positioned so that the deflectors are within 4 to 6 in. from the ceiling.

Exception: Special residential sprinklers shall be installed in accordance with the listing limitations.

4-2.3*   Sprinklers shall be positioned so that the response time and discharge are not unduly affected by obstructions such as ceiling slope, beams or light fixtures.

Location of sprinklers in violation of Table A-4.2.3 and Figure A-4.2.3 will cause obstruction of the discharge and may prevent the sprinkler from controlling a fire.
Sprinklers designed to have different discharge characteristics may be positioned closer to a beam than is permitted by Table A-4-2.3. However, it requires that test data be supplied to support such different positioning.

A-4-2.3   Fire testing has indicated the need to wet walls in the area protected by residential sprinklers at a level closer to the ceiling than that accomplished by standard sprinkler distribution. Where beams, light fixtures, sloped ceilings and other obstructions occur, additional residential sprinklers may be necessary to achieve proper response and distribution. Guidance may be obtained from the manufacturer.

Table A-4.2.3 and Figure A-4.2.3 provide guidance for location of sprinklers near ceiling obstructions.

**Table A-4-2.3   Maximum Distance from
Sprinkler Deflector to Bottom of Ceiling Obstruction**

Distance from Sprinkler to Side of Ceiling Obstruction	Maximum Distance from Sprinkler Deflector to Bottom of Ceiling Obstruction
6 in. or less	Not permitted
6 in. to less than 1 ft	0 in.
1 ft to less than 2 ft	1 in.
2 ft to less than 2 ft 6 in.	2 in.
2 ft 6 in. to less than 3 ft	3 in.
3 ft to less than 3 ft 6 in.	4 in.
3 ft 6 in. to less than 4 ft	6 in.
4 ft to less than 4 ft 6 in.	7 in.
4 ft 6 in. to less than 5 ft	9 in.
5 ft to less than 5 ft 6 in.	11 in.
5 ft 6 in. to less than 6 ft	14 in.

For SI Units: 1 in. = 25.4 mm; 1 ft = 0.3048 m.

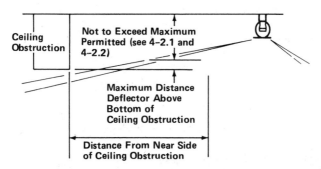

Figure A-4-2.3   Position of Deflector, Upright or Pendent, When Located Above Bottom of Ceiling Obstruction.

4-3 System Types.

4-3.1 Wet-Pipe Systems.   A wet-pipe system shall be used when all piping is installed in areas not subject to freezing.

4-3.2 Dry-Pipe Systems.   Where system piping is located in unheated areas subject to freezing, a dry-pipe system shall be used.

4-4 Pipe Sizing.

4-4.1   Piping shall be sized in accordance with 4-4.3 and 4-4.4. If more than one design discharge is required (*see 4-1.1*), the pipe sizing procedure shall be repeated for each design discharge.

Exception: When piping is sized hydraulically, calculations shall be made in accordance with the methods described in NFPA 13, Standard for the Installation of Sprinkler Systems.

4-4.2   Minimum pipe size shall be ¾ in. for copper; 1 in. for steel.

The requirements for the minimum pipe size are the same as those specified in NFPA 13, *Standard for the Installation of Sprinkler Systems*. This results in less friction loss than if smaller pipe and tubing sizes were permitted. Also, it lessens the possibility of pipe clogging from scale and rust.

4-4.3*   To size piping for systems connected to a city water supply, the following approximate method is acceptable.

(a) Establish system flow rate in accordance with Section 4-1.

(b) Determine water pressure in the street.

(c) Select pipe sizes.

(d) Deduct meter pressure losses if any. [*See Table 4-4.3(d).*]

(e) Deduct pressure loss for elevation. (Building height above street in ft x 0.434 = psi.) (Building height above street in meters x 0.098 = bars.)

(f) Deduct pressure losses from city main to control valve by multiplying the factor from Table 4-4.3(a) or (b) by the total length(s) of pipe in ft (m).

(g) Deduct pressure losses for piping within building by multiplying factor from Table 4-4.3(a) or (b) by the total length in ft (m) of each size of pipe between the control valve and the farthest sprinkler.

(h) Deduct valve and fitting pressure losses. Count the valves and fittings from the control valve to the farthest sprinkler. Determine the equivalent length for each valve and fitting as shown in Table 4-4.3(c) and add these values to obtain the total equivalent length for each pipe size. Multiply the equivalent length for each size by the factor from Table 4-4.3(a) or (b) and total these values.

(i) In multilevel buildings, steps (a) through (g) shall be repeated to size piping for each floor.

(j) If the remaining pressure is less than the operating pressure established by the testing laboratory for the sprinkler being used, a redesign is necessary. If this pressure is higher than required, smaller piping may be used when justified by calculations.

(k) The remaining piping shall be sized the same as the piping to the farthest sprinkler unless smaller sizes are justified by calculations.

Exception: For sprinklered dwellings manufactured off-site, the minimum pressure needed to satisfy the system design criteria on the system side of the meter shall be specified on a data plate by the manufacturer. (See Section 2-3.)

Table 4-4.3(a)   Pressure Losses (psi/ft) Schedule 40 Steel Pipe. C = 120

Pipe Size in.					Flow Rate	GPM						
	10	12	14	16	18	20	25	30	35	40	45	50
1	0.04	0.05	0.07	0.09	0.11	0.13	0.20	0.28	0.37	0.47	0.58	0.71
1¼	0.01	0.01	0.02	0.02	0.03	0.03	0.05	0.07	0.10	0.12	0.15	0.19
1½	0.01	0.01	0.01	0.01	0.01	0.02	0.02	0.03	0.05	0.06	0.07	0.09
2						0.01	0.01	0.01	0.01	0.02	0.02	0.03

For SI Units: 1 gal = 3.785 L; 1 psi = 0.0689 bar; 1 ft = 0.3048 m.

Table 4-4.3(b)   Pressure Losses (psi/ft) Copper Tubing — Types K, L, & M. C = 150

Tubing Size in.	Type	10	12	14	16	Flow Rate	GPM	25	30	35	40	45	50
						18	20						
¾	M	0.08	0.12	0.16	0.20	0.25	0.30	0.46	0.64	0.85	—	—	—
	L	0.10	0.14	0.18	0.23	0.29	0.35	0.53	0.75	1.00	—	—	—
	K	0.13	0.18	0.24	0.30	0.38	0.46	0.69	0.97	1.28	—	—	—
1	M	0.02	0.03	0.04	0.06	0.07	0.08	0.13	0.18	0.24	0.30	0.38	0.46
	L	0.03	0.04	0.05	0.06	0.08	0.10	0.15	0.20	0.27	0.35	0.45	0.53
	K	0.03	0.04	0.06	0.07	0.09	0.11	0.17	0.24	0.31	0.40	0.50	0.61
1¼	M	0.01	0.01	0.02	0.02	0.03	0.03	0.05	0.07	0.09	0.11	0.14	0.17
	L	0.01	0.01	0.02	0.02	0.03	0.03	0.05	0.07	0.10	0.12	0.16	0.19
	K	0.01	0.01	0.02	0.02	0.03	0.04	0.06	0.08	0.11	0.13	0.17	0.20
1½	M	—	0.01	0.01	0.01	0.01	0.01	0.02	0.03	0.04	0.05	0.06	0.08
	L	—	0.01	0.01	0.01	0.01	0.01	0.02	0.03	0.04	0.05	0.07	0.08
	K	—	0.01	0.01	0.01	0.01	0.02	0.02	0.03	0.05	0.06	0.07	0.09
2	M	—	—	—	—	—	—	0.01	0.01	0.01	0.01	0.02	0.02
	L	—	—	—	—	—	—	0.01	0.01	0.01	0.01	0.02	0.02
	K	—	—	—	—	—	—	0.01	0.01	0.01	0.01	0.02	0.02

For SI Units: 1 gal = 3.785 L; 1 psi = 0.0689 bar; 1 ft = 0.3048 m.

**Table 4-4.3(c)   Equivalent Length of Pipe
in Feet for Steel and Copper Fittings and Valves**

Fitting/Valve Diameter in.	45 Degrees	Elbows 90 Degrees	Long Radius	Tees Flow Thru Branch	Flow Thru Run	Gate	Angle	Globe	Valves Globe "Y" Pattern	Cock	Check
¾	1	2	1	4	1	1	10	21	11	3	3
1	1	3	2	5	2	1	12	28	15	4	4
1¼	2	3	2	6	2	2	15	35	18	5	5
1½	2	4	3	8	3	2	18	43	22	6	6
2	3	5	3	10	3	2	24	57	28	7	8

Based on Crane Technical Paper No. 410.

For SI Units: 1 ft = 0.3048 m.

Table 4-4.3(d)   Pressure Losses in Water Meters

Meter (Inches)	Pressure Loss (psi) Flow (gpm)					
	18	23	26	31	39	52
⅜	9	14	18	26	*	*
¾	4	8	9	13	*	*
1	2	3	3	4	6	10
1½	**	1	2	2	4	7
2	**	**	**	1	2	3

NOTE: Lower pressure losses may be used when supporting data is provided by the meter manufacturer.

* Above maximum rated flow of commonly available meters.

** Less than 1 psi.

For SI Units: 1 gpm = 3.785 L/min; 1 in. = 25.4 mm.

A-4-4.3   Determination of public water supply pressure should take into account probable minimum pressure condition prevailing during such periods as at night, or during summer months when heavy usage may occur; also, possibility of interruption by floods, or ice conditions in winter.

This section of the standard describes a simplified method of sizing pipe in a residence to ensure that the flow rate from the sprinklers in the design area will meet the design discharge criteria of 4-1.1.

In performing the calculation for a specific system, it is necessary to go through the calculation procedure twice. The first calculation is made for a single sprinkler flowing at 18 gpm, and the second calculation is made for two sprinklers flowing at 26 gpm [13 gpm each]. The pressure required to produce this flow at the sprinkler will depend on the coefficient of discharge ("K" factor), which must be obtained from the manufacturer of the particular residential sprinkler that will be used. In multilevel dwellings calculations will need to be done for all levels of the building to be certain that pipe is sized for the variations in flow which can occur due to elevation differences between floors and due to differences in piping geometry between floors.

The second step is to determine the water pressure in the street in front of the property. This is done by obtaining information on the public water supply pressure from the local water company or by arranging with a local plumber to place a pressure gage on the water supply inlet connection in the dwelling.

Next, pipe sizes are arbitrarily selected and valves and fittings laid out to meet the piping arrangements described in Figures A-2-2, A-2-3(a), A-2-3(b), A-2-3(c), A-2-3(d), and A-2-3(e). Losses in piping, fittings, meters, valves and for elevation are then determined in accordance with Steps d, e, f, g and h, and these are deducted from the city pressure to determine the pressure at the sprinkler design area on each floor of the building. If this pressure is less than the operating pressure determined in the calculation to meet flow demands (see Step 1 above), a redesign is necessary and pipe must be upsized or the water supply improved. If the remaining pressure is higher than required to supply the demand rate at the sprinkler, the designer can either stop and accept the calculation or downsize the pipe to achieve optimum sizing.

A sample calculation for a typical residence follows:

This example is based on the dwelling illustrated by Figure 4.1, which is a plot plan also showing an existing city main. Figures 4.2 and 4.3 show the sprinkler and piping layout for the basement and first floor of this dwelling.

Assume the following:

• That the sprinklers used are listed for application rates, design area, and area of coverage as indicated in paragraphs 4-1.1, 4-1.2 and 4-1.4.

• That the flow from one sprinkler is 18 gpm and from two sprinklers is 13 gpm each [26 gpm total] per 4-1.1 of this standard.

• That the sprinkler used has a flow coefficient ("K" factor) of 3.3. Thus, the pressure needed for the system at the flowing sprinkler is calculated as follows:

$Q = K \sqrt{P}$ $P = (Q/K)^2$
For 18 gpm = $P = (18/3.3)^2 = (5.45)^2 = 29.8$ psi
For 13 gpm = $P = (13/3.3)^2 = (3.94) = 15.5$ psi

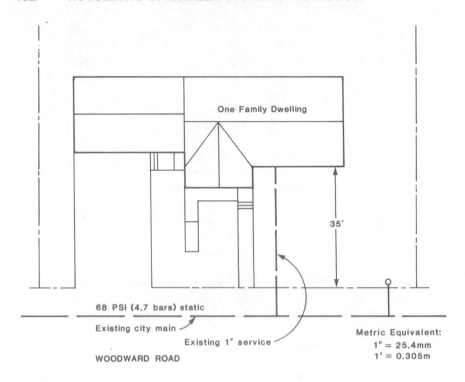

PLOT PLAN
Plan 1 of 3

Figure 4.1.   Typical residence for sample calculation.

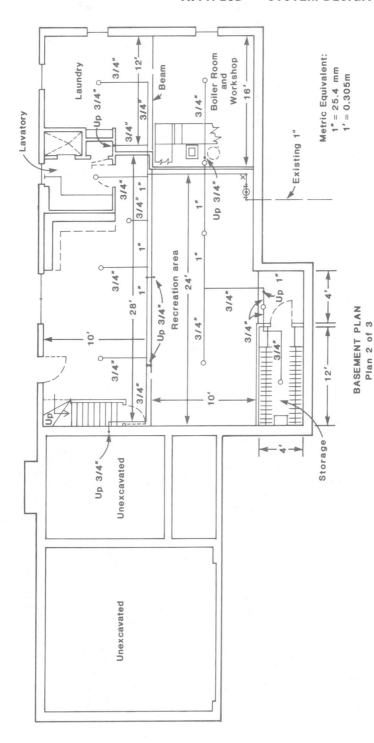

Figure 4.2.   Basement floor plan for sample sprinkler calculation.

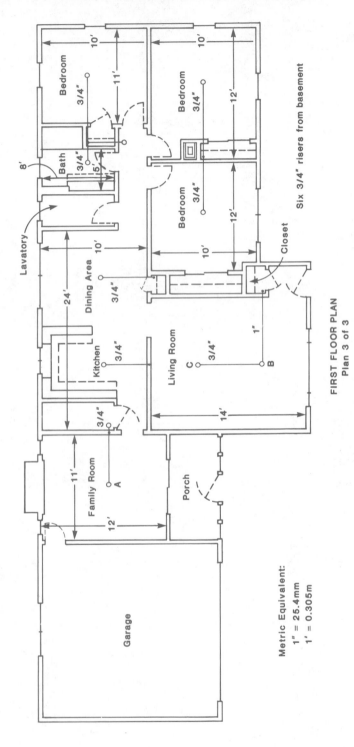

Figure 4.3.   First floor plan for sample sprinkler calculation.

Calculation #1 at 18 gpm (29.8 psi) from Sprinkler "A"

1st Floor: Steel pipe				Individual loss	Net total

a. Water pressure in street   _____   *68*

b. Arbitrarily selected pipe size *1" + ¾"*

c. Deduct meter loss (*¾* size)   *6.0*   *62*

d. Deduct loss for elevation

   System control valve*   ,

   (*15* ft X 0.434)   *6.5*   *55.5*

e. Deduct loss from city main

   to control valve

1"	pipe	*48* ft	
3	valves	*3* ft	
3	elbows	*9* ft	
1	tee	*5* ft	
	Total	*65* ft	X *0.11* *7.2* *48.3*

f. Deduct loss for piping: control

   valve to farthest sprinkler*

   3/4-in. pipe: 19 ft   X   *0.35*   *6.7*

   1-in. pipe: 35 ft   X   *0.11*   *3.9*   *37.7*

Size	Quan.	Description	Total equiv. ft		
	4	90° elbow	*8*		
¾"	_____	45° elbow	_____		
	_____	tee	_____		
	_____	check valve	_____		
	_____	valve (___)	_____		
		Total	*8*	ft X *0.4* = *3.2*	*34.5*

Size	Quan.	Description	Total equiv. ft		
	1	90° elbow	*3*		
1"	_____	45° elbow	_____		
	1	tee	*5*		
	1	check valve	*4*		
	_____	valve (___)	_____		
		Total	*12*	ft X *0.11* = *1.3*	*33.2*

This exceeds 29.8 psi and design is acceptable.

Note: Repeat calculation for basement to verify adequacy of basement piping size.

Remaining pressure for sprinkler operations.

* Factors from Table 4.4.3(a), (c), and (d).

For SI Units: 1 ft = 0.3018 m; 1 psi = 0.0689 bar.

Figure 4.4.   Calculation Example No. 1.

Calculation #2 at 26gpm (15.5 psi) from Sprinklers "B" and "C"

1st Floor: Steel pipe	Individual loss	Net total
a. Water pressure in street		_68_
b. Arbitrarily selected pipe size _1" + 3/4"_		
c. Deduct meter loss (_3/4_ size)	_11.0_	_57.0_
d. Deduct loss for elevation		
System control valve*		
(_15_ ft X 0.434)	_6.5_	_50.5_

e. Deduct loss from city main

to control valve

_1_   pipe   _48_ ft
_3_   valves   _3_ ft
_3_   elbows   _9_ ft
_1_   tee   _5_ ft

Total _65_ ft X _0.21_	_13.7_	_36.8_

f. Deduct loss for piping: control

valve to farthest sprinkler*

3/4-in. pipe: 12 ft X _0.19_	_2.3_	
1-in. pipe: 14 ft X _0.69_	_9.7_	_24.8_

Size	Quan.	Description	Total equiv. ft		
_3/4"_	_____	90° elbow	_____		
	_____	45° elbow	_____		
	_____	tee	_____		
	_____	check valve	_____		
	_____	valve (____)	_____		
		Total	_____ ft X _____ = _____		_____

Size	Quan.	Description	Total equiv. ft		
_1"_	_1_	90° elbow	_3_		
	_____	45° elbow	_____		
	_1_	tee	_5_		
	_1_	check valve	_4_		
	_____	valve (____)	_____		
		Total	_12_ ft X _0.21_ = _2.5_		_17.5_

This exceeds 15.5 psi and design is acceptable.

Note: Repeat calculation for basement to verify adequacy of basement piping size.

Remaining pressure for sprinkler operations.

* Factors from Table 4.4.3(a), (c), and (d).

For SI Units: 1 ft = 0.3018 m; 1 psi = 0.0689 bar.

Figure 4.5   Calculation Example No. 2. The farthest sprinkler in this case is the farthest sprinkler in the only room having more than one sprinkler.

	Individual Loss	Net Total

a. Water Pressure in Street _____   _____   _____

b. Arbitrarily Select Pipe Size _____

c. Deduct Meter Loss (_____Size) _____   _____   _____

d. Deduct Head Loss for Elevation System Control
Valve*
(_____ft × 0.434) _____

e. Deduct Pressure Loss from City Main to
Sprinkler _____   _____   _____

    _____Pipe — _____ft

    _____Valves — _____ft

    _____Elbows — _____ft

    _____Tee — _____ft

    _____Total — _____ft × _____   _____   _____

f. Deduct Pressure Loss for Piping — Control
Valve to Farthest Sprinkler*

Size	Quan.	Description	Total Equiv. Feet
	_____90° Elbow		_____
	_____45° Elbow		_____
	_____Tee		_____
	_____Check Valve		_____
	_____Valve (_____)		_____
	_____Total		Ft × _____ = _____ _____

Size	Quan.	Description	Total Equiv. Feet
	_____90° Elbow		_____
	_____45° Elbow		_____
	_____Tee		_____
	_____Check Valve		_____
	_____Valve (_____)		_____
	_____Total		Ft × _____ = _____ _____

Remaining Pressure for Sprinkler Operations

*Factors from Table 4-4.3(a), (b), (c) and (d)

For SI Units: 1 ft = 0.3048 m; 1 psi = 0.0689 bar.

Figure A-4-4.3(1)   Calculation Sheet.

	Individual Loss	Net Total
Water Pressure at Supply Outlet	_____	_____
a. Deduct Head Loss for Elevation (____ft × 0.434)	_____	_____
b. Deduct Pressure Loss from Piping Within Building*	_____	_____
Remaining Pressure for Sprinkler Operation	_____	_____

*Factors from Table 4-4.3(a), (b), (c) and (d)

For SI Units: 1 ft = 0.3048 m; 1 psi = 0.0689 bar.

Figure A-4-4.3(2)   Calculation Sheet — Elevated Tank, Booster Pump, Pump Tank Supply.

4-4.4   To size piping for systems with an elevated tank, pump or pump-tank combination, determine the pressure at the water supply outlet and proceed through steps (c), (e), (g), (h), (i), (j) and (k) of 4-4.3.

4-5 Piping Configurations.   Piping configurations may be looped, gridded, straight run, or combinations thereof.

When piping configurations are looped or gridded, calculations must be conducted in accordance with the exception to Section 4-4.1 of this standard. This means that hydraulic calculations to size the piping must be done in accordance with the method described in NFPA 13.

4-6 Location of Sprinklers.   Sprinklers shall be installed in all areas.

Exception No. 1:   Sprinklers may be omitted from bathrooms not exceeding 55 sq ft (5.1 m²) with noncombustible plumbing fixtures.

Exception No. 2:   Sprinklers may be omitted from small closets where the least dimension does not exceed 3 ft (0.9 m) and the area does not exceed 24 sq ft (2.2 m²) and the walls and ceiling are surfaced with noncombustible materials.

Exception No. 3:   Sprinklers may be omitted from open attached porches, garages, carports and similar structures.

Exception No. 4:   Sprinklers may be omitted from attics and crawl spaces which are not used or intended for living purposes or storage.

Exception No. 5:   Sprinklers may be omitted from entrance foyers which are not the only means of egress.

The basis for omission of sprinklers in Exceptions 1 through 6 has been previously described under Section A-2-2 and is based on NFPA statistics in Table A-1-2. These areas are the ones shown to have a low incidence of life loss from fires in dwellings. In addition, the committee recognized the following.

• With respect to Exception No. 1, the combustible load in bathrooms is normally extremely low, especially with noncombustible plumbing fixtures.

• With respect to Exceptions Nos. 3, 4 and 5, mandatory sprinklering of these areas would necessitate the use of dry-pipe systems in areas where freezing weather is encountered. This would detract from the rapid response of the system within the occupied areas of the dwelling and detract from, rather than enhance, life safety. Further, most building codes require a 1-hour fire rated separation between garages and other portions of the dwelling.

5

MANDATORY REFERENCED PUBLICATIONS

5-1 This chapter lists publications referenced within this document which, in whole or in part, are part of the requirements of this document.

5-1.1 NFPA Publications. The following publications are available from the National Fire Protection Association, Batterymarch Park, Quincy, MA 02269.

NFPA 13-1983, *Standard for the Installation of Sprinkler Systems*

NFPA 13A-1981, *Recommended Practice for the Inspection, Testing and Maintenance of Sprinkler Systems*

NFPA 20-1983, *Standard for the Installation of Centrifugal Fire Pumps*

NFPA 22-1984, *Standard for Water Tanks for Private Fire Protection*

NFPA 74-1984, *Standard for the Installation, Maintenance, and Use of Household Fire Warning Equipment*

NFPA 101®-1981, *Code for Safety to Life from Fire in Buildings and Structures*

NFPA 220-1979, *Standard on Types of Building Construction*

5-1.2 Other Codes and Standards. This publication makes reference to the following codes and standards and the year dates shown indicate the latest editions available.

Publications designated ANSI are available from the American National Standard Institute, Inc., 1450 Broadway, New York, NY 10018; those designated ASTM are available from the American Society for Testing and Materials, 1916 Race Street, Philadelphia, PA 19105; those designated AWS are available from the American Welding Society, 550 NW Le Jeune Road, Miami, FL 33135.

ANSI B16.1-1975, *Cast Iron Pipe Flanges and Flanged Fittings, Class 25, 125, 250 and 800*

ANSI B16.11-1980, *Forged Steel Fittings, Socket Welding and Threaded*

ANSI B16.18-1978, *Cast Copper Alloy Solder Joint Pressure Fittings*

ANSI B16.22-1980, *Wrought Copper and Copper Alloy Solder Joint Pressure Fittings*

ANSI B16.25-1979, *Buttwelding Ends*

ANSI B16.3-1977, *Malleable Iron Threaded Fittings, Class 150 and 300*

ANSI B16.4-1977, *Cast Iron Threaded Fittings, Class 125 and 250*

ANSI B16.5-1981, *Pipe Flanges and Flanged Fittings*

ANSI B16.9-1978, *Factory-Made Wrought Steel Buttwelding Fittings*

ANSI B36.10-1979, *Welded and Seamless Wrought Steel Pipe*

ASTM B32-1976, *Specifications for Solder Metal*

ASTM A53-83, *Specifications for Pipe, Steel Black, and Hot-Dipped, Zinc-Coated Welded and Seamless Steel Pipe*

ASTM B75-81, *Specifications for Seamless Copper Tube*

ASTM B88-83, *Specifications for Seamless Copper Water Tube*

ASTM A120-83, *Specification for Black and Hot-Dipped Zinc Coated (Galvanized), Welded and Seamless Steel Pipe for Ordinary Uses*

ASTM A135-83, *Specifications for Electric Resistance Welded Steel Pipe*

ASTM A234-82, *Specifications for Piping Fittings of Wrought Carbon Steel and Alloy for Moderate and Elevated Temperatures*

ASTM E380-82, *Standard for Metric Practice*

ASTM B251-81, *Specifications for General Requirements for Wrought Seamless Copper and Copper-Alloy Tube*

ASTM A795-83, *Specification for Black and Hot-Dipped Zinc-Coated Welded and Seamless Steel Pipe for Fire Protection Use*

AWS A5.8-81, *Specification for Brazing Filler Metal*

APPENDIX A

This Appendix is not a part of the requirements of this NFPA document . . . but is included for information purposes only.

The material contained in Appendix A of this standard is included within the text of this Handbook, and therefore is not repeated here.

REFERENCES CITED IN COMMENTARY

[1]Halpin, B.M., Dinan, J.J., and Deters, O.J., *Assessment of the Potential Impact of Fire Protection Systems on Actual Fire Incidents*, John Hopkins University-Applied Physics Laboratory (JHU/APL), Laurel, Maryland, October 1978

This study describes an in-depth analysis of fires involving fatalities and includes an assessment of how use of detectors, sprinklers or remote alarms would have changed the results.

[2]Yurkonis, Peter, *Study to Establish the Existing Automatic Fire Suppression Technology for Use in Residential Occupancies*, Rolf Jensen & Associates, Inc., Deerfield, Illinois, December 1978.

This study identified suppression systems and evaluated design and cost factors affecting practical usage and user acceptance.

[3]Kung, H., Haines, D., Green, R. Jr., *Development of Low-Cost Residential Sprinkler Protection*: Factory Mutual Research Corporation (FMRC), Norwood, Massachusetts, February 1978.

This study is aimed *primarily* at developmet of low-cost residential sprinkler systems having minimal water discharge rates providing adequate life and property protection from smoldering fires.

[4]Henderson, N.C., Riegel, P.S., Patton, R.M., Larcomb, D.B., *Investigation of Low-Cost Residential Sprinkler Systems*, Battelle Columbus Laboratories (BCL), Columbus, Ohio, June 1978.

Fire tests of commercial nozzles and evaluation of piping methods to achieve low cost systems based on propane burner, wood crib and furniture fire tests.

[5]Clarke, Graham, *Performance Specifications for Low-Cost Residential Sprinkler System*, Factory Mutual Research, Norwood, Massachusetts, January 1978

A review of data from the listed USFA research efforts as a basis to prepare a proposed standard to install residential sprinkler systems and test residential sprinklers.

[6]Kung, H. C., Spaulding, R. D., Hill, E.E. Jr., *Sprinkler Performance in Residential Fire Tests*, Factory Mutual Research, Norwood, Massachusetts, December 1980.

Fire tests on representative living room and bedroom residential configuration with combinations of furnishings and with open and closed windows and with variations in sprinkler application rates and response sensitivity of sprinklers. This work has been closely evaluated by the NFPA 13D Subcommittee during its conduct, and the Subcommittee was responsible for directing many of the test conditions which were evaluated.

[7]Cote, A., Moore, D., *Field Test and Evaluation of Residential Sprinkler Systems, Los Angeles Test Series*, National Fire Protection Association, Boston, Massachusetts, April 1980. (A report for the NFPA 13D Subcommittee)

A series of fire tests in actual dwellings with sprinkler systems installed in accordance with the data resulting from all prior test work and the criteria included in the proposed 1980 revisions to NFPA 13D so as to evaluate the effectiveness of the system under conditions approaching actual use.

[8]Moore, D., *Data Summary of the North Carolina Test series of USFA Grant 79027 Field Test and Evaluation of Residential Sprinkler Systems*. National Fire Protection Association, Boston, Mass., September 1980. (A report for the NFPA 13D Subcommittee)

A series of fire tests in actual mobile homes with sprinkler systems installed in accordance with the data resulting from prior test work and following criteria described in NFPA 13D-1980 in order to evaluate the effectiveness of the system under conditions approaching actual use.

[9]Kung, H. C., Spaulding, R.D., Hill, E.E. Jr., Symonds, A.P., *Technical Report, Field Evaluation of Residential Prototype Sprinkler Los Angeles Fire Test Program*, Factory Mutual Research, Norwood, Massachusetts, February 1982.

[10]Cote, A. E., *Final Report on Field Test and Evaluation of Residential Sprinkler Systems*, National Fire Protection Association, Quincy, Massachusetts, July 1982.

[11]Factory Mutual Research Corporation/Underwriters Laboratories Inc., *Standard for Residential Automatic Sprinklers for Fire Protection Service*, Northbrook, Illinois, April 1980.

A study to develop a performance standard for testing residential sprinklers under room fire conditions that are in harmony with the requirements of NFPA 13D.

[12]Underwriters Laboratories Inc., *Proposed Standard for Residential Automatic Sprinklers for Fire Protection Service*, UL-1626, Northbrook, Illinois, October 1980.

This standard describes the requirements for the testing of residential sprinklers by a testing laboratory. Factory Mutual Engineering Association has a similar standard.

[13]Evans, D.D., Madrzykowski, D., *Characterizing the Thermal Response of Fusible-Link Sprinklers*, U.S. Department of Commerce, National Bureau of Standards., Washington DC, August 1981.

Errata

NFPA

Automatic Sprinkler Systems Handbook
Second Edition

In the printing of the second edition of the NFPA *Automatic Sprinkler Systems Handbook*, an error occurred and should be corrected as follows:

1. *In the portion dealing with NFPA 13, Standard for the Installation of Sprinkler Systems, 1985 edition, delete paragraph 3-10.3.2.2.*

5M-2-86-FP